Microbial Control of Weeds

Microbial
Control
of
Weeds

Edited by
David O. TeBeest

Chapman and Hall
New York.
London

First published in 1991 by
Chapman and Hall
an imprint of
Routledge, Chapman & Hall, Inc.
29 West 35 Street
New York, NY 10001

Published in Great Britain by

Chapman and Hall
2-6 Boundary Row
London SE1 8HN

Library of Congress Cataloging in Publication Data

Microbial control of weeds / edited by David O. TeBeest.
      p. cm.
   Includes bibliographical references.
   ISBN 0-412-01861-6
   1. Weeds—Biological control.   2. Phytopathogenic microorganisms.
3. Weeds—Diseases and pests.   4. Biological pest control agents.
I. TeBeest, David O.
SB611.5.M53   1990
632'.58—dc20   90-1457

British Library cataloguing publication data also available.

Contents

Preface

It is appropriate at this time to reflect on two decades of research in biological control of weeds with fungal plant pathogens. Some remarkable events have occurred in the last 20 years that represent a flurry of activity far beyond what could reasonably have been predicted.

In 1969 a special topics review article by C. L. Wilson was published in *Annual Reviews of Phytopathology* that examined the literature and the potential for biological control of weeds with plant pathogens. In that same year, experiments were conducted in Arkansas that determined whether a fungal plant pathogen could reduce the infestation of a single weed species in rice fields. In Florida a project was under way to determine the potential use of a soil-borne plant pathogen as a means for controlling a single weed species in citrus groves.

Work in Australia was published that described experiments that sought to determine whether a pathogen could safely and deliberately be imported and released into a country to control a weed of agricultural importance. All three projects were successful in the sense that *Puccinia chondrillina* was released into Australia to control rush skeletonweed and was released later into the United States as well, and that *Colletotrichum gloeosporioides* f.sp. *aeschynomene* and *Phytophthora palmivora* were later both marketed for the specific purpose of controlling specific weed species.

These three successes sparked interest worldwide, and today similar projects are found on six of the seven continents, Antarctica being the exception. Several world conferences on biological control of weeds have been held in the last 20 years that included reports of work with plant pathogens. Several professional societies now have standing committees whose members include those working on biological control of weeds with plant pathogens.

The emergence of genetic engineering captured the attention of the world during these two decades. Genetic manipulation of filamentous fungi, including fungal plant pathogens, with recombinant DNA techniques was demonstrated. And these techniques became part of several projects on biological control of weeds. Several recent papers have already discussed mechanisms for "improvement" of fungi as biological control agents. It would seem that biological control of weeds and genetic manipulation or engineering could be natural partners in biotechnology to develop this area of study further. The applied side of this area of study is obvious. The basic side is enormous and full of agriculturally relevant fundamental research problems.

This book attempts to discuss and focus attention on the progress made in several areas of research in the last 20 years relevant to biological control of weeds.

As the editor, I want to express my appreciation to each of the contributors who labored to fulfill this mission by summarizing their own subject areas. Each, in his own personal way, has contributed significantly to this effort.

David O. TeBeest,
University of Arkansas

Biological Control
of Weeds

1

The Classical Approach with Plant Pathogens

Alan K. Watson

I. Introduction

Biological weed control is an approach utilizing living organisms to control or reduce the population of an undesirable weed species. The classical or inoculative approach involves the importation and release of one or more natural enemies that attack the target weed in its native range, into areas where the weed is introduced and is troublesome and where its natural enemies are absent. The objective of classical biological weed control is generally not eradication of the weed species, but the self-perpetuating regulation of the weed population at acceptable low levels. The classical approach differs from the inundative or bioherbicide approach primarily in its ecological rather than technological response to a weed problem (86). The bioherbicide approach artificially increases the effectiveness of a candidate organism, whereas the classical approach relies on the innate ability of the introduced biocontrol organism to become established and flourish in its new habitat.

Biological control is based on the premise that biotic factors significantly affect the distribution and abundance of plant species. Despite the emphasis placed on the role of climatic and edaphic factors on vegetation composition, the biotic component has major effects on plant communities (22,35,58,70,75,76).

Many of the most noxious weeds throughout the world have been accidentally or intentionally introduced from one region of the world to another. In North America the most important weed species are not indigenous, with approximately one half of the weeds in the United States and 13 of the top 15 weeds being introduced species (6). Similarly, the majority (78 out of 107) of noxious Canadian weed species have been introduced from Europe or Asia (98). These introduced species tend to be far more aggressive in their new habitats than in their native range. Possible explanations for this increased aggressiveness include the following: (1) The new area has more suitable climatic or edaphic conditions, (2) plant competition from associated vegetation is less intense in new habitats, and (3) natural enemies are absent in the new habitat (97). The increased aggressiveness of introduced weeds is often difficult or impossible to explain by either of the first two options presented. In most instances exotic weed species have become established in their new environment in the absence of their natural enemies. Subsequently, within its new habitat, the population of the exotic weed species is not regulated by natural enemies and thrives. Hence, the basis of classical biological weed control is the introduction of selected host-specific natural from which the exotic weed species has escaped. Biotic agents that have been used in biological weed control programs include insects, mites, fungi, nematodes, and aquatic and terrestrial vertebrate herbivores.

Most of the biological weed control projects prior to 1970 utilized insects as the biotic agents, but subsequently plant pathogens have become increasingly more prevalent as biocontrol agents of weeds. Numerous excellent reviews highlight the fundamentals, progress, and prospects of biological weed control (e.g., 1,6,36,39,43,44,45,47,52,54,58,77,81,86,96,99), with a few emphasizing the use of plant pathogens (1,47,52,81,96).

II. Elements of Plant Disease and Restraints to Classical Biological Control

A susceptible host, a virulent pathogen, and favorable environmental conditions are essential components for plant disease to occur. The concepts of plant disease and epidemiology are well documented with articles by Quimby (75), Holcomb (56), Barrett (9), Leonard (62), and Shrum (79) reviewing the fundamentals of plant pathology as related to biological control of weeds with plant pathogens. In the bioherbicide approach opportunities exist for man to overcome restraints of disease development by increasing the quantity of inoculum, by timing inoculation to coincide with increased host plant susceptibility, by formulating infective propagules, or by applying irrigation to overcome environmental limitations. In the classical approach, however, opportunities to overcome restraints of disease development are fewer.

Genetic variability and spatial distribution of the target weed population are the key host-related factors in disease development. Spatial distribution may restrict disease spread if the target weed has a patchy distribution or populations are separated by geographical barriers. Because the usual target weed of classical biological control is a dominant, widespread weed forming dense infestations in pasture or rangeland habitats, spatial distribution may not be a limitation to disease in some host–pathogen systems. Similarly, in cultivated habitats, targeted weeds for biological control are usually dominant and relatively uniformly dispersed. If populations of the target weed are grouped in favorable habitats or widely separated from one another, spread of the disease could be aided by man.

Genetic variation does occur within natural plant populations, and weed populations are often described as highly variable. Barrett (9) reviewed genetic variation within weed populations and implications for biological control. He cautioned us in making generalizations but noted that weed species frequently have limited genetic variability. Populations of introduced weed species are usually genetically less variable than naturally occurring populations of the same species in its native range (9, 20, 21). Similarly, asexual reproduction, with its associated genetic homogeneity, has been positively linked to successful biological weed control (21). Little is known about the host–pathogen interactions in natural plant communities and less is known of weed–pathogen interactions. Most of the published information related to classical biological weed control with plant pathogens is on the *Chondrilla–Puccinia* association of an apomictic weed population and an "excessively" specific rust pathogen (20,23,65). Much has been learned from this association, but we need to be cautious in extrapolating to other weed–pathogen systems. Successful biological control will be favored by limited amounts of genetic variability, and Barrett (9) suggests that actual amounts of genetic variation in weed populations would be a poor predictor of whether or not successful biological control would be achieved. More important, the nature and level of genetic variability in both the weed and the pathogen populations will influence disease development. For example, Burden and Groves (23) suggested that partial resistance, common in many individuals in a plant population, may be an obstacle to classical biological control with plant pathogens, since it could affect the rate of development of disease epidemics. These concerns of resistance within the host population are appropriate when considering the *Chondrilla–Puccinia* systems (24,48,51) but may be less important in other weed–pathogen systems. For example, in eastern Europe basal leaves widely scattered diffuse knapweed (*Centaurea diffusa* Lam.) plants were infected with the rust *Puccinia jaceae* Otth ranging from a few small uredia (moderately resistant) to numerous large uredia (very susceptible) (95). Not all the European isolates of *P. jaceae* were virulent on North American diffuse knapweed, but when virulent isolates were inoculated onto plants from 15 different North American diffuse knapweed populations, all were highly susceptible to the pathogen (91).

Similarly, questions of whether or not resistance to the pathogen would develop or of when resistance will develop within the weed population are difficult to answer definitively. Since the objective of classical biological control is to reduce the target weed population to tolerable, manageable levels and not to eradicate the weed, selection pressure would be much less than that for chemical herbicides. Evolution of genetic resistance to herbicides has been relatively limited and restricted to those chemicals with target-specific mode of action sites (e.g., specific enzyme system). Resistance to biocontrol plant pathogens, with their usual broad spectrum of mechanism of action, will not likely be a major restraint to classical biological weed control.

Leonard (62) discussed genetic heterogeneity in plant pathogens as related to classical biological weed control. Much of his discussion has important implications in host specificity or crop safety concerns, but he also stressed the importance of pathogen virulence using the skeletonweed rust example. In addition to the pathogen being virulent on the target weed, it must be aggressive under local environmental conditions (79). Shrum (79) suggested that once a virulent pathogen has been selected, the efforts to maximize reproduction of the pathogen are paramount if a successful biocontrol epidemic is to develop. Successful classical biocontrol is dependent upon the fundamentals of epidemiology, including initial inoculum (x_o), rate of reproduction (r), and time (t) (79). Disease epidemics are often prevented because of lack of inoculum; therefore, mass rearing and release of additional inoculum early in the season coinciding with an infection window could help create epidemics and result in successful biocontrol.

Favorable environmental conditions are also essential for disease development, since environmental deficiencies (e.g., absence of free moisture, inhibitory temperatures), along with lack of sufficient inoculum and the innate resistance of the host population, constrain disease development. Unlike the bioherbicide strategy, the classical approach relies on establishing a restricted number of infection foci. The broadcast, often aqueous, application of pathogen propagules in the bioherbicide approach, coupled with the possible opportunity to apply irrigation to alleviate dry conditions, is usually not applicable to the classical approach. As a consequence, a thorough understanding of the etiology of the disease, specifically the environmental optimum for infection coupled with availability of susceptible host plant material, is essential to optimize the release of initial inoculum in the classical approach.

III. Procedures

The procedures followed in a classical biological weed control program with plant pathogens are essentially identical to those outlined by Harris (41) for insects. The steps followed are (1) determination of the suitability of the weed for classical biological control, (2) conducting surveys for suitable plant pathogens in the target weed's native range, (3) studying the ecology of potentially suitable (effective) plant pathogen(s), (4) evaluating the host specificity of selected plant pathogens, (5) introduction and establishment of selected plant pathogens into the new habitat, and (6) evaluation of the effect of the biocontrol agent on the target weed population. Specific procedures with plant pathogenic agents in classical biological weed control have been reviewed by Hasan (47) and Schroeder (77).

Suitability of Target Weed

Most biological control programs in the past were undertaken as a last resort because conventional control methods could not be applied, had failed, or would be uneconomical (77). However, prior to the initiation of a classical biocontrol project, a careful analysis of the suitability of the target weed should be conducted (52,77). In general, the ideal target weed for classical biological control is an aggressive, introduced weed species that infests large areas of marginal lands, such as rangelands or pastures. In addition, the weed should not be highly valued by other segments of society.

Classical biological weed control has been reported to have limited application for weeds common in intensive (cultivated) crop production systems (42). However, the success of the skeleton weed project in Australia (32), the successful control of ragweed (*Ambrosia artemisiifolia* L.) in cultivated fields in the Soviet Union (61), and the suppression of groundsel (*Senecio vulgaris* L.) by a rust fungus in lettuce fields (72) suggest that the classical approach also has application as a component of integrated weed management systems in intensive crop production (92).

Suitable Agents

The initial aspect of the discovery phase of a classical biological weed control program is a comprehensive literature survey to gather information on the occurrence and distribution of the target weed and its associated diseases in its native and introduced ranges. These literature surveys provide the basis for selection of possible candidate biocontrol agents and selection of areas for foreign exploration to enhance opportunities to collect pathogens.

As Hasan (52) and Templeton and Trujillo (81) have stated, theoretically all kinds of plant pathogens, including viruses, bacteria, fungi, nematodes, and others that adversely affect the growth and/or reproduction of a weed, could be considered as classical biological weed control agents. To date, rust fungi have been the agent of choice almost exclusively in most classical biological weed control programs (Figures 1.1 and 1.2). Rust fungi are often damaging to their hosts. They are wind disseminated, usually have a high degree of host specialization, and are relatively well known and documented in the literature.

Potential Effectiveness

The choice or selection of biocontrol agents is theoretically based on two criteria: efficacy and safety (77). However, in most of the classical biological weed projects, especially with insects, the emphasis has been on the selection of "safe" biocontrol agents with little or no evaluation of potential effectiveness of a prospective biocontrol agent prior to its introduction. Harris (42) and Goeden (40) have proposed various mechanisms, including scoring systems, to determine the relative potential effectiveness of insects prior to introduction. These approaches are based primarily on the biology and ecology of the agent and not on the impact (real or potential) of the agent on the target weed. Wapshere (87) has also discussed the difficulty of making realistic experimental estimates of effectiveness.

It is tempting to follow the lead of the entomologists and develop a modified

Figure 1.1.   Weed Targets and Exotic Pathogens Intentionally Released for Biological Weed Control Projects

Target Weed	Pathogen	Country of Introduction	References
Asteraceae *Acroptilon repens* (L.) DC. (=*Centaurea repens* L.) (Russian knapweed)	*Subanguina picridis* (Kirj.) Brzeski	Canada and U.S. (continental)	89, 90
Ageratina riparia (Regel) *K. & R.* (=*Eupatorium riparium* *Regel)* (hamakua pa-makani, mistflower)	*Entyloma ageratinae* sp. now.	U.S. (Hawaii)	8, 83, 84
Carduus nutans L. (musk thistle, nodding thistle)	*Puccinia carduorum* Jacky	U.S. (continental)	12, 13, 73, 74
Chondrilla juncea L. (skeletonweed)	*Puccinia chondrillina* Bubak & Syd.	Australia U.S. (continental)	1, 2, 3, 10, 24, 32, 37, 46, 48, 51, 53, 80
Eupatorium adenophorum Spreng. (crofton weed)	*Phaeoramularia* sp.	South Africa	66
Fabaceae (Leguminosae) *Acacia saligna (Labill.)* Wendl. (Acacia)	*Uromycladium tepperianum* (Sacc.) Mac Alp.	South Africa	66
Galega officinalis L. (galega, goat's rue)	*Uromyces galegae (Opic)* Sacc.	Chile	69
Rosaceae *Rubus* spp. (blackberry)	*Phragmidium violaceum* (Schultz) Winter	Chile	67, 68

scoring system to assess arbitrarily and rank the potential efficacy of different plant pathogens as classical biological control agents. This approach, however, has not been particularly useful in selecting more effective insects for the biocontrol of weeds, and the regulatory authorities in most countries of proposed introductions of plant pathogens require clear indication that not only is the agent host specific (safe), but the pathogen has an adverse effect on the target weed population.

Prior to the introduction of *Puccinia chondrillina* Bubak and Syd. into Australia for the control of skeletonweed (*Chondrilla juncea* L.), the role of the pathogen in reducing skeletonweed populations was clearly demonstrated in its native habitat (53,87). Specially designed field experiments in the native range of the weed, in areas with climates analogous to those existing in the proposed areas of

Figure 1.2. Weed Targets and Exotic Pathogens Under Investigation in Classical Biological Weed Control Programs

Target Weed	Pathogen	Country of Intended Introduction	References
Asclepiadaceae			
Crptostegia grandiflora R.Br.	*Maravalia cryptostegiae* (Cumm.) Ono	Australia	82
Morrenia odorata (H. & A.) Lindl. (strangler vine, milkweed vine)	*Aecidium asclepiadinum* Speg.	U.S. (continental)	29
	Araujia mosaic virus	U.S. (continental)	28
	Puccinia araujae Lév.	U.S. (continental)	29
Asteraceae			
Centaurea diffusa Lam. (diffuse knapweed)	*Puccinia jaceae* Otth	Canada	54, 91, 93, 95
Centaurea maculosa Lam. (spotted knapweed)	*Puccinia centaureae* DC.	Canada	93
Centaurea solstitialis L. (yellow starthistle)	*Puccinia jaceae* Otth	U.S. (continental)	11, 12
Chondrilla juncea L. (skeletonweed)	*Erysiphe cichoracearum* DC. ex Marat	Australia	47, 49
	Leveillula taurica (Lév.) Arnaud f. sp. *chondrillae*	Australia	47, 49
Parthenium hysterophorus L. (parthenium)	*Puccinia abrupta* Diet. & Holw. var. *partheniicola* (Jackson) Parmelee	Australia	38, 63, 71
Senecia jacobaea L. (tansy ragwort)	*Puccinia expansa* Link (=*P. glomerata* Grev.)	Australia	4, 5
Xanthium spp. (cocklebur)	*Puccinia xanthii* Schw.	Australia	47, 49
Boraginaceae			
Heliotropium europaeum L. (common heliotrope)	*Cercospora* sp.	Australia	50
	Uromyces heliotropii Sredinski	Australia	50
Euphorbiaceae			
Euphorbia esula L. complex (leafy spurge)	*Melampsora euphorbiae* (Schub.) Cast.	U.S. (continental)	14
	Uromyces scutellatus (Pers.) Lév.	U.S. (continental)	34

(continued)

Figure 1.2.   (continued)

Target Weed	Pathogen	Country of Intended Introduction	References
Hydrocharitaceae			
Hydrilla verticillata (L.f.) Royle (hydrilla)	*Fusarium roseum* "Culmorum" (Link ex Fr.) Snyder & Hans.	U.S. (continental)	30
Liliaceae			
Asphodelus fistulosus L. (onion weed)	*Puccinia barbeyi* (Roum.) P. Magn.	Australia	7
Polygonaceae			
Emex spp.	*Cercopora tripolitana* Sacc. et Trott.	Australia	7
	Peronospora rumicis Cdla.	Australia	7
Rumex crispus L. (curled dock)	*Uromyces rumicis* (Schum.) Winter	U.S. (continental)	59, 78
Pontederiaceae			
Eichornia crassipes (Mart.) Solms (water hyacinth)	*Uredo eichorniae* Gonz.-Frag. & Cif.	U.S. (continental)	26, 27
Rosaceae			
Rubus fruticosus L. (European blackberry)	*Phragmidium violaceum* (Schultz) Winter	Australia	15–19

introduction, might provide the best estimates of potential effectiveness of selected plant pathogens as biocontrol agents (49). However, access and/or availability of resources in the native range of the target weed often limit the possibility of conducting meaningful field experiments. As a result, detailed efficacy evaluation of selected plant pathogens will of necessity most often be conducted under quarantined controlled laboratory conditions.

Research within controlled environmental facilities has certain limitations (e.g., plant morphology and anatomy of controlled-environment-grown plants may differ from field-grown plants), but the advantages of being able to evaluate effects of temperature, fertility, soil moisture, humidity, light, and plant (crop) competition on the potential of the biocontrol pathogen to affect the target weed adversely are expanded when compared to the usual single opportunity each year to conduct field experiments.

The success of most classical biological weed control insect agents has been due not to direct mortality of the target weed caused by the insect, but to the increased relative competitiveness of associated plant species or increased susceptibility of the stressed weed to adverse environmental conditions (42,44,57). Similarly *P. chondrillina* was more injurious to skeletonweed in mixtures when the population of the susceptible form was greater than that of the resistant form (24). Paul and Ayres (72) have demonstrated that rust infection reduced the impact of groundsel in lettuce without causing increased mortality in groundsel. The effect of the rust was expressed in decreased competitive ability of rust-infected groundsel and the impact of the rust was most pronounced at high population densities where resources were most limiting. Paul and Ayres (72) correctly suggested that efficacy evaluations of rust pathogens in biological weed control programs should be on the basis of crop yield rather than on that of injury or mortality of the weed host. As a consequence, future efforts in our classical biological weed control programs with plant pathogens would be strengthened by the design and implementation of suitable experiments to evaluate the potential efficacy of pathogens as classical biological weed control agents.

Host Specificity

In classical biological control the importance of specificity cannot be overemphasized. Rigorous host range testing is essential to ensure that a prospective exotic pathogen will not damage beneficial plants in the country of proposed introduction and neighboring countries (49,62,86,88). Extensive preintroduction investigations are conducted with the candidate biotic agent to ensure that the organisms will not attack or harm any desirable plant species. For North American projects these studies with plant pathogens are conducted in Europe or within specialized quarantine facilities at the Agriculture Canada Research Station in Regina, Saskatchewan; at the United States Department of Agriculture (USDA) quarantine facility at Fort Detrick in Frederick, Maryland; at the quarantine facility at the University of Florida, Gainesville; or within the containment facility of the Biopesticide Research Laboratory of Macdonald College of McGill University, Ste-Anne-de-Bellevue, Quebec.

All studies with exotic organisms in domestic facilities must be conducted under strict quarantine conditions within a government-approved facility. The quarantine laboratory is specifically designed to prevent the escape of plant pathogens of unconfirmed host range and virulence during scientific study (64,94). The Plant Protection and Quarantine section of the USDA Animal and Plant Health Inspection Service (APHIS) has the legal authority for issuance of federal permits for introduction, movement, and release of exotic organisms for weed control in the United States (25). In Canada, importation of exotic organisms is under the legal authority of the Plant Products and Quarantine Directorate of Agriculture Canada in Ottawa and the Plant Quarantine Section of the Commonwealth Department of Health is the regulating authority in Australia.

In North America a Working Group on Biological Control of Weeds (WGBCW) was established in 1957 to advise researchers and the U.S. and Canadian government authorities on importation, scientific study, and release of exotic weed control organisms (60). Guidelines have been published by this working group to assist in the planning of a biological weed control program involving the importation of foreign organisms (60).

Procedures for these host specificity studies have been reviewed by Zwölfer and Harris (100), Wapshere (85), and Watson (88). Most host specificity testing follows the centrifugal phylogenetic method proposed by Wapshere (85), which attempts to delimit the extent of the biocontrol agent's host range and includes the testing of other plant species "at risk." Plant species "at risk" that should be tested include species (1) related to the target weed; (2) not previously exposed to the biocontrol agent; (3) having limited information on their natural enemies; (4) having similar secondary compounds and/or morphological similarities with the target weeds; (5) attacked by related organisms; and (6) recorded as hosts of the candidate agent (77,86).

To test the host range of a candidate pathogen, infectivity tests are suggested to be conducted under controlled conditions that are optimum for infection and disease development (49,77,81). Results of pathogenicity tests obtained from controlled-environment chambers or greenhouses are often slightly different from the actual field host range. Host ranges of pathogens have been artificially expanded under artificial growth conditions, which suggests more realistic testing procedures (31) and care in the interpretation of results (88). The reported host range of some of the rust fungi being investigated as possible biocontrol agents for *Carduus* and *Centaurea* weed species in North America have been extended in greenhouse and controlled-environment-chamber experiments (74,91,93). These results do not conclusively demonstrate that the host range of the test pathogen is broader than reported, since possible explanations for the extended or broader host ranges of pathogens tested under growth chamber and greenhouse conditions include the following: (1) The pathogen and host have not come into effective contact in nature; (2) field observations have not been sufficiently intense; or (3) the results are artifacts of the testing program (88).

The potential risk of damage to desirable plant species from an introduced biological weed control agent has been the topic of numerous research papers and review articles (e.g., 1,6,45,52,57,86,88). It is clear that the major concern surrounding the biological control of weeds with plant pathogens from all sectors—including the scientist, the regulatory authorities, and the public—is host specificity. Classical biological control must be conducted in the public interest (45), and host plant specificity is fundamental to this concern.

Release and Establishment

After an exotic pathogen has been found to be a potentially safe and effective biocontrol agent, a detailed proposal is prepared by the researcher and submitted to the WGBCW. The WGBCW reviews and makes recommendations, whereas the approval or authorization for release of an exotic plant pathogen is made by federal (Animal Plant Health Inspection Service, APHIS or Agriculture Canada) and state or provincial Department of Agriculture officials.

Once approval for release has been obtained, liberation of the natural enemy occurs at preselected, relatively undisturbed sites where the target weed occurs at relatively high densities (39). Since climate exerts a profound effect on disease development, location of release sites and timing of releases to coincide with suitable environmental conditions are extremely important. Similarly, care must be taken to ensure that the agents released are free of their own natural enemies, such as hyperparasites (47,79,96). The release phase of the program requires the cooperation of extension agents, research scientists, and landowners in areas where the target weed is a problem. Infestations of the target weed need not be located and the release site must be maintained for some time with minimal disturbance. Many attempts to establish insects as biocontrol agents of weeds have failed because of disturbance and/or a slow rate of establishment of the agent. Likewise, this stage is critical in a program with exotic plant pathogens, but fortunately the few examples of intentionally released pathogens as biocontrol agents of weeds have generally resulted in relatively rapid establishment and dispersal of the agent.

Evaluation

Once the agent is established, the final step is to determine if the population of the released biocontrol agent is increasing and to determine if the agent is having an effect on the target weed population. Evaluation studies are not essential to the success of a project, but they certainly have bearing on the implementation and success of future biocontrol projects.

Qualitative and quantitative records of the status of the target weed population before and after introduction of a biocontrol agent are essential for the evaluation of biocontrol projects. Different methods can be used to monitor the density of both the target weed and the introduced pathogen. In addition to actual counts, life table analysis, computer simulation models, and photographic records can contribute to the accurate assessment of the success of a biocontrol project.

IV. Progress in Classical Biological Weed Control with Plant Pathogens

Classical biological weed control projects using plant pathogen are listed in Figures 1.1 and 1.2. Figure 1.1 lists pathogens that have been intentionally introduced into a particular country. Figure 1.2 lists pathogens that have been or are currently being evaluated as biocontrol agents of specific weeds. The progress of some of these projects is summarized later.

Chondrilla Rust

The most dramatic demonstration of the utility of an exotic plant pathogen for the biological control of a weed is the intentional introduction of *Puccinia chondrillina* into Australia and the successful control of rush skeletonweed (*Chondrilla juncea*). Skeletonweed is an apomictic, herbaceous perennial species native to Eurasia that was introduced into other regions of the world. It has become an invasive, troublesome weed, particularly in the southeastern cereal cropping regions of Australia and the grasslands of the western United States.

The steps involved in the *Chondrilla* project, from the initial surveys to the decline of the weed population, have been chronicled in detail by Hasan (46–49,51,52), Hasan and Wapshere (53), and others (77,80). Several strains of *Puccinia chondrillina*, a macrocyclic, autoecious rust, were collected in the Mediterranean region, and one from southern Italy was most virulent on the most abundant form (narrow-leaf) of skeletonweed in Australia. Extensive host specificity studies demonstrated the extreme host specialization of *P. chondrillina*, and the rust was imported and released in southern Australia in 1971. The rust was readily established, rapidly dispersed, and quickly reduced populations of the narrow-leaf form of skeletonweed. Because of the strict host specialization of the rust, populations of the other two forms of skeletonweed (intermediate- and broad-leaf) were unaffected and generally increased (24). Subsequently, surveys to discover other strains of *P. chondrillina* were conducted in the Mediterranean regions. Two additional strains have been released in Australia for the control of the intermediate-leaf form of skeletonweed, and attempts to locate a virulent strain against the broad-leaf form are continuing (48,51).

P. chondrillina was also intentionally introduced into the United States in October 1976 in Placer and El Dorado counties of California (80). This was the first intentional introduction of an exotic plant pathogen into North America for the biological weed control. Aspects of the biology and epidemiology of the strains released into the western United States have been examined (2,3,10,37), and recently Supkoff et al. (80) have demonstrated the significant impact of *P. chondrillina* in the reduction of skeletonweed populations in the western United States.

Blackberry Rust

The dramatic success of the *Chondrilla* rust is closely followed by that of the blackberry rust, *Phragmidium violaceum* (Schultz) Winter. Introduced European blackberries (*Rubus fruticosus* L. aggregate) have become excessively weedy in Chile, Australia, and other parts of the world. Interestingly, *P. violaceum* was first considered for the biological control of European blackberry in New Zealand by Cunningham (cited in Bruzzese and Hasan [16]) in 1927, but the project did not progress, because the spores of the fungus did not germinate. Subsequently, after limited host specificity testing, the rust was successfully introduced into southern Chile in 1973 (67,68) and has provided good control of *Rubus constrictus* Lefèvre et R. J. Mueller, but limited control of *R. ulmifolius* Schott (68).

This rust is also being evaluated for possible release into Australia for the control of European blackberry in the southeastern regions of that country. Extensive studies on the biology, potential efficacy, and host specificity have been completed (16–19), and specific strains of the fungus are being considered for introduction into Australia. This well-planned blackberry rust program in Australia received a setback in 1984, when it was discovered that the rust had been apparently intentionally, and illegally, introduced prior to official government approval. The rust spread rapidly, and four species of blackberry—*R. procerus* R. J. Mueller, *R. polyanthemus* Lindeb., *R. laciniatus* Willd., and some *R. ulmifolius* hybrids—were attacked in the field (15). Laboratory inoculations showed that *R. ulmifolius* and *R. vestitus* Weihe & Nees were susceptible to the introduced strain, but *R. cissburiensis* Barton & Riddelsd., *R. rosaceus* Weihe & Nees, and other *R. ulmifolius* hybrids were very resistant (15). The illegally introduced strain (or strains) appears to be less virulent on certain weedy *Rubus* species than the pool of isolates used in the host specificity testing (17). Fortunately, the illegally introduced strain was not virulent on desirable *Rubus* species, and additional strains of *P. violaceum* are being evaluated for importation into Australia (52).

Pamakani Leaf Spot

Hamakua pamakani, or mistflower [*Ageratina riparia* (Regel) K. & R.], was introduced into Hawaii in the 1920s as an ornamental, but has since become a major weed in the cool, moist forest floors and upcountry grazing lands of the Hawaiian islands (83). The white smut pathogen *Entyloma ageratinae* sp. nov. (8) was imported from Jamaica, demonstrated to be host specific, and released at three sites on Oahu on November 4, 1975 (83,84). This project represents the first effort in Hawaii to use exotic plant pathogens for biological weed control. The pathogen provided more than 95% control of the weed in less than 1 year after inoculation in areas with optimum temperatures (18°–20°C) and high rainfall; 95% control after 3–4 years in areas with temperatures less than or greater than optimum, but with high rainfall; and less than 80% control after 8 years in areas with low rainfall (83).

Carduus Rust

Several introduced *Carduus* species, including *C. nutans* L. (musk or nodding thistle), are aggressive introduced weeds of pastures and rangelands in the United States, Canada, and other parts of the world. There has been an active biological control program against nodding thistle in North America, with the seed-head weevil *Rhinocyllus conicus* Froelich providing excellent control in many areas. The insect, however, does not provide adequate control in many areas. The insect, however, does not provide adequate control in all habitats and under all management systems; therefore, additional biocontrol agents were considered necessary. Among those additional prospective agents is the Eurasian rust *Puccinia carduorum* Jacky.

The *Carduus* rust project began in 1978 with the collection of *P. carduorum* isolates in Bulgaria, Romania, and Turkey by R. G. Emge. Studies on the biology, host specificity, and effect on the target weed were conducted within the USDA Plant Pathogen Containment Facility at Frederick, Maryland, with some collaborative projects conducted within the quarantine facility at Macdonald College, McGill University in Ste-Anne-de-Bellevue, Quebec. The pathogen is virulent and aggressive on nodding thistle, causing reduction in biomass under controlled conditions (73), but host specificity studies demonstrated that other species, including globe artichoke (*Cynara scolymus* L.) and some *Cirsium* species within the Carduinae subtribe of the Cynareae tribe, were susceptible to the rust (74). Further evaluation of the rust on artichoke (13) and on native *Cirsium* species (W. L. Bruckart, personal communication) demonstrated that these species are very to moderately resistant, and plant reaction ratings declined when assessed after reinoculation with the rust. Although the rust was able to infect globe artichoke and some native *Cirsium* species under greenhouse conditions, it was suggested that these species were not at risk if *P. carduorum* was introduced into the United States to control nodding thistle. Proposals were submitted to the WGBCW, and close collaboration among USDA, APHIS, Virginia Department of Agriculture, and Virginia Polytechnic Institute culminated in the approval and granting of permits and the actual release of *P. carduorum* in a contained field plot on October 29, 1987, in Virginia. The rust has become established and is continuing to be evaluated in 1989 within the restricted field plot area prior to consideration for possible widespread release of the rust on nodding thistle populations in the United States and Canada (W. L. Bruckart, personal communication).

Centaurea Projects

Four *Centaurea* species, *C. diffusa* Lam. (diffuse knapweed), *C. maculosa* Lam. (spotted knapweed), *C. solstitatilis* L. (yellow starthistle), and *C. repens* L. syn. *Acroptilon repens* (L.) DC. (Russian knapweed) are troublesome introduced weeds of pastures and rangelands in the United States and Canada. There are active biological control programs directed toward each of these four weeds (93). The progress of the nematode *Subanguina picridis* (Kirj.) Brzeski (89,90) for the control of russian knapweed is discussed in detail in the following chapter by P. E. Parker. In addition to the nematode, rust pathogens, including *Puccinia jaceae* Otth and *P. centaureae* DC. are being actively evaluated as potential biocontrol agents of *Centaurea* species in North America (11,12,54,91,93,95). The progress of these projects is hampered to some degree by the taxonomic complexities of the autoecious *Puccinia* species associated with *Centaurea*. Evaluation of these rusts in controlled-environment experiments within containment facilities has demonstrated the rusts to be virulent and aggressive pathogens of their respective hosts, but they are also pathogenic on related plant species, including safflower (*Carthamus tinctorius* L.) (54,91,93). Within controlled-environment conditions, cotyledons of safflower cultivars inoculated with *P. jaceae* or *P. centaureae* were fully susceptible, but the first true leaves were resistant and upper leaves were immune to infection (12,91,93). Field experiments with the European knapweed rust *P. jaceae* conducted in the Mediterranean provided similar results, with safflower cotyledon being susceptible and upper leaves being immune, but rust infections were observed on flower parts of all safflower cultivars tested (54). Although the effect of the rust on plant height and seed production appeared minimal, Hasan et al. (54) do not favor release of *P. jaceae* in North America. Interestingly, *P. jaceae* has recently (fall 1988) been discovered on diffuse knapweed in southern British Columbia (K. Mortensen, personal communication). The origin and significance of this discovery are not known. Safflower does not appear to be at risk from the *Centaurea* rusts, but many questions are still unanswered. Proposals are being prepared for submission to the WGBCW for consideration of possible restricted field trials with one or more of these rusts in the future.

V. Limitations and Future Prospects

The major perceived constraint to the classical biological weed control with plant pathogens is the strict host specificity requirement of imported exotic plant pathogens. The specificity requirement has generally limited research to obligate such parasites as rust fungi, which are often thought of as causing relatively limited damage to their host (75). However, examples of successful classical biocontrol with plant pathogens of skeletonweed, blackberry, and mistflower certainly demonstrate that introduced obligate pathogens can provide excellent long-term control of introduced troublesome weeds (8,32,67). Interpretation of the expanded host ranges or European and Mediterranean rust fungi *P. carduorum, P. jaceae,* and *P. centaureae* observed under controlled environment conditions (11,13,54,74,88,91,93) further challenges researchers and government authorities in rendering decisions concerning possible introduction of these pathogens.

Progress in classical biological weed control with plant pathogen has been relatively slow. Some of the reasons for this slow progress include limited funding, limited personnel, less incentive when compared to the bioherbicide approach, higher infrastructure requirements (quarantine laboratories, foreign collaboration and exploration, etc.), and host plant specificity concerns. Despite these difficulties, classical biological control with plant pathogens along with the bioherbicide approach will become increasingly more important as the real and perceived problems associated with chemical herbicides intensify. Classical biological weed control should not be considered to be limited to pasture or rangeland habitats, but it also has application in intensive, cultivated cropping systems. Biological control is normally thought of as an alternative to the other methods of control when these other methods have failed to control a particularly troublesome, dominant weed. However, excellent opportunities exist to use classical biocontrol as a complementary approach and to integrate biological control agents with other methods of control (92). Biocontrol workers are challenged to understand the mechanisms and systems involved and to develop acceptable, reliable, and efficacious biological weed control strategies.

Literature Cited

1. Adams, E. B. 1988. Fungi in classical biocontrol of weeds. Pages 111–124, *in* M. N. Burge (ed.), Fungi in Biological Control Systems. Manchester University Press, New York.

2. Adams, E. B., and R. F. Line. 1984. Biology of *Puccinia chondrillina* in Washington. Phytopathology 74:742–745.

3. Adams, E. B., and R. F. Line. 1984. Epidemiology and host morphology in the parasitism of rush skeleton weed by *Puccinia chondrillina*. Phytopathology 74:745–748.

4. Alber, G., C. Défago, H. Kern, and L. Sedlar. 1986. Host range of *Puccinia expansa* Link (= P. glomerata Grev)., a possible fungal biocontrol agent against *Senecio* weeds. Weed Res. 26:69–74.

5. Alber, G., G. Défago, L. Sedlar, and H. Kern. 1985. Damage to *Senecio jacobaea* by the rust fungus *Puccinia expansa*. Pages 587–592, in: E. S. Delfosse (ed.), Proc. VI Int. Symp. Biol. Contr. Weeds. Vancouver, Aug. 19–25, 1984. Agriculture Canada, Ottawa.

6. Andres, L. A., and R. D. Goeden. 1971. The biological control of weeds by introduced natural enemies. Pages 143–164, in: C. B. Huffaker (ed.), Biological Control. Plenum Press, New York.

7. Anonymous. 1988. Biennial Report 1985–87. Division of Entomology, CSIRO, Canberra.

8. Barreto, R. W., and H. C. Evans. 1988. Taxonomy of a fungus introduced into Hawaii for biological control of *Ageratina riparia* (Eupatorieae; Compositae), with observations on related weed pathogens. Trans. Br. Mycol. Soc. 91:81–97.

9. Barrett, S. C. H. 1982. Genetic variation in weeds. Pages 73–98, in: R. Charudattan and H. L. Walker (eds.), Biological Control of Weeds with Plant Pathogens. Wiley, New York.

10. Blanchette, B. L., and G. A. Lee. 1981. The influence of environmental factors on infection of rush skeleton weed (*Chondrilla juncea*) by *Puccinia chondrillina*. Weed Sci. 29:364–367.

11. Bruckart, W. L. 1989. Host range determination of *Puccinia jaceae* from yellow starthistle. Plant Dis. 73:155–160.

12. Bruckart, W. L., and W. M. Dowler. 1986. Evaluation of exotic rust fungi in the United States for classical biological control of weeds. Weed Sci. 34 (Suppl. 1): 11–14.

13. Bruckart, W. L., D. J. Politis, and E. M. Sutker. 1985. Susceptibility of *Cynara scolymus* (artichoke) to *Puccinia carduorum* observed under greenhouse conditions. Pages 603–607, in:

E. S. Delfosse (ed.), Proc. VI Int. Symp. Biol. Contr. Weeds. Vancouver, Aug. 19–25, 1984. Agriculture Canada, Ottawa.

14. Bruckart, W. L., S. K. Turner, E. M. Sutker, R. Vonmoos, L. Sedlar, and G. Défago. 1986. Relative virulence of *Melampsora euphorbiae* from Central Europe toward North America and European spurges. Plant Dis. 70:847–850.

15. Bruzzese, E., and R. P. Field. 1985. Occurrence and spread of *Phragmidium violaceum* on blackberry (*Rubus fruticosus*) in Victoria, Australia. Pages 609–612, *in* E. S. Delfosse (ed.), Proc. VI Int. Symp. Biol. Contr. Weeds. Vancouver, Aug. 19–25, 1984. Agriculture Canada, Ottawa.

16. Bruzzese, E., and S. Hasan. 1986. Host specificity of the rust *Phragmidium violaceum,* a potential biological control agent of European blackberry. Ann. Appl. Biol. 108:585–596.

17. Bruzzese, E., and S. Hasan. 1986. Infection of Australian and New Zealand *Rubus* subgenera *Dalebarda* and *Lampobatus* by the European blackberry rust fungus *Phragmidium violaceum.* Plant Pathol. 35:413–416.

18. Bruzzese, E., and S. Hasan. 1986. The collection and selection in Europe of isolates of *Phragmidium violaceum* (Uredinales) pathogenic to species of European blackberry naturalized in Australia. Ann. Appl. Biol. 108:527–533.

19. Bruzzese, E., and S. Hasan. 1987. Infection of blackberry cultivars by the European blackberry rust fungus, *Phragmidium violaceum.* J. Horticult. Sci. 64:475–479.

20. Burdon, J. J. 1985. Pathogens and genetic structure of plant populations. Pages 313–325, *in*: J. White (ed.), Studies on Plant Demography: a Festschrift for John L. Harper. Academic Press, London.

21. Burdon, J. J., and D. R. Marshall. 1981. Biological control and the reproductive mode of weeds. J. Appl. Ecol. 18:649–658.

22. Burdon, J. J., and G. A. Chilvers. 1974. Fungal and insect parasites contributing to niche differentiation in mixed species stands of Eucalypt saplings. Aust. J. Bot. 22:103–114.

23. Burdon, J. J., and R. H. Groves. 1986. Ecological aspects of plant-pathogen interactions. Pages 37–40, *in*: Proc. Workshop on Potential for Mycoherbicides in Australia, May 1986, Agric. Res & Vet. Centre, Orange.

24. Burdon, J. J., R. H. Groves, P. E. Kaye, and S. S. Speer, 1984. Competition in mixtures of susceptible and resistant genotypes of *Chondrilla juncea* differentially infected with rust. Oecologia (Berlin) 64:199–203.

25. Charudattan, R. 1982. Regulation of microbial weed control agents. Pages 175–188, *in*: R. Charudattan and H. L. Walker (eds.), Biological Control of Weeds with Plant Pathogens. Wiley, New York.

26. Charudattan, R., and K. E. Conway. 1975. Comparison of *Uredo eichhorniae,* the waterhyacinth rust, and *Uromyces pontederiae.* Mycologia 67:653–657.

27. Charudattan, R., D. E. McKinney, H. A. Cordo, and A. Silveira-Guido. 1978. *Uredo eichhorniae,* a potential biocontrol agent for waterhyacinth. Pages 210–213, *in*: T. E. Freeman (ed.), Proc. IV Int. Symp. Biol. Contr. Weeds. Gainesville, Aug. 30–Sept. 2, 1976. University of Florida, Gainesville.

28. Charudattan, R., F. W. Zettler, H. A. Cordo, and R. G. Christie. 1980. Partial characterization of a polyvirus infecting milkweed vine, *Morrenia odorata.* Phytopathology 70:909–913.

29. Charudattan, R., H. A. Cordo, A. Silveira-Guido, and F. W. Zettler. 1978. Obligate pathogens of the milkweed vine *Morrenia odorata,* as biological control agents (Abstract). Page 241, *in*:

T. E. Freeman (ed.), Proc. IV Int. Symp. Biol. Contr. Weeds. Gainesville, Aug. 30–Sept. 2, 1976. University of Florida, Gainesville.

30. Charudattan, R., T. E. Freeman, R. E. Cullen, and F. M. Hofmeister. 1981. Evaluation of *Fusarium roseum* "Culmorum" as a biological control for *Hydrilla verticillata*: safety. Pages 307–323, *in*: E. S. Delfosse, (ed.), Proc. V Int. Symp. Biol. Contr. Weeds. Brisbane, July 22–27, 1980. CSIRO, Melbourne.

31. Cother, E. J. 1975. *Phytophthora dreschsleri*: pathogenicity testing and determination of effective host range. Aust. J. Bot. 23:87–94.

32. Cullen, J. M. 1978. Evaluating the success of the programme for the biological control of *Chondrilla juncea* L. Pages 117–121, *in*: T. E. Freeman (ed.), Proc. IV Int. Symp. Int. Symp. Biol. Contr. Weeds. Gainesville, Aug. 30–Sept. 2, 1976, University of Florida, Gainesville.

33. Cullen, J. M., P. F. Kable, and M. Catt. 1973. Epidemic spread of a rust imported for biological control. Nature 244:462–464.

34. Défago, G. H. Kern, and L. Sedlar. 1985. Potential control of weedy spurges by the rust *Uromyces scutellatus* s.l. Weed Sci. 33:857–860.

35. Dinoor, A., and N. Eshed. 1984. The role and importance of pathogens in natural plant communities. Ann. Rev. Phytopathol. 22:443–466.

36. Ehler, L. E., and L. A. Andres. 1983. Biological control: exotic natural enemies to control exotic pests. Pages 395–418, *in*: C. L. Wilson and C. L. Graham (eds.), Exotic Plant Pests and North American Agriculture. Academic Press, New York.

37. Emge, R. G., J. S. Melching, and C. H. Kingsolver. 1981. Epidemiology of *Puccinia chondrillina*, a rust pathogen for the biological control of rush skeleton weed in the United States. Phytopathology 71:839–843.

38. Evans, H. C. 1986. The life cycle of *Puccinia abrupta* var. *partheniicola*, a potential biological control agent of *Parthenium hysterophorus*. Trans. Br. Mycol. Soc. 88:105–111.

39. Goeden, R. D. 1977. Biological control of weeds. Pages 43–47, in: B. Truelove (ed.), Research Methods in Weed Science. Southern Weed Science Society of America, Auburn, AL.

40. Goeden, R. D. 1983. Critique and revision of Harris' scoring system for selection of insect agents in biological control of weeds. Prot. Ecol. 5:287–301.

41. Harris, P. 1971. Current approaches to biological control of weeds. Pages 67–76, *in*: Biological Control Programmes Against Insects and Weeds in Canada 1959–1968. Commonw. Inst. Biol. Contr., Tech. Comm. No. 4.

42. Harris, P. 1973. The selection of effective agents for the biological control of weeds. Can. Ent. 105:1495–1503.

43. Harris, P. 1986. Biological control of weeds. Fortsch. Zool. 32:123–138.

44. Harris, P. 1986. Biological control of weeds. Pages 123–138, *in*: J. M. Franz (ed.), Biological Plant and Health Protection, Biological Control of Plant Pests and Vectors of Human and Animal Diseases. Gustav Fischer Verlag, New York.

45. Harris, P. 1988. Environmental impact of weed-control insects. BioScience 38:542–548.

46. Hasan, S. 1972. Specificity and host specialization of *Puccinia chondrillina*. Ann. Appl. Biol. 72:257–263.

47. Hasan, S. 1980. Plant pathogens and biological control of weeds. Rev. Plant Pathol. 59:349–356.

48. Hasan, S. 1981. A new strain of the rust fungus *Puccinia chondrillina* for biological control of skeleton weed in Australia. Ann. Appl. Biol 99:119–124.

49. Hasan, S. 1983. Biological control of weeds with plant pathogens—status and prospects. Proc. 10th Int. Congress Plant Prot. 2:759–776.

50. Hasan, S. 1985. Prospects for biological control of *Heliotropium europeaeum* by fungal pathogens. Pages 617–623, in: E. S. Delfosse, (ed.), Proc. VI Int. Symp. Biol. Contr. Weeds. Vancouver, Aug. 19–25, 1984, Agriculture Canada, Ottawa.

51. Hasan, S. 1985. Search in Greece and Turkey for *Puccinia chondrillina* strains suitable to Australian forms of skeleton weed. Pages 625–632, in: E. S. Delfosse (ed.), Proc. VI Int. Symp. Biol. Contr. Weeds. Vancouver, Aug. 19–25, 1984. Agriculture Canada, Ottawa.

52. Hasan, S. 1988. Biocontrol of weeds with microbes. Pages 129–151, in K. G. Mukerji and K. L. Garg (eds.), Biological Control of Plant Disease, Vol. 1. CRC Press, Boca Raton, FL.

53. Hasan, S., and A. J. Wapshere. 1973. The biology of *Puccinia chondrillina* a potential biological control agent of skeletonweed. Ann. App. Biol. 74:325–332.

54. Hasan, S., P. Chaboudey, and K. Mortensen. 1989. Field experiment with the European knapweed rust (*Puccinia jaceae*) on safflower, sweet sultan and bachelor's button, in: E. S. Delfosse (ed.), Proc. VII Int. Symp. Biol. Contr. Weeds. Rome, March 6–11, 1988, 1st Sper. Patol. Veg. (MPAF)P (in press).

55. Hokkanen, H. M. T. 1985. Success in classical biological control. CRC Crit. Rev. Plant Sci. 3:35–72.

56. Holcomb, G. E. 1982. Constraints on disease development. Pages 61–71, in: R. Charudattan and H. L. Walker (eds.), Biological Control of Weeds with Plant Pathogens. Wiley, New York.

57. Huffaker, C. B. 1957. Fundamentals of biological control of weeds. Hilgardia 27:101–157.

58. Huffaker, C. B., D. L. Dahlsten, D. H. Janzen, and G. G. Kennedy. 1984. Insect influences in the regulation of plant populations and communities. Pages 659–691, in: C. B. Huffaker and R. L. Robb (eds.), Ecological Entomology. Wiley, New York.

59. Inman, R. E. 1971. A preliminary evaluation of *Rumex* rust as a biological control agent for curly dock. Phytopathology 61:102–107.

60. Klingman, D. L., and J. R. Coulson. 1982. Guidelines for introducing foreign organisms into the U.S. for biological control of weeds. Plant Dis. 66:1205–1209.

61. Kovalev, O. V. 1989. New factors of effectiveness of phytophages: a solitary population wave (SPW) and succession process, in: E. S. Delfosse (ed.), Proc. VII Int. Symp. Biol. Contr. Weeds. Rome, March 6–11, 1988, 1st Sper. Patol. Veg. (MPAF)P (in press).

62. Leonard, K. J. 1982. The benefits and potential hazards of genetic heterogeneity in plant pathogens. Pages 99–112, in: R. Charudattan and H. L. Walker (eds.), Biological Control of Weeds with Plant Pathogens. Wiley, New York.

63. McClay, A. S. 1985. Biocontrol agents for *Parthenium hysterophorus* from Mexico. Pages 771–778, in: E. S. Delfosse (ed.), Proc. VI Int. Symp. Biol. Contr. Weeds. Vancouver, Aug. 19–25, 1984. Agriculture Canada, Ottawa.

64. Melching, J. S., K. R. Bromfield, and C. H. Kingsolver. 1983. The plant pathogen containment facility at Frederick, Maryland. Plant Dis. 67:717–722.

65. Miles, J. W. and J. M. Lenné. 1984. Genetic variation within a natural *Stylosanthes guianensis, Colletotrichum gloeosporioides* host–pathogen population. Aust. J. Agric. Res. 35:211–218.

66. Morris, M. J. 1989. The use of plant pathogens as control agents for South Africn weeds, in: E. S. Delfosse (ed.), Proc. VII Int. Symp. Biol. Contr. Weeds. Rome, March 6–11, 1988, 1st Sper. Patol Veg. (MPAF)P (in press).

67. Oehrens, E. 1977. Biological control of the blackberry through the introduction of rust, *Phragmidium violaceum* in Chile. FAO Plant Prot. Bull. 25:26–28.

68. Oehrens, E. B., and S. M. Gonzáles. 1974. Induccion de *Phragmidium violaceum* (Schulz) Winter como factor de control biologico de zarzamora (*Rubus constrictus* Lef. et *M. y R. ulmifolius* Schott.). Agro Sur 2:30–33.

69. Oehrens, E. B., and S. M. Gonzáles. 1975. Introduccion de *Uromyces galegae* (Opiz) Saccardo como factor de control biologico de galega (*Galega officinalis* L.) Agro Sur 3:87–91.

70. Ohr, H. D. 1974. Plant disease impacts on weeds in the natural ecosystem. Proc. Am. Phytopathol. Soc. 1:181–184.

71. Parker, A. 1989. Biological control of parthenium weed using two rust fungi, *in*: E. S. Delfosse (ed.), Proc. VII Int. Symp. Biol. Contr. Weeds. Rome, March 6–11, 1988, 1st Sper. Patol. Veg. (MPAF)P (in press).

72. Paul, N. D., and P. G. Ayres. 1987. Effects of rust infection of *Senecio vulgaris* on competition with lettuce. Weed Res. 37:431–441.

73. Politis, D. J., and W. L. Bruckart, 1986. Infection of musk thistle by *Puccinia carduorum* influenced by conditions of dew and plant age. Plant Dis. 70:288–290.

74. Politis, D. J., A. K. Watson, and W. L. Bruckart. 1984. Susceptibility of musk thistle and related composites to *Puccinia carduorum*. Phytopathology 74:687–691.

75. Quimby, P. C., Jr. 1982. Impact of diseases on plant populations. Pages 47–60, *in*: R. Charudattan and H. L. Walker (eds.), Biological Control of Weeds with Plant Pathogens. Wiley, New York.

76. Rausher, M. D., and P. Feeny. 1980. Herbivory, plant density, and plant reproductive success: the effect of *Battus philenor* on *Uristolochia reticulata*. Ecology 61:905–917.

77. Schroeder, D. 1983. Biological control of weeds. Pages 41–78, *in*: W. W. Fletcher (ed.), Recent Advances in Weed Research, Commonwealth Agricultural Bureaux, Slough.

78. Schubiger, F. X., G. Défago, L. Sedlar, and H. Kern. 1985. Host range of the haplontic phase of *Uromyces rumicis*. Pages 653–659, *in*: E. S. Delfosse (ed.), Proc. VI Int. Symp. Biol. Contr. Weeds. Vancouver, Aug. 19–25, 1984. Agriculture Canada, Ottawa.

79. Shrum, R. D. 1982. Creating epiphytotics. Pages 113–136, *in*: R. Charudattan and H. L. Walker (eds.), Biological Control of Weeds with Plant Pathogens. Wiley, New York.

80. Supkoff, D. M., D. B. Joley, and J. J. Marois. 1988. Effect of introduced biological control organisms on the density of *Chondrilla juncea* in California. J. Appl. Ecol. 25:1089–1095.

81. Templeton, G. E., and E. E. Trujillo. 1981. The use of plant pathogens in the biological control of weeds. Pages 345–350, *in*: D. Pimental (ed.), CRC Handbook of Pest Management in Agriculture, Vol. 2. CRC Press, Boca Raton, FL.

82. Thomley, A. J. 1989. The biological control program for *Cryptostegia grandiflora* in Australia, *in*: E. S. Delfosse (ed.), Proc. VII Int. Symp. Biol. Contr. Weeds. Rome, March 6–11, 1988, 1st Sper. Patol. Veg. (MPAF)P (in press).

83. Trujillo, E. E. 1985. Biological control of hamakua pa-makani with *Cercosporella* sp. in Hawaii. Pages 661–671, *in*: E. S. Delfosse (ed.), Proc. VI Int. Symp. Biol. Contr. Weeds. Vancouver, Aug. 19–25, 1984. Agriculture Canada, Ottawa.

84. Trujillo, E. E., M. Aragaki, and R. A. Shoemaker. 1988. Infection, disease development, and axenic cultures of *Entyloma compositarum,* the cause of hamakua pamakani blight in Hawaii. Plant Dis. 72:355–357.

85. Wapshere, A. J. 1974. A strategy for evaluating the safety of organisms for biological weed control. Ann. Appl. Biol. 77:201–211.

86. Wapshere, A. J. 1982. Biological control of weeds. Pages 47–56, *in*: W. Holzner and N. Numata (eds.), Biology and Ecology of Weeds. Junk Publisher, The Hague.

87. Wapshere, A. J. 1985. Effectiveness of biological control agents for weeds: present quandaries. Agric. Ecol. Envir. 13:261–280.

88. Watson, A. K. 1985. Host specificity of plant pathogens in biological weed control. Pages 577–586, *in*: E. S. Delfosse (ed.), Proc. VI Int. Symp. Biol. Contr. Weeds. Vancouver, Aug. 19–25, 1984. Agriculture Canada, Ottawa.

89. Watson, A. K. 1986. Biology of *Subanguina picridis,* a potential biological control agent of Russian knapweed. J. Nematol. 18:149–154.

90. Watson, A. K. 1986. Host range of, and plant reaction to, *Subanguina picridis*. J. Nematol. 18:112–120.

91. Watson, A. K., and I. Alkhoury. 1981. Response of safflower cultivars to *Puccinia jaceae* collected from diffuse knapweed in eastern Europe. Pages 301–305, *in*: E. S. Delfosse (ed.), Proc. II Int. Symp. Biol. Cont. Weeds, Brisbane, July 22–29, 1980. CSIRO, Melbourne.

92. Watson, A. K., and L. A. Wymore. 1989. Biological control, a component of integrated weed management, *in* E. S. Delfosse (ed.), Proc. VII Int. Symp. Biol. Contr. Weeds. Rome, March 6–11, 1988, 1st Sper. Patol. Veg. (MPAF)P (in press).

93. Watson, A. K., and M. Clément. 1986. Evaluation of rust fungi as biological control agents of weedy *Centaurea* in North America. Weed Sci. 34(Suppl. 1):7–10.

94. Watson, A. K., and W. E. Sackston. 1985. Plant pathogen containment (quarantine) facility at Macdonald College. Can. J. Plant Pathol. 7:177–180.

95. Watson, A. K., D. Schroeder, and I. Alkhoury. 1981. Collection of *Puccinia* species from diffuse knapweed in eastern Europe. Can. J. Plant Pathol. 3:6–8.

96. Wilson, C. L. 1969. The use of plant pathogens in weed control. Ann. Rev. Phytopathol. 7:411–434.

97. Wilson, F. 1950. The biological control of weeds. New Biol. 8:51–74.

98. Zwölfer, H. 1968. Some aspects of biological weed control in Europe and North America. Proc. 9th Br. Weed Contr. Conf., pp. 1147–1156.

99. Zwölfer, H. 1973. Possibilities and limitations in biological control of weeds. OEPP/EPPO Bull. 3:19–30.

100. Zwölfer, H., and P. Harris. 1971. Host specificity determination of insects for biological control of weeds. Ann. Rev. Entomol. 16:159–178.

2

The Mycoherbicide Approach with Plant Pathogens

R. Charudattan

It is nearly a decade since the first of fungal herbicides, DeVine and COLLEGO, were introduced for commercial use, establishing mycoherbicides as a practical means of weed management. In 1982 Templeton (156) reviewed the status of weed control with plant pathogens, providing a comprehensive list of mycoherbicide candidates (Figure 2.1). Since then numerous other reviews have appeared detailing the mycoherbicide tactic and mycoherbicide agents (8,14,40,43,46,78, 81,84,85,123,133,154,157–164,182), attesting to the intense scientific and commercial interest in this field. What has been accomplished during this time? What have we gained in the conceptual and practical fronts? What does the future hold for mycoherbicides? These are the questions I attempt to discuss in this chapter and to show that the field has indeed grown significantly.

The Mycoherbicide Tactic

The concept of mycoherbicide was first introduced by Daniel et al. (69), who demonstrated that an endemic (i.e., native) pathogen might be rendered completely destructive to its weed host by applying a massive dose of inoculum at a particularly susceptible stage of weed growth. The application of an inundative dose of inoculum and its proper timing would shorten the lag period for inoculum buildup and pathogen distribution, essential for natural epiphytotics. This would also make it possible to employ any natural or artificial conditions favorable for disease development (69). To be successful in this approach, it must be possible to produce abundant and durable inoculum in artificial culture, the pathogen must be genetically stable and specific to the target weed, and it must be possible to infect and kill the weed in environments of reasonably wide latitude (69). The fungus would be applied annually shortly after the weed's emergence when conditions for disease were favorable. The spore inoculum for this purpose was

I thank Dr. George Templeton and Dr. Gerald Van Dyke for critically reviewing this chapter and offering helpful suggestions for improvement.

Figure 2.1. Status of Biological Weed Control Projects

	In 1982*	In 1983†
Number of weed species targeted for control by fungal pathogens	45	NC
Of the above, number targeted for control by rust fungi only	10	NC
Number of weeds targeted for mycoherbicidal control	35	67
Number of fungal taxa studied as potential mycoherbicidal agents	54	107

*Based on data summarized by Templeton (156).
†Summary from Figure 2.2. NC = not considered for this discussion.

to be raised in artificial media, harvested, prepared in a manner to withstand storage and handling, and applied like a chemical herbicide. Thus, the in vitro culturing of the pathogen to obtain large quantities of inoculum and the inundative application of inoculum to achieve rapid epidemic buildup and a high level of disease are two distinctive aspects of the mycoherbicide concept (43).

Since this initial definition of the concept, the term *mycoherbicide* has been redefined as "plant pathogenic fungi developed and used in the inundative strategy to control weeds in the way chemical herbicides are used" (154), or as "living products that control specific weeds in agriculture as effectively as chemicals" (161). The use of the pathogen in a "product form" and an "application technique similar to the chemical tactic" are salient features distinguishing the mycoherbicides from classical agents. These features require that mycoherbicides be treated as pesticides and therefore be subject to regulations governing pesticides rather than to those concerning classical introductions (37,154). The regulatory requirements also dictate that the product conform to certain performance and safety standards and therefore be standardized at least with respect to each inoculum batch (37).

The type of parasitism exhibited by a fungus affects its ability to serve as a mycoherbicide (164). Since obligate parasites are typically less damaging to their hosts than facultative parasites or facultative saprophytes, they have less potential for use as mycoherbicides in situations requiring rapid and complete control (i.e., weed kill). The technical difficulties in producing obligate parasites are also a deterrent to their use, but advances in ex planta culturing of obligate parasites may create new opportunities in the future for mycoherbicidal use of obligate fungi.

Whether a pathogen is native ("endemic" sensu Daniel et al. [69]) or exotic is not the point that distinguishes mycoherbicides from classical agents. An exotic facultative parasite could very well be developed as a mycoherbicide if it possesses the epidemiological attributes exploitable for mycoherbicidal purposes (39,43). However, a pathogen that normally incites an endemic disease (i.e., a disease that is more or less constantly present from year to year in a moderate to severe form [181]), whether it is native to the area of intended use or not, is an ideal candidate for a mycoherbicide. Because endemic diseases evolve in regions of host–parasite coadaptation, it is logical to find them in native or naturalized ranges of weeds. Once developed as mycoherbicides, these pathogens may prove to be effective outside the region of coevolution, provided a favorable environment and host susceptibility are prevalent (40,43).

Clearly, a case could be made for searching for mycoherbicidal candidates in the adventive (=exotic) ranges of weeds, not merely in the native homes. This point and the related aspect of surveying congeneric and conspecific species of weeds should not be overlooked. In fact, it has been argued that new host–pathogen associations (namely, associations of newly encountered host and pathogen), as constrasted with long-evolved associations, may in fact yield a greater number of virulent and highly effective biocontrol agents than old associations that typically form the basis of classical agents (88). The lack of evolved homeostasis in new host–pathogen associations may explain the greater destructive ability of the pathogen and the consequent higher probability of biocontrol success with such a pathogen. Many of the facultative parasites of weeds under consideration for mycoherbicidal use may represent new host–pathogen assocations. For example, the same facultative parasites are found on water hyacinth (*Eichhornia crassipes*) in several widely separated regions of the world (38), and this pattern of distribution is not explainable on the basis of parasite comigration with the weed. For if comigration had happened, we should expect to find the obligate parasite *Uredo eichhorniae* beyond the original range of water hyacinth along with the facultatives. However, this is not the case (45); the rust is found only in the South American home of water hyacinth, whereas many facultative parasites, such as *Acremonium zonatum*, *Cercospora piaropi*, and *Rhizoctonia solani*, occur on this host in several continents. Also, certain facultative parasites such as *Alternaria eichhorniae* and *Myrothecium roridum* have been found in the adventive range but not in the weed's center of origin (37).

Augmentation Tactic and Necrogenic Microorganisms as Mycoherbicides

Since Daniel et al. (69) described the mycoherbicide tactic, two related approaches to weed control have since been attempted, In the first, *Puccinia canaliculata*, a rust fungus native to the United States has been used effectively for controlling yellow nut sedge, *Cyperus esculentus* (127,128). Uredospore inoculum of this fungus is gathered from infected plants, stockpiled, and applied with suitable carriers. No in vitro mass production of inoculum is possible because of the obligate nature of the parasite. Although it is possible to stockpile large amounts of inoculum for applications over several hectares, there will be obvious limitations to the amount of inoculum that can be gathered and applied. In practice, only a few grams of inoculum per hectare are needed to initiate a high level of disease sustainable throughout the growing season—similar to the inoculative strategy (Phatak, personal communication). Epidemiologically, this rust fungus is typically host–density dependent for multiplication and dissemination. It is normally present at an endemic level in various parts of the United States. By applying a surplus of inoculum over localized areas, an augmentive effect is produced, comparable to the augmentive strategy with insect biocontrol agents. Therefore, this approach should rightly be termed the augmentation tactic rather than the mycoherbicide tactic (39,43). The augmentation strategy is similar to the mycoherbicide strategy in that in both cases human intervention is involved in inoculum distribution. But in the former, inoculum is neither mass produced nor applied as an inundative dose over a large proportion of the weed population. Other examples in this respect include the use or the attempted use of *Puccinia obtegens* to control Canada thistle, *Cirsium arvense* (73,169), the smut fungus *Sphacelotheca holci* to control Johnson grass, *Sorghum halepense* (113); and *Puccinia punctiformis* to control Canada thistle (170).

The use of necrogenic microorganisms as mycoherbicides is illustrated by the work of Howell and Stiponovic (92,93), and Jones et al. (97) have shown that by incorporating into soil dried and ground preparations of *Gliocladium virens*, which produces the phytotoxin viridiol, a number of plant species, including several important weeds, could be controlled. When *G. virens* culture on a rice medium was worked into pigweed (*Amaranthus retroflexus*)–infested soil supporting cotton (*Gossypium hirsutum*) seedlings, viridiol was apparently produced in sufficient quantity and duration to prevent pigweed emergence without harm to the emerging cotton seedlings. The longevity of *G. virens* in the dried rice preparation and the ease with which it is stored and applied suggested that it is possible to use this fungus as a phytotoxin-generating mycoherbicide (93). Jones et al. (97) confirmed the ability of the fungus to cause root necrosis and to control weeds. The herbicidal activity was correlated with the production of viridiol, which was particularly active against redroot pigweed and annual composite species and to a lesser extent against monocots. Toxicity to crop species was avoided by applying the fungal preparation out of the crop's root zone—for example, between the seed and the soil surface. It is conceivable that phytotoxins produced by a mycoherbicide candidate could be used as adjuvants to improve weed control ability of the candidate. Likewise, phytotoxin-producing, nonpathogenic microbes could be utilized as necrogenic agents to control weeds, as has been shown with hydrilla, *Hydrilla verticillata* (35,48,130), and eurasian water milfoil, *Myriophyllum spicatum* (83). Unlike fungi whose primary mode of weed control action is through parasitism, the preceding fungi are active mainly through toxigenicity. These necrogenic fungi usually are not principal pathogens. Nevertheless, if weed control is accomplished in these cases through the deployment of cultured fungal products that release herbicidal metabolites in situ, the approach could be included under the mycoherbicide tactic. If a microbially derived chemical is applied in pure form or in preponderance over the live propagules, then it should fall under chemical herbicide technology despite the possibility that some of these microbial chemicals may fall under the category of biorational pesticides (37).

Accomplishments

Presently two mycoherbicides, DeVine and COLLEGO, are used commercially in the United States to control, respectively, milkweed vine (stranglervine), *Morrenia odorata,* in citrus groves of Florida and northern joint vetch, *Aeschynomene virginica,* in rice and soybean fields of Arkansas and neighboring states (19,20,100,137,145,146,157,163).

DeVine, marketed by Abbott Laboratories, is the first registered mycoherbicide (31,100,137,138,193). The mycoherbicidal product consists of a liquid concentrate of chlamydospores of a pathotype of *Phytophthora palmivora* native to Florida. It is capable of killing seedlings and adult stranglervine, a weed of South American origin that is troublesome in Florida's citrus groves. Extensive host range and efficacy data from laboratory and field trials supported the safety and biocontrol potential of this fungus (137). DeVine is applied as a postemergent, directed spray. Typically, more than 90% control is obtained, and control lasts for at least 2 years after initial application. Despite concerns about susceptibility of some nontarget plants to this pathogen, it was deemed acceptable as a mycoherbicide, and registration was granted in 1981. Safety to nontarget susceptible plants was addressed by label restrictions, which prescribe selective, site-specific application of the mycoherbicide. DeVine is not to be used where susceptible nontarget plants are grown or occur (4), and to date no adverse nontarget effects have occurred. The market for DeVine is quite small, specialized, and concentrated in the citrus-growing areas of Florida.

COLLEGO, based on *Colletotrichum gloesporioides* f.sp. *aeschynomene,* an anthracnose-inciting pathogen of northern joint vetch discovered in Arkansas, was developed by scientists of the University of Arkansas, the U.S. Department of Agriculture, and the Upjohn Company (5,69,145,146,154,162). The commercial product, a wettable powder formulation of dried spores produced by liquid fermentation, was registered in 1982 (19,20,57,145,146,154). Upjohn continues to produce the product, which is now owned and sold by Ecogen, Inc. COLLEGO is applied postemergence, aerially or with land-based sprayers (146). The history, development, registration, integrated use, and postregistration status of COLLEGO have been reviewed (19,20,101,103,145,146,154,157). It is capable of killing seedlings as well as mature northern joint vetch, a hard-seeded, leguminous weed in rice and soybean crops. COLLEGO has provided consistently high levels of weed control, typically above 85%, and its acceptance by rice and soybean growers has been very good (20). Although the fungus has a wider host range than originally thought (152) and was found to be capable of infecting but not killing several economically important legumes (152), under field use it has not posed any danger to nontarget plants. It spreads poorly because of the sticky nature of its spores, and the chance of uncontrollable epidemic is nonexistent. Following inundative applications, the inoculum declines to very low levels in soil and irrigation water, but the pathogen may persist on infected host tissue and overwinter (151,153). The lack of high survival rate, which may be typical for inundatively applied inoculum, is an added safety feature of mycoherbicides. However, this necessitates annual applications for effective weed control. During nearly two decades of use, no environmental or user hazards have ever been found with COLLEGO (154). Grower acceptance of COLLEGO has been good.

Two other mycoherbicides, CASST (*Alternaria cassiae* against sicklepod, *Cassia obtusifolia*) and BioMal (*Colletotrichum gloeosporioides* f.sp. *malvae* against round-leaf mallow, *Malva pusilla* [=M. rotundifolia]), are under advanced stages of development (10,53,124,179). *Cephalosporium diospyri,* a wilt-causing fungus, has been used for a number of years in Oklahoma as a selective silvicide against weedy persimmon, *Diospyros virginiana* (67,82,188,189). *Colletotrichum gloeosporioides* f.sp. *cuscutae,* under the name of LUBOA II, is used against dodders (*Cuscuta* spp). in the People's Republic of China (160). Several other candidates have undergone extensive testing for commercial development. These include *Cercospora rodmanii* for water hyacinth, *Eichhornia crassipes* (38) and *Colletotrichum coccodes* for velvet leaf, *Abutilon theophrasti* (194). Others, notably *Bipolaris sorghicola* for Johnson grass, *Sorghum halepense* (192); *Chondrostereum purpureum* for black cherry, *Prunus serotina* (71,141); *Colletotrichum orbiculare* for spiny cocklebur, *Xanthium spinosum* (9,116,117); *Colletotrichum gloeosporioides* f.sp. *jussiaeae* for winged water primrose, *Jussiaea decurrens* (24); *Colletotrichum malvarum* for prickly sida, *Sida spinosa* (102); *Fusarium solani* f. sp. *cucurbitae* for Texas gourd, *Cucurbita texana* (23,184–186); and *Phomopsis convolvulus* for field bindweed, *Convolvulus arvensis* (126), have undergone or are undergoing precommercial evaluations. Many other pathogens are under various stages of research and development. Figure 2.2 represents an attempt at a comprehensive, worldwide listing of mycoherbicide projects based on a literature survey completed in April 1989. It appears that mycoherbicide research is being conducted in 16 countries and 44 locations, including the United States. In the latter, 18 locations or research groups are involved.

Alternaria cassiae, a foliar blight-inducing pathogen, was discovered in Mississippi and shown by Walker and co-workers (174–177,179) to be a safe and efficacious mycoherbicide for sicklepod. Subsequently; it was demonstrated to be highly efficacious in controlling the weed in soybean under field conditions in a regionwide trial involving five Southern states (53). It has a narrow host range and is capable of controlling three economically important leguminous weeds—sicklepod, coffee senna (*Cassia occidentalis*), and showy crotalaria (*Crotalaria*

Figure 2.2. A List of Mycoherbicide Projects

Weed*	Weed Type	Pathogen	Fungal Class	1982 Status	1989 Status	Commercial/ Practical Prospects (Product Name)	Reference(s)
Abutilon theophrasti (velvetleaf)	A/H/CW	*Colletotrichum coccodes*	Coelomycetes	NL	E	2 (VELGO)	86, 194, 195
		Fusarium lateritium	Hyphomycetes	NL	D	2	26,173
Aeschynomene virginica (northern joint vetch)	A/H/CW	*Colletotrichum gloeosporioides* f.sp. *aeschynomene*	Coelomycetes	E	F	4 (COLLEGO)	5, 19, 20, 57, 69, 101, 103, 145, 146, 151–155, 157, 162, 163
Albizzia julibrissin (silktree albizzia)	P/T/FW	*Fusarium oxysporum* f.sp. *perniciosum*	Hyphomycetes	E	G	0	36
Alternanthera philoxeroides (alligatorweed)	P/H/AW	*Alternaria alternantherae*	Hyphomycetes	C	G	0	89, 90
Ambrosia trifida (giant ragweed)	A/H/CW/ RW	*Fusarium lateritium*	Hyphomycetes	NL	C	2	7
		Protomyces gravidus	Hemiascomycetes	NL	C	0	34
Anoda cristata (spurred anoda)	A/H/CW	*Altenaria macrospora*	Hyphomycetes	D	C	1	68, 172, 180
		Fusarium lateritium	Hyphomycetes	NL	D	2	68 173
Arceuthobium spp. (dwarf mistletoes)	P/H/PW	*Colletotrichum gloeosporioides*	Coelomycetes	NL	B	0	125
		Nectria fuckeliana var. *macrospora*	Pyrenomycetes	D	G	0	156
		Various fungi		D	G	0	156
Asclepias seriaca (common milkweed)	P/H/CW/ RW	*Colletotrichum fusarioides*	Coelomycetes	NL	C	1	7
Avena fatua (wild oat)	A/G/CW/ RW	*Septoria tritici* f.sp. *avenae*	Coelomycetes	NL	C	1	110
Brachiaria platyphylla (broadleaf signalgrass)	A/G/CW	*Bipolaris setariae*	Hyphomycetes	NL	C	2	135
Bresenia schreberi (water shield)	P/H/AW	*Dichotomophthoropsis nymphaearum*	Hyphomycetes	C	G	0	96
Cannabis sativa (marijuana)	A/H/IC	*Fusarium oxysporum* f.sp. *cannabis*	Hyphomycetes	G	D	1	114
		Phomopsis ganjae sp. nov.	Coelomycetes	NL	G	1	115

				NL	B	1	
Caperonia palustris (texasweed)	A/H/CW	Amphobotrys ricini	Hyphomycetes	NL	B	1	187
Carduus pycnocephalus (Italian thistle)	P/H/RW	Alternaria sp.	Hyphomycetes	NL	D	2	2
Cassia obtusifolia (sicklepod)	A/H/CW	Alternaria cassiae	Hyphomycetes	NL	E	4 (CASST, pending registration)	10–12, 22, 53, 171, 174–179
		Pseudocercospora nigricans	Hyphomycetes	NL	D	1	87
Cassia surratensis (brushweed)	P/T/FW	Cephalosporium sp.	Hyphomycetes	D	G	0	167
Chenopodium album (common lambs-quarter)	A/H/CW	Ascochyta caulina	Coelomycetes	NL	C	1	142
		Cercospora chenopodii	Hyphomycetes	NL	C	1	142
		C. dubia	Hyphomycetes	NL	B	1	190
Cirsium arvense (Canada thistle)	P/H/RW	Alternaria spp.	Hyphomycetes	D	G	0	156
		Fusarium roseum	Hyphomycetes	C	G	0	134
		Sclerotinia sclerotiorum	Discomycetes	NL	D	1	28
Clidemia hirta (Koster's curse)	A/H/FW	Colletotrichum gloeosporioides f.sp. clidemiae	Coelomycetes	NL	F	4 (Inoculum distributed by hikers)	166
Convolvulus arvensis (field bindweed)	P/V/CW	Phomopsis convolvulus	Coelomycetes	NL	D	3	126
		Various fungi		B	G	0	156
Crotalaria spectabilis (showy crotalaria)	A/H/CW/ RW/FW	Colletotrichum dematium f.sp. crotalariae	Coelomycetes	NL	D	2	42
		Fusarium udum f.sp. crotalariae	Hyphomycetes	NL	D	2	42
		Alternaria cassiae	Hyphomycetes	NL	D	2	22
Cucurbita texana (Texas gourd)	A/V/CW	Fusarium solani f.sp. cucurbitae	Hyphomycetes	NL	D	3	23,184–186
Cuscuta spp. (dodders)	A/V/PW	Alternaria sp.	Hyphomycetes	NL	D	3	17,18
		Fusarium tricinctum	Hyphomycetes	NL	D	1	17
		Colletotrichum gloeosporioides f.sp. cuscutae	Coelomycetes	NL	F	4 (LUBOA 2)	160

(continued)

31

Figure 2.2. (continued)   A List of Mycoherbicide Projects

Weed*	Weed Type	Pathogen	Fungal Class	1982 Status	1989 Status	Commercial/ Practical Prospects (Product Name)	Reference(s)
Cyperus esculentus (yellow nut sedge)	P/Sg/CW	Cercospora caricis	Hyphomycetes	NL	B	1	7
		Phyllachora cyperis	Pyrenomycetes	NL	B	1	7
		Sclerotinia homoeocarpa	Discomycetes	D	G	0	36
Cyperus rotundus (purple nut sedge)	P/Sg/CW	Balansia cyperi	Pyrenomycetes	NL	C	1	56, 109, 148
Datura stramonium (jimsonweed)	A/H/CW	Alternaria crassa	Hyphomycetes	NL	D	3	21, 25, 27
Desmodium tortuosum (Florida beggarweed)	A/H/CW	Colletotrichum truncatum	Coelomycetes	NL	D	1	32
Digitaria sanguinalis (large crabgrass)	A/G/CW	Pyricularia grisea	Hyphomycetes	NL	B	1	7
Diospyros virginiana (persimmon)	P/T/RW	Cephalosporium diospyri	Hyphomycetes	E	F	4 (Under practical use)	160
Echinochloa crusgalli (barnyardgrass)	A/G/CW	Cochliobolus lunatus	Pyrenomycetes	NL	C	2	140
Eichhornia crassipes (water hyacinth)	P/H/AW	Acremonium zonatum	Hyphomycetes	D	G	0	35, 38, 111, 139
		Alternaria eichhorniae	Hyphomycetes	G	D	1	35, 38, 51, 143
		Cercospora piaropi	Hyphomycetes	C	G	0	76, 112
		Cercospora rodmanii	Hyphomycetes	E	D	3 (ABG 5003)	38, 41, 49, 60–65, 77
		Myrothecium roridum	Hyphomycetes	C	G	0	38, 129
		Rhizoctonia sp. (Aquathanatephorus pendulus)	Hyphomycetes	C	G	0	38, 99, 168
Eleocharis kuroguwai (water chestnut)	P/Sr/AW/CW	Epicoccosorus nematosporus gen. et. sp. nov.	Hyphomycetes	NL	D	3	149
Eleucine indica (goosegrass)	A/G/CW	Bipolaris setariae	Hyphomycetes	NL	C	1	75
		Pyricularia grisea	Hyphomycetes	NL	C	2	7
Eupatorium adenophorum (Crofton weed or pamakani)	P/H/CW	Cercospora eupatorii	Hyphomycetes	F	F	4 (of limited effectiveness under noninundative incidence)	72

Weed (common name)		Pathogen	Class				Ref.
Euphorbia esula (leafy spurge)	P/H/RW	*Alternaria tenuissima* f.sp. *euphorbiae*	Hyphomycetes	NL	D	1	107
E. heterophylla (milk weed)	A/H/CW	*Alternaria* sp.	Hyphomycetes	NL	C	1	197
		Helminthosporium sp.	Hyphomycetes	NL	D	2	198
E. supina (prostrate spurge)	A/H/CW	*Amphobotrys ricini*	Hyphomycetes	NL	C	0	91
Hakea sericea (hakea)	P/S/RW/FW	*Colletotrichum gloeosporioides*	Coelomycetes	NL	D	2	122
Hydrilla verticillata (hydrilla)	P/H/AW	*Fusarium roseum* 'Culmorum'	Hyphomycetes	D	D	3	47, 50
		Fusarium solani	Hyphomycetes	D	G	0	36
		Sclerotium sp.	Hyphomycetes	NL	D	2	98
		Various fungi		NL	G	0	16
Ipomoea purpurea (tall morning glory)	A/V/CW	*Colletotrichum dematium*	Coelomycetes	NL	C	1	7
Jussiaea decurrens (winged water primrose)	P/S/AW	*Colletotrichum gloeosporiodes* f. sp. *jussiaeae*	Coelomycetes	E	D	2	24
Lemna spp. (duckweed)	A/H/AW	*Pythium aphanidermatum*	Oomycetes	NL	D	2	58
		P. myriotylum	Oomycetes	NL	C	2	136
Malva pusilla (round-leaf mallow)	A/H/CW	*Colletotrichum gloeosporioides* f. sp. *malvae*	Coelomycetes	NL	E	4 (BioMal, awaiting registration)	124
Molluccella (=*Mollucella*) sp. (molluccabalm)	A/H/CW	*Cercospora mollucellae*	Hyphomycetes	C	G	0	36
Morrenia odorata (strangler vine or milkweed vine)	P/V/CW	*Phytophthora citrophthora* (later identified as *P. palmivora*)	Oomycetes	E	F	4 (DeVine)	4, 31, 100, 137, 138, 193
Myriophyllum brasiliense (parrotfeather)	P/H/AW	*Pythium carolinianum*	Oomycetes	NL	D	0	15
M. spicatum (eurasian watermilfoil)	P/H/AW	*Colletotrichum gloeosporioides*	Coelomycetes	NL	D	0	144, 147
		Mycoleptodiscus terrestris	Coelomycetes	NL	D	3	83

(continued)

Figure 2.2. (continued) A List of Mycoherbicide Projects

Weed*	Weed Type	Pathogen	Fungal Class	1982 Status	1989 Status	Commercial/ Practical Prospects (Product Name)	Reference(s)
Nymphaea odorata (fragrant waterlily)	P/H/AW	Dichotomophthoropsis nymphaearum	Hyphomycetes	C	G	0	96
Nymphaea tuberosa (white waterlily)	P/H/AW	Dichotomophthoropsis nymphaearum	Hyphomycetes	C	G	0	96
Nymphoides orbiculata (waterlily)	P/H/AW	Various fungi		C	G	0	96
Orobanche spp. (broomrape)	P/H/PW	Fusarium oxysporum var. orthoceras	Hyphomycetes	F	F	4 (presumably under practical use in USSR)	106
Panicum dichotomiflorum (fall panicum)	A/G/CW	Sorosporium cenchri	Teliomycetes	D	G	0	156
Portulaca oleracea (common purslane)	A/H/CW	Dichotomophthora indica	Hyphomycetes	NL	B	1	13
		D. portulacae	Hyphomycetes	NL	C	1	104, 119
Populus alba (aspen or white poplar)	P/T/FW	Valsa spp.	Pyrenomycetes	D	G	0	156
Prunus serotina (black cherry)	P/T/FW	Chondrostereum purpureum	Hymenomycetes	NL	D	3	71, 141
Pteridium aquilinum (bracken)	P/F/RW	Ascochyta pteridis	Coelomycetes	NL	C	1	30, 95, 183
		Cryptomycina pteridis	Discomycetes	NL	C	1	6
		Phoma aquilina	Coelomycetes	NL	C	1	30, 95
Pueraria lobata	P/V/FW/ RS	Cercospora pueraricola	Hyphomycetes	NL	G	0	7
Quercus spp. (red and bur oaks)	P/T/FW	Ceratocystis fagacearum	Plectomycetes	D	G	0	79
Roettbollia cochin-chinensis (itchgrass)	A/G/CW	Curvularia sp.,	Hyphomycetes	NL	B	1	74
		Phaeoseptoria sp.	Coelomycetes	NL	B	1	74
		Phoma sp.	Coelomycetes	NL	B	1	74

Weed species	A, B, or P/H/CW	Pathogen	Class				Ref.
Rumex spp. (curly docks)	A, B, or P/H/CW	Ramularia rubella	Hyphomycetes	NL	B	1	94
Sesbania exaltata (hemp sesbania)	A/H/CW	Colletotrichum truncatum	Coelomycetes	NL	C	1	7
Sida spinosa (prickly sida)	A/H/CW	Colletotrichum malvarum	Coelomycetes	C	D	2	102
		Fusarium lateritium	Hyphomycetes	NL	D	2	26, 173
Silybum marianum (milk thistle)	P/H/RW	Septoria silybi	Coelomycetes	NL	C	1	7
Solanum ptycanthum	A/H/CW	Colletotrichum coccodes	Coelomycetes	NL	C	1	3
Sorghum halepense	P/G/CW	Sphacelotheca cruenta (also S. holci)	Teliomycetes	C	D	1	113
(Johnson grass)		Bipolaris halepense	Hyphomycetes	NL	C	2	54, 55
		B. sorghicola	Hyphomycetes	NL	E	2	191, 192
		Colletotrichum graminicola	Hyphomycetes	NL	C	2	55, 121
		Gloeocercospora sorghi	Hyphomycetes	NL	C	2	55, 121
Trianthema portulacas-trum (horse purslane)	A/H/CW	Gibbago trianthemae	Hyphomycetes	NL	C	1	120
		Drechslea (=Exserohilum) indica	Hyphomycetes	NL	B	1	150
Xanthium spinosum (spiny cocklebur)	A/H/CW	Colletotrichum orbiculare	Coelomycetes	NL	D	3	9, 116, 117
Xanthium strumarium (heartleaf cocklebur)	A/H/CW	Alternaria helianthi	Hyphomycetes	NL	C	1	131
Various aquatic weeds	A/P/AW	Various fungi	Various	NL	B	1	105

*Some of the weed species from Templeton's 1982 list (156) have been grouped under "spp." NL=Not in the previous list. Weed type: A=Annual; B=biennial; P=perennial; F=fern; G=grass; H=herb; S=shrub; Sg=sedge; Sr=spike rush; T=tree; and V=vine; AW=aquatic weed; CW=crop weed; FW=forest weed; IC=illegal crop; PW=parasitic weed; RS=roadside weed; and RW=rangeland weed. The most important situation or weedy nature by which the weed is most commonly known is listed first; a weed may be problematic in more than one situation. Status: A=planning stage; B=Surveys completed; C=completed or under laboratory/greenhouse tests; D=completed or under small-scale field tests; E=completed or under large-scale field tests comparable to commercial evaluation; F=in commercial or public use; and G=status unknown or project inactive. Commercial/Practical Prospects: 0=not good; 1=unknown; 2=fair; 3=good; and 4=excellent. Current Status and Commercial/Practical Prospects are based on this author's best estimate considering data on host specificity, efficacy, regulatory aspects, and technological feasibility. All with Current Status of G=Prospect 0 at this time. Similarly, all F=4.

spectabilis) (22,175). The fungus is currently under commercial development by Mycogen Corporation as a wettable powder formulation, to be marketed under the trade name CASST (10–12).

Colletotrichum gloeosporioides f.sp. *malvae,* an anthracnose-causing pathogen of round-leaf mallow, is being developed as a mycoherbicide by PhilomBios, a biotechnology company based in Saskatchewan, Canada (124; Gantotti and Mortensen, personal communications). The fungus, native to midwestern Canada, is highly virulent, efficacious, easily cultured, and easily applied with conventional equipment (Mortensen, personal communication). No chemical herbicides give satisfactory control of this weed, so the market outlook for this mycoherbicide is good. However, safflower is a nontarget host of this pathogen. The risk to safflower is being addressed through extensive host range and pathogenicity trials (Mortensen, personal communication) to decide the question of risks versus benefit.

Many other prospective candidates, although successful in research trials, have failed to gain registration and commercial use due to one or more of the following reasons: lack of acceptable level of efficacy, technological difficulties in production and marketing of a commercially acceptable formulation, competition from chemical herbicides, and unprofitable markets. The case of *C. rodmanii* is perhaps typical. This fungus incites a debilitating leaf spot disease, causing the leaf to die back from the tip. Severely infected plants become chlorotic and stressed. In advanced stages of disease, root deterioration occurs (60,64,77). With the spread of the fungus, the plant population begins to decline, and open water appears where previously there had been dense stands of water hyacinth. Small clusters of heavily diseased plants float away from the mat, and finally the entire clusters of plants gradually sink to the bottom. This type of disease progression, which has been observed under controlled conditions, may take several weeks to months and will occur only under severe and sustained disease pressure (77).

The fungus is host specific and safe for use (61,63). Between 1974 and 1984 *C. rodmanii* was field tested in several locations in Florida and the southeastern United States using industrial and laboratory-grown fungal inocula. It was established that *C. rodmanii* can severely affect water hyacinth growth, especially under conditions that caused a slow rate of plant growth. Although the greatest effect of *C. rodmanii* was in reducing plant height and biomass, plant death and plant elimination can occur under severe disease pressure (60). The use of *C. rodmanii* as a biological control agent for water hyacinth was patented by the University of Florida, and Abbott Laboratories was licensed to develop the fungus as a microbial herbicide for commercial use. Abbott developed wettable powder formulations of *C. rodmanii* and obtained U.S. Environmental Protection Agency Experimental Use Permit to evaluate their herbicidal use. The possibility of registration and commercialization of *C. rodmanii* appeared good at first. Nonetheless, in 1984 Abbott decided not to register, because of technical difficulties in assuring efficacy of the microbial herbicide and economic uncertainties of the marketplace.

Water hyacinth is one of the most productive plants, and the biological control efficacy of *C. rodmanii* is related to the growth rate of its host. Conway et al. (64) stated that under conditions favorable for growth, water hyacinth was able to produce one new leaf every 5–6 days and was thus capable of outgrowing the cercospora disease. When conditions are present that favor disease development and limit leaf production to less than one leaf for every 3 weeks the pathogen could kill leaves faster than the plant could produce new leaves. The plant would then become debilitated and die unless conditions changed to stimulate its regrowth or conditions became less favorable for the disease. Charudattan et al. (49) tested this hypothesis by determining the relationship between the disease and host growth rate at different nutrient levels. This was accomplished by measuring plant growth, disease incidence, and disease severity, and calculating the level of disease stress and rate of disease progress required to kill water hyacinth. Although the amount of disease and the speed of disease development obtained in this study were insufficient to kill water hyacinth, there were reductions of 20–90% in host growth rates, as measured by weekly increments of green leaves due to the disease (49). Higher reductions in growth rates occurred on the lower nutrient levels, but when the nutrient concentration was most favorable for growth, water hyacinth grew at a rate faster than the rate at which the disease progressed to newer leaves.

These results indicated that for practical levels of control of water hyacinth by *C. rodmanii,* the fungus should be used under conditions that support only low to moderate host growth rates, which is not feasible under field conditions. Alternatively, the biocontrol efficacy of the fungus should be somehow improved. For example, the disease could be established on host populations at severe levels by multiple applications of inoculum when water hyacinth is in the early phase of seasonal growth. Or the fungus could be combined with other biotic or abiotic agents capable of retarding host growth, such as with insects or sublethal rates of chemical herbicides. A combination of insect (water hyacinth weevils) and *C. rodmanii* infections yielded 98% control of water hyacinth in an experimental field study, confirming the potential of this approach (41). However, the marketplace for aquatic weed control agents is dominated by chemical herbicides that provide fast, economical, and predictable control. Thus, the incentive for developing and registering *C. rodmanii* has been limited due to competition from the alternative control options.

Current Status

Figure 2.1 is an analysis of information from Templeton's 1982 list (156), and Figure 2.2 is a compilation of projects currently active. Comparing these two lists, it is clear that considerable progress has indeed been made. In 1982, 45 weeds were identified as targets for control by fungal pathogens. Of these, 10 were targets for control solely by rust or other nonculturable fungi that most probably were intended as classical or augmentative agents. The remainder, 35 weed species, were targets for control by 54 different facultative parasites fitting the criteria of mycoherbicide agents. Of the projects dealing with these 54 agents, the majority, 36, were reported to be active (status B through F), whereas 18 were inactive or of unknown status. The active projects represented a significant number, and the weeds in question were, and continue to be, of considerable economic significance. Yet there is no new information or research on 23 of these projects relating to 19 weed targets; these must be considered inactive or unlikely to yield usable mycoherbicides. The remaining 13 projects relating to 12 weed targets are still active, and the mycoherbicide candidates involved have progressed beyond the previous situation. In some cases, new research efforts or new pathogens have been added for these erstwhile active weed targets. Since 1982 there has been an increase in the number of weed targets and candidate pathogens studied (69 and 109, respectively), including 73 host–agent combinations not mentioned in 1982, confirming a worldwide trend in seeking mycoherbicides.

Figure 2.2 is intended to provide a comprehensive, worldwide, and up-to-date list of mycoherbicide projects (excluding augmentative and necrogenic agents). It also includes information on any supplemental research on the projects reported to be active in 1982 (156) and a bibliography. The literature survey for this was completed in April 1989.

A numerical analysis of the data in Figure 2.2 is provided in Figure 2.3. The data bear out the following facts. Most of the current efforts are directed nearly equally at controlling annual and perennial weeds; the majority are aimed at herbaceous angiosperms and weeds in crops. Grasses, trees, and vines follow. After crop weeds, the second most studied are aquatic weeds, attesting to the importance of water resources as sites for mycoherbicides.

Many of the current projects are undergoing or have completed small-scale field tests (29.36%) or laboratory or greenhouse testing (26.61%). However, a substantial number of projects are under unknown or inactive status (22.02%), implying that the failure rate or the lack of sustained initiative on the part of the researcher may be high in this field. This also creates a sense of uncertainty about these projects, many of which have initially claimed successful results.

The prospects for commercial or practical success are judged to be not good for about 25.69% of the projects. The prospects of another 35.78% are unknown. Assuming that researchers will continue to champion their mycoherbicide candidates, the remaining projects are considered to have fair to excellent potential for commercial or practical success; the following agents are included in these categories (see Figure 2.2 for references). Agents rated Fair face more biotic, technological, regulatory constraints to their further development than those rated

Figure 2.3. A Numerical Analysis of Data in Figure 2.2

Category	Number of	Percentage
Type of weeds studied (1989 data)	Weeds	
Annual	35	50.72
Biennial	1	1.45
Perennial	33	47.83
Total	69	100.00
Fern	1	1.45
Grass	8	11.59
Herb	41	59.42
Parasitic	3	4.35
Sedge	2	2.90
Shrub	2	2.90
Spikerush	1	1.45
Tree	6	8.70
Vine	5	7.25
Total	69	100.01
Target weed is in (1989 data)		
Aquatic sites	12	17.39
Crops	36	52.17
Forests	7	10.14
Rangelands	6	8.70
Multiple sites	7	10.14
Weed is an illegal crop	1	1.45
Total	69	99.99
1982 Status	Projects	
A–Planning stage	0	0
B–Survey under way or completed	1	0.92
C–Lab./greenhouse studies in progress or completed	12	11.01
D–Small-scale field tests in progress or completed	12	11.01
E–Large-scale field trials in progress or completed	6	5.50
F–Under commercial/practical use	2	1.83
G–Status unknown/inactive project	3	2.75
NL–Not listed in 1982	73	66.97
Total	109	99.99
1989 Status	Projects	
A–Planning stage	0	0
B–Survey underway or completed	13	11.93
C–Lab./greenhouse studies in progress or completed	29	26.61
D–Small-scale field tests in progress or completed	32	29.36
E–Large-scale field trials in progress or completed	4	3.67
F–Under commercial/practical use	7	6.42
G–Status unknown/inactive project	24	22.02
Total	109	100.01

(continued)

Figure 2.3. (continued)   A Numerical Analysis of Data in Figure 2.2

Category	Number of	Percentage
Prospects for commercial/practical use (1989 data)	Agents	
0–Not good	28	25.69
1–Unknown	39	35.78
2–Fair	23	21.10
3–Good	10	9.17
4–Excellent	9	8.26
Total	109	100.00
Agent studied belongs to (1989 data)	Taxa	
Coelomycetes	27	24.77
Discomycetes	3	2.75
Hemiascomycetes	1	0.92
Hymenomycetes	1	0.92
Hyphomycetes	61	55.96
Oomycetes	4	3.67
Plectomycetes	1	0.92
Pyrenomycetes	5	4.59
Teliomycetes	2	1.83
Unspecified	4	3.67
Total	109	100.00

Good. Those rated Excellent are already in commercial/practical use or such use is considered imminent.

Prospect level 2, Fair: *Alternaria* spp. (against Italian thistle); *Bipolaris halepense* and *B. sorghicola* (Johnson grass); *B. setariae* (broadleaf signal grass); *Cochliobolus lunatus* (barnyard grass); *Colletotrichum coccodes* (velvetleaf); *C. dematium* f. sp. *crotalariae, Fusarium udum* f.sp. *crotalariae,* and *Alternaria cassiae* (showy crotalaria); *Colletotrichum gloeosporioides* f.sp. *jussiaeae* (winged water primrose); *C. graminicola* (Johnson grass); *C. malvarum* (prickly sida); *Fusarium lateritium* (giant ragweed, prickly sida, spurred anoda, and velvetleaf); *Gloeocercospora sorghi* (Johnson grass); *Pythium aphanidermatum* and *P. myriotylum* (duckweeds); and *Sclerotium* sp. (hydrilla).

Level 3, Good: *Alternaria crassa* (jimsonweed); *Alternaria* sp. (dodder); *Cercospora rodmanii* (water hyacinth); *Chondrostereum purpureum* (black cherry); *Colletotrichum orbiculare* (spiny cocklebur); *Epicoccosorus nematosporus (water chestnut); Fusarium roseum* "Culmorum" (hydrilla); *Fusarium solani* f.sp. *cucurbitae* (Texas gourd); *Mycoleptodiscus terrestris* (eurasian watermilfoil); and *Phomopsis convolvulus* (field bindweed).

Level 4, Excellent: *Alternaria cassiae* (sicklepod), *Cephalosporium diospyri* (persimmon), *Cercospora eupatorii* (Crofton weed); *Colletotrichum gloeosporioides* f.sp. *aeschynomene* (northern joint vetch); *C. gloeosporioides* f.sp. *clidemiae* (Koster's curse); *C. gloeosporioides* f.sp. *cuscutae* (dodder); *C. gloeosporioides* f.sp. *malvae* (round-leaf mallow); *Fusarium oxysporum* var. *orthoceras* (broomrapes); and *Phytophthora palmivora* (milkweed vine). With-

out exception all these have undergone some measure of commercial development, commercial use, or practical use through user distributors. The number
of colletotrichums in this group is noteworthy. This is no doubt due to the
success of *C. gloeosporioides* f.sp. *aeschynomene* (COLLEGO) as a commercial mycoherbicide and the technical feasibility of developing this species into
efficacious mycoherbicidal products (19,20,57,157).

More than half of the mycoherbicide candidates tested belong to Hyphomycetes
(56.96%), a large and varied class of conidial and nonsporulating fungi. The
Coelomycetes, which include *Colletotrichum* species, come second (24.77%).
Seven other classes claim the remaining 19.27% of the candidates. On a generic
basis, 18 of the candidates belong to *Colletotrichum* spp., 13 to *Fusarium* spp.,
12 to *Alternaria* spp., and 8 to *Cercospora* spp. In all, 41 genera of pathogenic
fungi are being considered. Thus, a wide choice of candidates is being explored,
although preferred pathogens appear to be those capable of causing some of the
most destructive diseases, such as anthracnoses, wilts, blights, and foliar spots.
This is to be expected because the efficacy and performance standards for mycoherbicides dictate a high capacity for plant kill or damage (44).

Major Conceptual and Technical Contributions

As with any emerging experimental science, the field of mycoherbicides has
grown with every practical and empirical attempt in the field. Several significant
developments have taken place in the last decade, starting from our understanding
of how to employ mycoherbicide under field conditions to complex issues of
epidemiology, integration, commercial development, and governmental regulations (52). A list of significant contributions in this field is presented in Figure
2.4.

Constraints

With our increasing knowledge of mycoherbicides has come a better understanding of the limitations facing this field. These include efficacy, extreme level of
host specificity, incompatibility with chemical pesticides, the need to produce
spores as inocula, technical difficulties in product development, competition from
chemical herbicides, economics based on market size, and potential buildup of
weeds resistant to the microbial herbicide. Some of these constraints are easily
solvable through research; others are more difficult to overcome.

Efficacy encompasses the level of weed control provided by the agent as well
as the speed and ease with which the control is accomplished (44). Unless the
mycoherbicide is fairly fast-acting, affords high levels of control similar to
chemical herbicides, and is easy to use and predictable in performance, it is
difficult to expect acceptance by industries and users. One of the ways to improve

Figure 2.4. Major Conceptual and Technical Developments in Mycoherbicides

Development	Weed–pathogen system involved	Key reference(s)
First registered mycoherbicide–DeVine	Milkweed vine–*Phytophthora palmivora*	100, 137
Liquid formulation of chlamydospores–DeVine	Milkweed vine–*Phytophthora palmivora*	100
Second registered mycoherbicide–COLLEGO	Northern joint vetch–*Colletotrichum gloeosporioides* f.sp. *aeschynomene*	69, 157
Wettable formulation of dried conidia–COLLEGO	Northern joint vetch–*C. gloeosporioides* f.sp. *aeschynomene*	154
First of the U.S. patents issued for mycoherbicidal use of fungi	Northern joint vetch–*C. gloeosporioides* f.sp. *aeschynomene*	5, 19, 20, 57
	Water hyacinth–*Cercospora rodmanii*	65, 70, 165
First test cases for EPA registration of mycoherbicides	Milkweed vine–*Phytophthora palmivora*	4, 100, 137
	Northern joint vetch–*C. gloeosporioides* f.sp. *aeschynomene*	5, 19, 20
Practical integration of a mycoherbicide with other crop protection chemicals and crop management practices	Northern joint vetch–*C. gloeosporioides* f. sp. *aeschynomene*	101, 103, 145, 146, 154
Practical and experimental application technology for mycoherbicides under grower's field conditions	Northern joint vetch–*C. gloeosporioides* f.sp. *aeschynomene*	132, 145, 146
Demonstration of improvement in mycoherbicide efficacy through combination with chemical herbicides or plant growth regulators	Water hyacinth–*C. rodmanii*	41
	Velvetleaf–*Colletotrichum coccodes*	86, 195
Feasibility of control of two weeds by combined application of two mycoherbicide agents	Northern joint vetch and winged water primrose–*C. Gloeosporioides* f.sp. *aeschynomene* and *C. gloeosporioides* f.sp. *jussiaeae*	24
Demonstration of combined use of two mycoherbicide candidates to control a difficult-to-control weed	Spurred anoda–*Fusarium lateritium* and *Alternaria macrospora*	68
	Showy crotalaria–*Colletotrichum dematium* f.sp. *crotalariae* and *Fusarium udum* f.sp.*crotalariae*	42
A broad-spectrum mycoherbicide capable of controlling more than one weed	Sicklepod, coffee senna, and showy crotalaria–*Alternaria cassiae*	22, 175
	Spurred anoda, prickly sida, and velvetleaf–*Fusarium lateritium*	173
Feasibility of integration of arthropods and a mycoherbicide	Water hyacinth–water hyacinth weevils and *C. rodmanii*	41
Demonstration of the use of a rust fungus in an augmentative strategy to control a crop weed	Yellow nut sedge–*Puccinia canaliculata*	127, 128

(continued)

42

Figure 2.4. (continued)   Major Conceptual and Technical Developments in Mycoherbicides

Development	Weed–pathogen system involved	Key reference(s)
Demonstration of the potential of necrogenic, toxin-producing fungi as mycoherbicides	Hydrilla–various toxin–producing fungi Various weeds–*Gliocladium virens*	48, 92, 93, 97, 130
First mutationally altered mycoherbicide agent and attempts at genetic improvement of mycoherbicides	Northern joint vetch–*C. gloeosporioides* f.sp. *aeschynomene*	27, 155
Development of a large-scale fungal spore production technique for mycoherbicidal use	Spurred anoda–*Alternaria macrospora*	172

efficacy is to combine the mycoherbicide with chemical herbicides or plant growth regulators. In some cases, chemical herbicides are known to aid the invasion of plants by microorganisms, possibly by lowering host resistance to infection (1,80). Attempts have been made to exploit the synergistic interaction of certain chemical herbicides, plant growth regulators, or surfactants and mycoherbicides such as *Alternaria cassiae, Cercospora rodmanii,* and *Colletotrichum coccodes* (11,41,194,195). It is also possible to improve efficacy through manipulation of virulence genes via genetic engineering or protoplast fusions (40,196). Although studies are under way to incorporate genes coding for phytotoxins and enzymes involved in pathogenesis, practical results are yet to be realized.

Efficacy is also related to the epidemiology, in particular the requirement of free moisture for the pathogen's germination and infection. Although the optimum duration of free moisture and the temperature–moisture relationships necessary for infection and disease development can be easily identified, it is often difficult to create these conditions on a field scale. Formulations that prevent drying or promote moisture retention are used to overcome this limitation (59,178).

Mycoherbicides are often specific to one or a small group of plants. It is rare to find a pathogen that will attack more than one economically significant weed and yet be safe for use. Therefore, users must rely on more than the given mycoherbicide to control a spectrum of weeds infesting a crop or a habitat. To overcome this problem, as has been demonstrated in the case of rice and soybean (101,145,146), combinations of mycoherbicides and chemical herbicides can be integrated effectively and economically into a weed management system.

In situations where mycoherbicides are to be used in combination with chemical pesticides, viability and efficacy of the biological agents may be adversely affected by the chemicals. For example, in cases where fungicides are used for disease control, the mycoherbicide has been adversely affected (101,145,146). A careful sequencing of fungicide and mycoherbicide applications, based on much research and extension efforts, can help overcome this problem.

Although mycelial inoculum can be infective in some cases (e.g., *Cercospora rodmanii* [77]), spores are often required for infection. Hence, spores are preferred for inoculum, but sporulation in culture could be a difficult step, requiring special techniques. This, coupled with technological problems in mass culturing, formulation, shelf life, and delivery could add further constrains. Any additional research needed to solve these problems may increase the cost and discourage industrial participation in mycoherbicides.

One of the most serious challenges to mycoherbicides is from chemical herbicides. The availability of economical and effective chemicals is a powerful disincentive of mycoherbicides. The availability of 2,4-D (2,4-dichlorophenoxy acetic acid), diquat (6,7-dihydrodipyrido[1,2-*a*:2′,1′-*c*]pyr azinediium ion), and other chemicals as broad-spectrum aquatic herbicides has discouraged industrial development of potential mycoherbicide candidates such as *Cercospora rodmanii* for water hyacinth and *Fusarium culmorum* for hydrilla. The cancellation of 2,4,5-T ([2,4,5-trichlorophenoxy] acetic acid) opened a noncompetitive market for COLLEGO. Similarly, the absence of suitable chemicals for controlling sicklepod has stimulated commercial development of *Alternaria cassiae*. In fact, it will be well to target for mycoherbicide research only those weeds for which there are no suitable chemical herbicides, where no herbicides are forthcoming, or where existing chemicals are likely to be lost.

The potential for buildup of weeds resistant to mycoherbicides has been debated (29,40,156), but so far it has not been of concern. DeVine and COLLEGO have been researched or in use for nearly two decades with no appearance of resistant weed biotypes. As Leonard (108) has pointed out, it should be possible to match any incidence of disease resistance among weeds with genetic selection among mycoherbicide pathogens for increased virulence.

Future Outlook and Needs

Experience with COLLEGO, DeVine, CASST, and BioMal (10,19,20,53,100, 124,137,138,145,146,156,179) leaves no doubt that mycoherbicides are effective and practical as weed control agents. It might be argued that despite the intense research efforts, only two commercial products are available in the United States, and perhaps worldwide only six or seven agents are practically used with any regularity. This may appear as a small number relative to the number of projects in Figure 2.3. However, the chemical industry is known to screen thousands of chemicals for every commercially feasible herbicide. When viewed in this light, mycoherbicides have had a remarkably high rate of return on scientific and monetary input. Moreover, with understanding comes deeper concerns, a greater need to answer those concerns, and the consequent slower pace in registering and using the agents. Experience with agents like *Alternaria cassiae, Cercospora rodmanii, Colletotrichum coccodes,* and *C. gloeosporioides* f.sp. *malvae* suggest that we are indeed witnessing this second phase of growth in mycoherbicides in which tougher challenges, both scientific and commercial, are being posed. It is hoped that the challenges can be met with thorough empirical knowledge, leading to speedier development and delivery of candidates to the users.

The future direction of mycoherbicides is being influenced by current scientific, practical, and governmental decisions. Major emphasis has been placed on biological pest control (66,118), but unless significant financial and personnel resources are expressly allocated, future growth of this field is bound to be limited. Private industries must also come forward to develop and market mycoherbicides in spite of their expected small market size. Since microbial pesticides are considerably cheaper to research and develop than chemical pesticides (33,156), a reasonable return on investment may be gained despite the market size and the risks in assuring the efficacy of a biological agent. On the research front, the following are emerging as major areas of importance.

1. *More mycoherbicide candidates of important weeds*. With each weed–pathogen system, new conceptual and practical problems are bound to come to light. These in turn will provide a deeper understanding of mycoherbicides. The acceptance of mycoherbicides and the strength of this field will increase with each new pathogen delivered for practical use.

2. *Integration of mycoherbicides with chemical pesticides*. As an ongoing effort, the compatibility–incompatibility of mycoherbicides and chemicals should continue. This will be mandated by the fact that each weed–mycoherbicide–pest management system will be different, and specific recommendations for the use of mycoherbicides will be needed.

3. *Integration of mycoherbicides and chemical plant growth regulators for improved weed control through decrease in weed growth and increase in mycoherbicide efficacy*. Weeds possessing high rates of vegetative growth and vegetative proliferation tend to be difficult to control with mycoherbicides. The ability to outgrow disease pressure is a characteristic of these weeds (49,192,195). In such cases, the integration of mycoherbicides with plant growth regulators, which by themselves may not afford weed control, offer a useful solution (41,195).

4. *Development of suitable formulations to improve viability, efficacy, and ease of application of mycoherbicides*. As stated earlier, the need for optimum moisture and specific temperature regimes for infection pose problems in assuring mycoherbicide efficacy. The lack of proper epidemiological conditions for infections and disease development and the adverse effect of solar radiation on fungal propagules can be countered to an extent through formulation technology. Substances that improve moisture retention, reduce drying and uv-irradiation, dilute and evenly disperse the inoculum, and provide better host–pathogen contact are being studied (59).

5. *Fermentation technology*. Current industrial preference favors submerged liquid fermentation to produce mycoherbicide products (57,162). Although successful, cost-effective, and readily available, this technique is not suit-

able for fungi that do not sporulate in submerged culture. Solid-substrate culturing and air-lift fermentation can offer solutions, but these methods are generally not preferred by industries with investment and expertise in liquid fermentation. Labor-intensive methods are also unsuitable because of high cost and problems in quality assurance. More exploration is therefore needed to find economical and practical methods of commercial inoculum production.

6. *Molecular genetic basis of virulence and host specificity.* Genetic improvement of mycoherbicide candidates through bioengineering for increased virulence and increased or decreased host specificity deserves research emphasis. With several mycoherbicide candidates the level of virulence is less than desirable. By incorporating genes for virulence factors such as host-specific toxins and phytotoxic metabolites or host receptors, it should be possible to improve weed control ability of these candidates. On the other hand, several highly virulent and destructive pathogens exist that are unsuitable as mycoherbicides on account of their broad host range. By genetically narrowing their host range, perhaps selecting strains specific only to the weed, these pathogens may be rendered useful (28). Mutation–selection, gene cloning, interspecific and intrageneric portoplast fusions, electroporation, and other methods can be useful for this purpose.

7. *Discovery of host-specific and nonspecific herbicidal metabolites of microbial origin that could be used as virulence and host specificity factors for genetic engineering.*

8. *Increased public and private funding as well as administrative support for research and development of mycoherbicides.* Current levels of funding and personnel resources are inadequate to meet future challenges. Involvement of private biotechnology companies is a welcome sign, but their continued involvement and support will be vital to this field.

9. *Education of scientists unfamiliar with mycoherbicides and the user public, which is required for technology transfer.* Mycoherbicides, like many other biological control agents, are sensitive to environmental conditions and need to be handled in strict accordance to the prescribed methods. Unlike chemical pesticides, they are usually slower in producing the desirable results. The users must therefore be educated about the use and performance features of mycoherbicides.

Summary

The level of scientific activity in mycoherbicides has increased tremendously since 1982. Both the number of weeds targeted for control and candidate pathogens studied have increased. Pratical registered or unregistered uses of mycoherbicides have also increased worldwide to about seven agents. Likewise, the number of U.S. patents issued for mycoherbicidal use of fungi and mycoherbicidal technology have increased, perhaps foretelling an increasing reliance on mycoherbicides in the future. Yet no new mycoherbicides have entered the markets as registered products since DeVine and COLLEGO were introduced in the early 1980s. This is largely due to lack of interest on the part of commercial industries, which apparently regard mycoherbicides as economically unattractive. It is clear from such examples as DeVine, COLLEGO, and other experimental mycoherbicides that the market size is small, profits are low, and the risks related to product efficacy and user acceptance are great. Besides, chemical herbicides are still preferred over other forms of weed control by western agrochemical industries and societies on account of the reliable efficacy and economic returns offered by chemicals. Only in cases where the market size is too small are the industries unwilling to develop or reregister chemical herbicides. Thus, chemicals pose a formidable challenge to mycoherbicides; either mycoherbicides must be developed for weeds that do not have chemical controls or we must rely on nonprofit agencies and public financial support for developing mycoherbicides in the future. However, the record of public agencies in funding and promoting biological controls is poor at best. Perhaps a solution will come from small venture companies that are interested in niche markets and that can operate efficiently and profitably despite the market size. Whether many of the mycoherbicide agents identified in Figure 2.2 will eventually reach the users, either through private industries or through public agencies, remains to be seen.

Literature Cited

1. Altman, J., and Campbell, L. C. 1977. Effect of herbicides on plant diseases. Annu. Rev. Phytopathology. 15:361–385.

2. Anderson, G. L., and Lindow, S. E. 1984. Biological control of *Carduus pycnocephalus* with *Alternaria* sp. Pages 593–600, *in*: Proc. VI Int. Symp. Biol. Control Weeds, E. S. Delfosse, ed. Agric. Canada, Canadian Govt. Publ. Centre, Ottawa.

3. Andersen, R. N., and Walker, H. L. 1985. *Colletotrichum coccodes*: a pathogen of eastern black nightshade (*Solanum ptycanthum*). Weed Sci. 33:902–905.

4. Anonymous. 1981. DeVine Brochure. Agricultural Products Division, Abbott Laboratory, Chicago, IL.

5. Anonymous. 1982. COLLEGO Technical Manual. Tuco, The Upjohn Company, Agricultural Division, Kalamazoo, MI.

6. Anonymous. 1985. Annual Report of the Cooperative Regional Research Project, S–136: Biological control of Weeds with Plant Pathogens. Available from R. Charudattan.

7. Anonymous. 1989. Discovery and Development of Plant Pathogens for Biological Control of Weeds. Regional Research Project S–136. Available from R. Charudattan.

8. Auld, B. A. 1986. Potential for Mycoherbicides in Australia. Workshop Proc. Agric. Res. Vet. Centre, Orange, NSW, Australia.

9. Auld, B. A., McRae, C. F., and Say, M. M. 1988. Possible control of *Xanthium spinosum* by a fungus. Agric. Ecosyst. Environ. 21:219–223.

10. Bannon, J. S. 1988. CASST™ herbicide (*Alternaria cassiae*): a case history of a mycoherbicide. Am. J. Alt. Agric. 3:73–76.

11. Bannon, J. S. 1988. Synergistic mycoherbicidal compositions. U.S. Patent No. 4,755,207, dated July 5, 1988.

12. Bannon, J. S., and Hudson, R. A. 1988. The effect of application timing and lighting intensity on efficacy of CASST™ herbicide (*Alternaria cassiae*) on sicklepod (*Cassia obtusifolia*). Weed Sci. Soc. Am. Abstr. 28:143.

13. Baudoin, A. B. A. M. 1986. First report on *Dichotomophthora indica* on common purslane in Virginia. Plant Dis. 70:352.

14. Bemmann, W. 1985. From weeds isolated fungi and their capability for weed control—a review of publications. Zentralbl. Mikrobiol. 140:111–148 (in German).

15. Bernhardt, E. A., and Duniway, J. M. 1984. Root and stem rot of parrotfeather (*Myriophyllum brasiliense*) caused by *Pythium carolinianum*. Plant Dis. 68:999–1003.

16. Bernhardt, E. A., and Duniway, J. M. 1986. Decay of pondweed and hydrilla hybernacula by fungi. J. Aquat. Plant Manag. 24:20–24.

17. Bewick, T. A., Binning, L. K., and Stevenson, W. R. 1986. Discovery of two fungal pathogens of swamp dodder (*Cuscuta gronovii* Willd.). Weed Sci. Soc. Am. Abstr. 26:27.

18. Bewick, T. A., Binning, L. K., Stevenson, W. R., and Stewart, J. 1987. A mycoherbicide for control of swamp dodder (*Cuscuta gronovii* Willd.). Pages 93–104, *in*: Parasitic Flowering Plants, H. Chr. Weber and W. Forstreuter, eds. Marburg, Federal Republic of Germany.

19. Bowers, R. C. 1982. Commercialization of microbial biological control agents. Pages 157–173, *in*: Biological Control of Weeds with Plant Pathogens, R. Charudattan and H. L. Walker, eds. Wiley, New York.

20. Bowers, R. C. 1986. Commercialization of Collego—an industrialist's view. Weed Sci. 34 (Suppl. 1):24–25.

21. Boyette, C. D. 1986. Evaluation of *Alternaria crassa* for biological control of jimsonweed: host range and virulence. Plant Sci. 45:223–228.

22. Boyette, C. D. 1988. Biocontrol of three leguminous weed species with *Alternaria cassiae*. Weed Technol. 2:414–417.

23. Boyette, C. D., Templeton, G. E., and Oliver, L. R. 1984. Texas gourd (*Cucurbita texana*) control with *Fusarium solani* f. sp. *cucurbitae*. Weed Sci. 32:649–655.

24. Boyette, C. D., Templeton, G. E., and Smith, R. J., Jr. 1979. Control of winged waterprimrose (*Jussiaea decurrens*) and northern jointvetch (*Aeschynomene virginica*) with fungal pathogens. Weed Sci. 27:497–501.

25. Boyette, C. D., and Turfitt, L. B. 1988. Factors influencing biocontrol of jimsonweed (*Datura stramonium* L.) with the leaf-spotting fungus *Alternaria crassa*. Plant Sci. 56:261–264.

26. Boyette, C. D., and Walker, H. L. 1985. Factors influencing biocontrol of velvetleaf (*Abutilon theophrasti*) and prickly sida (*Sida spinosa*) with *Fusarium lateritium*. Weed Sci. 33:209–211.

27. Brooker, N. L., TeBeest, D. O., and Spiegel, F. W. 1988. Nuclear number and pathogenic variability in *Alternaria crassa* protoplasts. Phytopathology 78:1519.

28. Brosten, B. S., and Sands, D. C. 1986. Field trials of *Sclerotinia sclerotiorum* to control Canada thistle (*Cirsium arvense*). Weed Sci. 34:377–380.

29. Burdon, J. J., and Marshall, D. R. 1986. Ecological aspects of plant–pathogen interactions. Pages 37–40, *in*: B. A. Auld, ed. Potential for Mycoherbicides in Australia. Workshop Proc. Agric. Res. Vet. Centre, Orange, NSW, Australia.

30. Burge, M. N., and Irvine, J. A. 1985. Recent studies on the potential for biological control of bracken using fungi. Proc. R. Soc. Edinb. 86B:187–194.

31. Burnett, H. C., Tucker, D. P. H., and Ridings, W. H. 1974. Phytophthora root and stem rot of milkweed vine. Plant Dis. Rep. 58:355–357.

32. Cardina, J., Littrell, R. H., and Hanlin, R. T. 1988. Anthracnose of Florida beggarweed (*Desmodium tortuosum*) caused by *Colletotrichum truncatum*. Weed Sci. 36:329–334.

33. Carlton, P. 1989. UCLA Symposium. Alan Liss, New York. In press.

34. Cartwright, R. D., and Templeton, G. E. 1988. Biological limitation of *Protomyces gravidus* as a mycoherbicide for giant ragweed, *Ambrosia trifida*. Plant Dis. 72:580–582.

35. Charudattan, R. 1973. Pathogenicity of fungi and bacteria from India to hydrilla and waterhyacinth in Florida. Hyacinth Control J. 11:44–48.

36. Charudattan, R. 1978. Biological Control Projects in Plant Pathology—A Directory, Misc. Publ. Plant Pathol. Dept., Univ. Florida, Gainesville, FL.

37. Charudattan, R. 1982. Regulation of microbial weed control agents. Pages 175–188, *in*: Biological Control of Weeds with Plant Pathogens, R. Charudattan and H. L. Walker, eds. Wiley, New York.

38. Charudattan, R. 1984. Role of *Cercospora rodmanii* and other pathogens in the biological and integrated controls of water hyacinth. Pages 834–859, *in*: Proc. Int. Conf. on Water Hyacinth, G. Thyagarajan, ed. U.N. Environ. Prog., Nairobi, Kenya.

39. Charudattan, R. 1984. Microbial control of plant pathogens and weeds. J. Georgia Entomol. Soc. 19 (Suppl. 2):40–62.

40. Charudattan, R. 1985. The use of natural and genetically altered strains of pathogens for weed control. Pages 347–372 *in*: Biological Control in Agricultural IPM Systems, M. A. Hoy and D. C. Herzog, eds. Academic Press, Orlando, FL.

41. Charudattan, R. 1986. Integrated control of waterhyacinth with a pathogen, insects, and herbicides. Weed Sci. 34 (Suppl. 1):26–30.

42. Charudattan, R. 1986. Biological control of showy crotalaria (*Crotalaria spectabilis*) with two fungal pathogens. Weed Sci. Soc. Am. Abstr. 26:51.

43. Charudattan, R. 1988. Inundative control of weeds with indigenous fungal pathogens. Pages 88–110, *in*: Fungi in Biological Control Systems, M. N. Burge, ed. Manchester University Press, Manchester, England.

44. Charudattan, R. 1989. Assessment of efficacy of mycoherbicide candidates, *in*: Proc. VII Int. Symp. Biol. Control Weeds, E. S. Delfosse, ed.

45. Charudattan, R., and Conway, K. E. 1975. Comparison of *Uredo eichhorniae*, the waterhyacinth rust with *Uromyces pontederiae*. Mycologia 67:653–657.

46. Charudattan, R., and DeLoach, C. J., Jr. 1988. Management of pathogens and insects for weed control in agroecosystems. Pages 246–264, *in*: Weed Management in Agroecosystems: Ecological Approaches, M. A. Altieri and M. Liebman, eds. CRC Press, Boca Raton, FL.

47. Charudattan, R., Freeman, T. E., Cullen, R. E., Hofmeister, F. M. 1980. Evaluation of *Fusarium roseum* "Culmorum" as a biological control agent for *Hydrilla verticillata*: safety. Pages 307–323, in: Proc. V Int. Symp. Biol. Control Weeds, E. S. Delfosse, ed. CSIRO, Canberra, Australia.

48. Charudattan, R., and Lin, C-Y. 1974. *Penicillium, Aspergillus* and *Trichoderma* isolates toxic to hydrilla and other aquatic plants. Hyacinth Control J. 12:70–73.

49. Charudattan, R., Linda, S. B., Kluepfel, M., and Osman, Y. A. 1985. Biocontrol efficacy of *Cercospora rodmanii* on waterhyacinth. Phytopathology 75:1263–1269.

50. Charudattan, R., and McKinney, D. E. 1978. A Dutch isolate of *Fusarium roseum* "Culmorum" may control *Hydrilla verticillata* in Florida. Pages 219–224, *in*: Proc. EWRS 5th Symp. on Aquatic Weeds. Wageningen, The Netherlands.

51. Charudattan, R., and Rao, K. V. 1982. Bostrycin and deoxybostrycin: two nonspecific phytotoxins produced by *Alternaria eichhorniae*. App. Environ. Microbiol. 43:846–849.

52. Charudattan, R., and Walker, H. L. 1982. Biological Control of Weeds with Plant Pathogens. Wiley, New York.

53. Charudattan, R., Walker, H. L., Boyette, C. D., Ridings, W. H., TeBeest, D. O., Van Dyke, C. G., and Worsham, A. D. 1986. Evaluation of *Alternaria cassiae* as a mycoherbicide for sicklepod (*Cassia obtusifolia*) in regional field tests. Southern Coop. Ser. Bull. 317. Alabama Agric. Exp. Sta., Auburn University, Auburn, AL.

54. Chiang, M-Y., Leonard, K. J., and Van Dyke, C. G. 1989. *Bipolaris halepense*: a new species from *Sorghum halepense* (johnsongrass). Mycologia 81:532–538.

55. Chiang, M-Y., Van Dyke, C. G., and Leonard, K. J. 1989. Evaluation of endemic foliar fungi for potential biological control of johnsongrass (*Sorghum halepense*): screening and host range tests. Plant Dis. 73:459–464.

56. Clay, K. 1984. New disease (*Balansia cyperi*) of purple nutsedge (*Cyperus rotundus*). Plant Dis. 70:597–599.

57. Churchill, B. W. 1982. Mass production of microorganisms for biological control. Pages 139–156, *in*: Biological Control of Weeds with Plant Pathogens, R. Charudattan and H. L. Walker, eds. Wiley, New York.

58. Colbaugh, P. F. 1988. Use of *Pythium aphanidermatum* as a biological control agent for duckweed and other floating aquatic weeds. *In*: Abstr. VII Int. Bymp. Biol. Control Weeds. Biol. Control Weeds Lab., USDA-ARS, Rome.

59. Connick, W. J. Jr., Lewis, J. A., and Quimby, P. C. Jr. 1989. Formulation of biocontrol agents for use in plant pathology, *in*: New Directions in Biological Control, R. R. Baker and P. E. Dunn, eds. UCLA Symposium. Alan Liss, New York.

60. Conway, K. E. 1976. Evaluation of *Cercospora rodmanii* as a biological control of waterhyacinth. Phytopathology 66:914–917.

61. Conway, K. E., and Cullen, R. 1978. The effect of *Cercospora rodmanii*, a biological control for waterhyacinth, on the fish, *Gambusia affinis*. Mycopathologia 66:113–116.

62. Conway, K. E., Cullen, R. E., Freeman, T. E., and Cornell, J. A. 1979. Field evaluation of *Cercospora rodmanii* as a biological control of waterhyacinth. Misc. Pap. A–79–6. U.S. Army Waterways Exp. Sta., Vicksburg, MS.

63. Conway, K. E., and Freeman, T. E. 1977. Host specificity of *Cercospora rodmanii*, a potential biological control of waterhyacinth. Plant Dis. Rep. 61:262–266.

64. Conway, K. E., Freeman, T. E., and Charudattan, R. 1978. Development of *Cercospora rodmanii* as a biological control for *Eichhornia crassipes*. Pages 225–230, *in*: Proc. EWRS 5th Symp. on Aquatic Weeds. Wageningen, The Netherlands.

65. Conway, K. E., Freeman, T. E., and Charudattan, R. 1978. Method and compositions for controlling waterhyacinth. U.S. Patent No. 4,097,261.

66. Cook, R. J., Andres, L., de Zoeten, G. A., Doane, C., Gwadz, R. W., Hardy, R., Hemming, B., Kuc, J., Mankau, R., Miller, D., Ryan, C. A. Jr., and Smith, S. 1987. Report of the Research Briefing Panel on Biological Control in Managed Ecosystems. National Academy Press, Washington, DC.

67. Crandall, B. S., and Baker, W. L. 1950. The wilt disease of American persimmon caused by *Cephalosporium diospyri*. Phytopathology 40:307–325.

68. Crawley, D. K., Walker, H. L., and Riley, J. A. 1986. Interaction of *Alternaria macrospora* and *Fusarium lateritium* on spurred anoda. Plant Dis. 69:977–979.

69. Daniel, J. T., Templeton, G. E., Smith, R. J., and Fox, W. T. 1973. Biological control of northern jointvetch in rice with an endemic fungal disease. Weed Sci. 21:303–307.

70. Daniel, J. T., Templeton, G. E., and Smith, R. J. Jr. 1974. Control of *Aeschynomene* species with *Colletotrichum gloeosporioides* f. sp. *aeschynomene*. U.S. Patent No. 3,849,104.

71. de Jong, M. D. 1988. Risk to Fruit Trees and Native Trees Due to Control of Black Cherry (*Prunus serotina*) by Silverleaf Fungus (*Chondrostereum purpureum*). Dissertation, Abstract in English, Landbouwuniversiteit te Wageningen, The Netherlands.

72. Dodd, A. P. 1961. Biological control of *Eupatorium adenophorum* in Queensland. Aust. J. Sci. 23:356–365.

73. Dyer, W. E., Turner, S. K., Fay, P. K., Sharp, E. L., and Sands, D. C. 1982. Control of Canada thistle by a rust, *Puccinia obtegens*. Pages 243–244, *in*: Biological Control of Weeds with Plant Pathogens, R. Charudattan and H. L. Walker, eds. Wiley, New York.

74. Evans, H., and Ellison, C. 1988. Preliminary work on the development of a mycoherbicide to control *Roettbollia cochinchinensis*. *In*: Abstr. VII Int. Symp. Biol. Control Weeds. Biol. Control Weeds Lab., USDA-ARS, Rome.

75. Figliola, S. S., Camper, N. D., and Ridings, W. H. 1988. Potential biological control agents for goosegrass (*Eleucine indica*). Weed Sci. 36:830–835.

76. Freeman, T. E., and Charudattan, R. 1974. Occurrence of *Cercospora piaropi* on waterhyacinth in Florida. Plant Dis. Rep. 58:277–278.

77. Freeman, T. E., and Charudattan, R. 1984. *Cercospora rodmanii* Conway, A Biocontrol Agent for Waterhyacinth. Florida Ag. Exp. Sta. Bull. 842. University of Florida, Gainesville.

78. Freeman, T. E., and Charudattan, R. 1984. Conflicts in the use of plant pathogens as biocontrol agents for weeds. Pages 351–357, in: Proc. VI Int. Sym. Biol. Control of Weeds, E. S. Delfosse, ed. Agric. Canada, Canadian Govt. Publ. Centre, Ottawa.

79. French, D. W., and Schroeder, D. B. 1969. The oak wilt fungus, *Ceratocystis fagacearum* as a selective silvicide. For Sci. 15:198–203.

80. Greaves, M. P., and Sargent, J. A. 1986. Herbicide-induced microbial invasion of plant roots. Weed Sci. 34 (Suppl. 1):50–53.

81. Greaves, M. P., Bailey, J. A., and Hargreaves, J. A. 1989. Mycoherbicides: opportunities for genetic manipulation. Pestic. Sci. 26:93–101.

82. Griffith, C. A. 1970. Persimmon Wilt Research. Anuual Report 1960–1970. Noble Foundation Agricultural Division, Ardmore, OK.

83. Gunner, H. B., Limpa-Amara, Y., and Weilerstein, P. J. 1988. Field evaluation of microbiological control agents on eurasion watermilfoil. Tech. Rep. A–88–1, Aquat. Plant Control Res. Prog., U.S. Army Waterways Exp. Sta., Vicksburg, MS.

84. Hasan, S. 1986. Industrial potential of plant pathogens as biocontrol agents of weeds. Symbiosis 2:151–163.

85. Hasan, S. 1988. Biocontrol of weeds with microbes. Pages 129–151, *in*: Biocontrol of Plant Diseases, K. G. Mukerji and K. L. Garg, eds. CRC Press, Boca Raton, FL.

86. Hodgson, R. H., and Snyder, R. H. 1989. Thidiazuron and *Colletotrichum coccodes* effects on ethylene production by velvetleaf (*Abutilon theophrasti*) and prickly sida (*Sida spinosa*). Weed Sci. 37:484–489.

87. Hofmeister, F. M., and Charudattan, R. 1987. *Pseudocercospora nigricans*, a pathogen of sicklepod (*Cassia obtusifolia*) with biocontrol potential. Plant Dis. 71:44–46.

88. Hokkanen, H. 1985. Exploiter–victim relationships of major plant diseases: implications for biological weed control. Agric. Ecosyst. Environ. 15:63–76.

89. Holcomb, G. E. 1978. *Alternaria alternantherae* from alligatorweed also is pathogenic to ornamental Amaranthaceae species. Phytopathology 68:265–266.

90. Holcomb, G. E., and Antonopoulos, A. A. 1976. *Alternaria alternantherae:* a new species found on alligatorweed. Mycologia 68:1125–1129.

91. Holcomb, G. E., Jones, J. P., and Wells, D. W. Blight of prostrate spurge and cultivated poinsettia caused by *Ambobotrys ricini*. Plant Dis. 73:74–75.

92. Howell, C. R. 1982. Effect of *Gliocladium virens* on *Pythium ultimum, Rhizoctonia solani,* and damping off of cotton seedlings. Phytopathology 72:496–498.

93. Howell, C. R., and Stiponovic, R. D. 1984. Phytotoxicity to crop plants and herbicidal effects on weeds of viridiol produced by *Gliocladium virens*. Phytopathology 74:1346–1349.

94. Huber-Meinicke, G., Defago, G., and Sedlar, L. 1988. *Ramularia rubella*, a potential mycoherbicide to control *Rumex* weeds. *In*: Abstr. VII Int. Symp. Biol. Control Weeds. Biol. Control Weeds Lab., USDA-ARS, Rome.

95. Irvine, J. I. M., Burge, M. N., and McElwee, M. 1987. Association of *Phoma aquilina* and *Ascochyta pteridis* with curl-tip disease of bracken. Ann. Appl. Biol. 110:25–31.

96. Johnson, D. A., and King, T. H. 1976. A leaf-spot disease of three genera of aquatic plants in Minnesota. Plant Dis. Rep. 60:726–730.

97. Jones, R. W., Lanini, W. T., and Hancock, J. G. 1988. Plant growth response to the phytotoxin viridiol produced by the fungus *Gliocladium virens*. Weed Sci. 36:683–687.

98. Joye, G. F. 1988. Biological control of *Hydrilla verticillata* (L. f.) Royle with an endemic fungal disease. Phytopathology 78:1593.

99. Joyner, B. G., and Freeman, T. E. 1973. Pathogenicity of *Rhizoctonia solani* to aquatic plants. Phytopathology 63:681–685.

100. Kenney, D. S. 1986. DeVine—the way it was developed—an industrialist's view. Weed Sci. 34 (Suppl. 1):15–16.

101. Khodayari, K., and Smith, R. J. Jr. A mycoherbicide integrated with fungicides in rice, *Oryza sativa*. Weed Technol. 2:282–285.

102. Kirkpatrick, T. L., Templeton, G. E., TeBeest, D. O., and Smith, R. J. 1982. Potential of *Colletotrichum malvarum* for biological control of prickly sida. Plant Dis. 66:323–325.

103. Klerk, R. A., Smith, R. J., and TeBeest, D. O. 1985. Integration of a microbial herbicide into weed and pest control programs in rice. Weed Sci. 33:95–99.

104. Klisiewicz, J. M. 1985. Growth and reproduction of *Dichotomophthora portulacae* and its biological activity on purslane. Plant Dis. 69:761–762.

105. Klokocar-Smit, Z., and Arsenovic, M. 1988. Survey of aquatic plants for possible biocontrol agents in Yugoslavia. *In*: Abstr. VII Int. Symp. Biol. Control Weeds. Biol. Control Weeds Lab., USDA-ARS, Rome.

106. Kovalev, O. V. 1977. Biological control of weeds (in Russian). Zashch. Past. 22:12–14.

107. Krupinsky, J. M. and Lorenz, R. J. 1983. An *Alternaria* sp. on leafy spurge (*Euphorbia esula*). Weed Sci. 31:86–88.

108. Leonard, K. J. 1982. The benefits and potential hazards of genetic heterogeneity in plant pathogens. Pages 99–112, *in*: Biological Control of Weeds with Plant Pathogens. R. Charudattan and H. L. Walker, eds. Wiley, New York.

109. Leuchtmann, A., and Clay, K. 1988. Experimental infection of host grasses and sedges with *Atkinsonella hypoxylon* and *Balansia cyperia* (Balansiae, Clavicipitaceae). Mycologia 80:291–297.

110. Madariaga, R. B., and Scharen, A. L. 1985. *Septoria tritici* blotch in Chilean wild oat. Plant Dis. 69:126–127.

111. Martyn, R. D., and Freeman, T. E. 1978. Evaluation of *Acremonium zonatum* as a potential biocontrol agent of waterhyacinth. Plant Dis. Rep. 62:604–608.

112. Martyn, R. D. 1985. Waterhyacinth decline in Texas caused by *Cercospora piaropi*. J. Aquat. Plant Manag. 23:29–32.

113. Massion, C. L., and Lindow, S. E. 1986. Effects of *Sphacelotheca holci* infection on morphology and competitiveness of johnsongrass (*Sorghum halepense*). Weed Sci. 34:883–888.

114. McCain, A. H., and Noviello, C. 1984. Biological control of *Cannabis sativa*. Pages 635–642, *in*: Proc. VI Int. Symp. Biol. Control Weeds, E. S. Delfosse, ed. Agric. Canada, Canadian Govt. Publ. Centre, Ottawa.

115. McPartland, J. M. 1983. *Phomopsis ganjae* sp. nov. on *Cannabis sativa*. Mycotaxon 18:527–530.

116. McRae, C. F., and Auld, B. A. 1988. The influence of environmental factors on anthracnose of *Xanthium spinosum*. Phytopathology 78:1182–1186.

117. McRae, C. F., Ridings, H. T., and Auld, B. A. 1988. Anthracnose of *Xanthium spinosum*— quantitative disease assessment and analysis. Aust. Plant Pathol. 17:11–13.

118. McWhorter, C. G. Future needs in weed science. 1984. Weed Sci. 32:850–855.

119. Mitchell, J. K. 1986. *Dichotomophthora portulacae* causing black stem rot on common purslane in Texas. Plant Dis. 70:603.

120. Mitchell, J. K. 1988. *Gibbago trianthemae*, a recently described hyphomycete with bioherbicide potential for control of horse purslane (*Trianthema portulacastrum*). Plant Dis. 72:354–355.

121. Mitchell, J. K. 1988. Evaluations of *Colletotrichum graminicola* (Cas.) Wils. and *Gloecercospora sorghi* D. Bain & Edg. as biological herbicides for controlling johnsongrass (*Sorghum halepense* [L.] Pers.). *In*: Abstr. VII Int. Symp. Biol. Control Weeds. Biol. Control Weeds Lab., USDA-ARS, Rome.

122. Morris, M. J. 1983. Evaluation of field trials with *Colletotrichum gloeosporioides* for the biological control of *Hakea sericea*. Phytophylactica 15:13–16.

123. Mortensen, K. 1986. Biological control of weeds with plant pathogens. Can. J. Plant Pathol. 8:229–234.

124. Mortensen, K. 1988. The potential of an endemic fungus, *Colletotrichum gloeosporioides*, for biological control of roundleaved mallow (*Malva pusilla*) and velvetleaf (*Abutilon theophrasti*). Weed Sci. 36:473–478.

125. Muir, J. A. 1977. Effects of the fungal hyperparasite *Colletotrichum gloeosporioides* of dwarf mistletoe (*Arceuthobium americanum*) on young lodgepole pine. Can. J. For. Res. 7:579–583.

126. Ormeno-Nunez, J., Reeleder, R. D., and Watson, A. K. 1988. A foliar disease of field bindweed (*Convolvulus arvensis*) caused by *Phomopsis convolvulus*. Plant Dis. 72:338–342.

127. Phatak, S. C., Callaway, M. B., and Vavrina, C. S. 1987. Biological control and its integration in weed management systems for purple and yellow nutsedge (*Cyperus rotundus* and *C. esculentus*). Weed Technol. 1:84–91.

128. Phatak, S. C., Sumner, D. R., Wells, H. D., Bell, D. K., and Glaze, N. C. 1983. Biological control of yellow nutsedge with the indigenous rust fungus *Puccinia canaliculata*. Science 219:1446–1447.

129. Ponnappa, K. M. 1970. On the pathogenicity of *Myrothecium roridum–Eichhornia crassipes* isolate. Hyacinth Control J. 8:18–20.

130. Prange, V. J., and Charudattan, R. 1988. Herbicidal metabolites as indicators of biological control efficacy of microorganisms against *Hydrilla verticillata*. *In*: Abstr. Annu. Meet. Aquat. Plant Manag. Soc. July 10–13, 1988, New Orleans, LA.

131. Quimby, P. C., Jr. 1989. Response of common cocklebur (*Xanthium strumarium*) to *Alternaria helianthi*. Weed Technol. 3:177–181.

132. Quimby, P. C., and Fulgam, F. E. 1984. A sprayer for mycoherbicide research. Weed Sci. Soc. Am. Abstr. 37:393.

133. Quimby, P. C., and Walker, H. L. 1982. Pathogens as mechanisms for integrated weed management. Weed Sci. 30 (Suppl. 1):30–34.

134. Rai, I. S., and Bridgmon, G. H. 1971. Studies on root rot of Canada thistle (*Cirsium arvense*) caused by *Fusarium roseum*. J. Colo. Wyo. Acad. Sci. 17:2.

135. Ravenell, D. I., and Van Dyke, C. G. 1986. Efficacy and host range of *Bipolaris setariae* as a potential biocontrol pathogen of broadleaf signalgrass (*Brachiaria platyphylla* [Griseb.] Nash). WSSA Abstr. 26:54.

136. Rejmankova, E., Blackwell, M., and Culley, D. D. 1986. Dynamics of fungal infection in duckweeds (Lemnaceae). Veroff. Geobot. Inst. ETH, Stiftung Rubel, Zurich 87:178–189.

137. Ridings, W. H. 1986. Biological control of stranglervine in citrus—a researcher's view. Weed Sci. 34 (Suppl. 1):31–32.

138. Ridings, W. H., Mitchell, D. J., Schoulties, C. L., and El-Gholl, N. E. 1976. Biological control of milkweed vine in Florida citrus groves with a pathotype of *Phytophthora citrophthora*. Pages 224–240, *in*: Proc. IV Int. Symp. Biol. Control Weeds, T. E. Freeman. ed. University of Florida, Gainesville, FL.

139. Rintz, R. E. 1973. A zonal leafspot of waterhyacinth. Hyacinth Control J. 11:41–44.

140. Scheepens, P. C. 1987. Joint action of *Cochliobolus lunatus* and atrazine on *Echinochloa crusgalli* (L.) Beauv. Weed Res. 27:43–47.

141. Scheepens, P. C., and Hoogerbrugge, A. 1988. Control of *Prunus serotina* in forests with the endemic fungus *Chondrostereum purpureum*. *In*: Abstr. VII Int. Symp. Biol. Control Weeds. Biol. Control Weeds Lab., USDA-ARS, Rome.

142. Scheepens, P. C., and van Zon, H. C. J. 1982. Microbial herbicides. Pages 623–641, *in*: Microbial and Viral Pesticides, E. Kurstak, ed. Dekker, New York.

143. Shabana, Y. M. N. El-Dean, 1987. Biological Control of Water Weeds by Using Plant Pathogens. Dissertation, Mansoura University, El-Mansoura, Egypt.

144. Smith, C. S., Slade, S. J., Andrews, J. H., and Harris, R. F. 1989. Pathogenicity of the fungus, *Colletotrichum gloeosporioides* (Penz.) Sacc. to eurasian watermilfoil (*Myriophyllum spicatum* L.). Aquat. Bot. 33:1–12.

145. Smith, R. J., Jr. 1982. Integration of microbial herbicides with existing pest management programs. Pages 189–203, in: Biological Control of Weeds with Plant Pathogens, R. Charudattan and H. L. Walker, eds. Wiley, New York.

146. Smith, R. J., Jr. 1986. Biological control of northern jointvetch in rice and soybeans—a researcher's view. Weed Sci. 34 (Suppl. 1):17–23.

147. Sorsa, K. K., Nordheim, E. V., and Andrews, J. H. 1988. Integrated control of eurasian water milfoil, *Myriophyllum spicatum,* by a fungal pathogen and a herbicide. J. Aquat. Plant Manag. 26:12–17.

148. Stovall, M. E., and Clay, K. 1988. The effect of the fungus, *Balansia cyperi* Edg., on growth and reproduction of purple nutsedge, *Cyperus rotundus* L. New Phytol. 109:351–359.

149. Suzuki, H. 1988. Trial of biocontrol of water chestnut (*Eleocharis kuroguwai*) in paddy field with fungal pathogen. Agric. Hortic. (Jpn) 63:741–744, 877–879, 969–974 (in Japanese).

150. Taber, R. A., Mitchell, J. K., and Brown, S. M. 1988. Potential for biological control of the weed *Trianthema* with *Drechslera* (*Exserohilum*) *indica*. Page 130, *in*: Abstr. Papers, 5th Int. Congr. of Plant Pathology, Aug. 20–27, 1988, Kyoto, Japan.

151. TeBeest, D. O. 1982. Survival of *Colletotrichum gloeosporioides* f. sp. *aeschynomene* in rice irrigation water and soil. Plant Dis. 66:469–472.

152. TeBeest, D. O. 1988. Additions to host range of *Colletotrichum gloeosporioides* f. sp. *aeschynomene*. Plant Dis. 72:16–18.

153. TeBeest, D. O., and Brumley, J. M. 1978. *Colletotrichum gloeosporioides* borne within the seed of *Aeschynomene virginica*. Plant Dis. Rep. 62:675–678.

154. TeBeest, D. O., and Templeton, G. E. 1985. Mycoherbicides: progress in the biological control of weeds. Plant Dis. 69:6–10.

155. TeBeest, D. O., and Weidemann, G. J. Preparation of protoplasts from conidia of *Colletotrichum gloeosporioides* f. sp. *aeschynomene*. Phytopathology 75:1361.

156. Templeton, G. E. 1982. Status of weed control with plant pathogens. Pages 29–44, *in*: Biological Control of Weeds with Plant Pathogens, R. Charudattan and H. L. Walker, eds. Wiley, New York.

157. Templeton, G. E. 1986. Mycoherbicide research at the University of Arkansas—Past, present, and future. Weed Sci. 34 (Suppl. 1):35–37.

158. Templeton, G. E. 1988. Biological control of weeds. Am. J. Alt. Agric. 3:69–72.

159. Templeton, G. E., and Greaves, M. P. 1984. Biological control of weeds with fungal pathogens. Trop. Pest Manag. 30:333–338.

160. Templeton, G. E., and Heiny, D. K. 1989. Improvement of fungi to enhance mycoherbicide potential.

161. Templeton, G. E., Smith, R. J., Jr., and TeBeest, D. O. 1986. Progress and potential of weed control with mycoherbicides. Rev. Weed Sci. 2:1–14.

162. Templeton, G. E., Smith, R. J., and Klomparens, W. 1980. Commercialization of fungi and bacteria for biological control, Biocontrol News Inf. 1:291–294.

163. Templeton, G. E., TeBeest, D. O., and Smith, R. J., Jr. 1984. Biological weed control in rice with a strain of *Colletotrichum gloeosporioides* (Penz.) Sacc. used as a mycoherbicide. Crop Prot. 3:409–422.

164. Templeton, G. E., TeBeest, D. O., and Smith, R. J., Jr. 1979. Biological weed control with mycoherbicides. Annu. Rev. Phytopathol. 17:301–310.

165. Templeton, G. E. 1976. *Colletotrichum malvarum* spore concentration and agricultural process. U.S. Patent No. 3,999,973.

166. Trujillo, E. E., Latterell, F. M., and Rossi, A. E. 1986. *Colletotrichum gloeosporioides,* a possible biological control agent for *Clidemia hirta* in Hawaiian forests. Plant Dis. 70:974–976.

167. Trujillo, E. E., and Obrero, F. P. 1976. Cephalosporium wilt of *Cassia surattensis* in Hawaii. Pages 217–220, *in:* Proc. IV Int. Symp. Biol. Control Weeds, T. E. Freeman, ed. University of Florida, Gainesville.

168. Tu, C. C., and Kimbrough, J. W. 1978. Systematics and phylogeny of fungi in the *Rhizoctonia* complex. Bot. Gaz. 139:454–466.

169. Turner, S. K., Fay, P. K., Sharp, E. L., and Sands, D. C. 1981. Resistance of Canada thistle (*Cirsium arvense*) ecotypes to a rust pathogen (*Puccinia obtegens*). Weed Sci. 29:623–624.

170. Van Den Ende, G., Frantzen, J., and Timmers, T. 1987. Teleutospores as origin of systemic infection of *Cirsium arvense* by *Puccinia punctiformis.* Neth. J. Plant Pathol. 93:233–239.

171. Van Dyke, C. G., and Trigiano, R. N. 1987. Light and scanning electron microscopy of the interaction of the biocontrol fungus *Alternaria cassiae* with sicklepod (*Cassia obtusifolia*). Can. J. Plant Pathol. 9:230–235.

172. Walker, H. L. 1980. *Alternaria macrospora* as a potential biocontrol agent for spurred anoda: Production of spores for field studies. Adv. Agric. Technol, AAT-S-12. USDA-SEA-AR, New Orleans, LA.

173. Walker, H. L. 1981. *Fusarium lateritium:* a pathogen of spurred anoda (*Anoda cristata*), prickly sida (*Sida spinosa*), and velvetleaf (*Abutilon theophrasti*). Weed Sci. 29:629–631.

174. Walker, H. L. 1982. A seedling blight of sicklepod caused by *Alternaria cassiae.* Plant Dis. 66:426–428.

175. Walker, H. L. 1983. Control of sicklepod, showy crotalaria, and coffee senna with a fungal pathogen. U.S. Patent No. 4,390,360.

176. Walker, H. L., and Boyette, C. D. 1985. Biological control of sicklepod (*Cassia obtusifolia*) in soybeans (*Glycine max*) with *Alternaria cassiae.* Weed Sci. 33:212–215.

177. Walker, H. L., and Boyette, C. D. 1986. Influence of sequential dew periods on biocontrol of sicklepod (*Cassia obtusifolia*) by *Alternaria cassiae.* Plant Dis. 70:962–963.

178. Walker, H. L., and Connick, W. J. 1983. Sodium alginate for production and formulation of mycoherbicides. Weed Sci. 31:333–338.

179. Walker, H. L., and Riley, J. A. 1982. Evaluation of *Alternaria cassiae* for the biocontrol of sicklepod (*Cassia obtusifolia*). Weed Sci. 30:651–654.

180. Walker, H. L., and Sciumbato, G. L. 1979. Evaluation of *Alternaria macrospora* as a potential biological control agent for spurred anoda (*Anoda cristata*): host range studies. Weed Sci. 27:612–614.

181. Walker, J. C. 1969. Plant Pathology, 3rd ed. McGraw-Hill, New York.

182. Watson, A. K. 1985. Host specificity of plant pathogens in biological weed control. Pages 577–586, in: Proc. VI Int. Symp. Biol. Control Weeds. E. S. Delfosse, ed. Agric. Canada, Canadian Govt. Publ. Centre, Ottawa.

183. Webb, R. R., and Lindow, S. E. 1987. Influence of environment and variation in host susceptibility on a disease of bracken fern caused by *Ascochyta pteridis.* Phytopathology 77:1144–1147.

184. Weidemann, G. J. 1988. Effects of nutritional amendments on conidial production of *Fusarium solani* f. sp. *cucurbitae* on sodium alginate granules and on control of Texas gourd. Plant Dis. 72:757–759.

185. Weidemann, G. J., and Templeton, G. E. 1988. Efficacy and soil persistence of *Fusarium solani* f. sp. *cucurbitae* for control of Texas gourd (*Cucurbita texana*). Plant Dis. 72:36–38.

186. Weidemann, G. J., and Templeton, G. E. 1988. Control of Texas gourd, *Cucurbita texana*, with *Fusarium solani* f. sp. *cucurbitae*. Weed Technol. 2:271–274.

187. Whitney, N. G., and Taber, R. A. 1986. First report of *Amphobotrys ricini* infecting *Caperonia palustris* in the United States. Plant Dis. 70:892.

188. Wilson, C. L. 1963. Wilting of persimmon caused by *Cephalosporium diospyri*. Phytopathology 53:1402–1406.

189. Wilson, C. L. 1965. Consideration of the use of persimmon wilt as a silvicide for weed persimmon. Plant Dis. Rep. 49:789–791.

190. Winder, R. S., and Van Dyke, C. G. 1986. *Cercospora dubia* (Reiss) Wint. on *Chenopodium album* L.: greenhouse efficacy and biocontrol potential. Proc. Southern Weed Sci. Soc. 39:387.

191. Winder, R. S., and Van Dyke, C. G. 1987. The effects of various adjuvants in biological control of johnsongrass (*Sorghum halepense* [L.] Pers.) with the fungus *Bipolaris sorghicola*. Weed Sci. Soc. Am. Abstracts 27:128.

192. Winder, R. S., and Van Dyke, C. G. 1989. The pathogenicity, virulence, and biocontrol potential of two *Bipolaris* species on johnsongrass (*Sorghum halepense*). Weed Sci. 37. 38:89–94.

193. Woodhead, S. H. 1981. Field efficacy of *Phytophthora palmivora* for control of milkweed vine. Phytopathology 71:913.

194. Wymore, L. A., and Watson, A. K. 1989. Interaction between velvetleaf isolate of *Colletotrichum coccodes* and thidiazuron for velvetleaf (*Abutilon theophrasti*) control in the field. Weed Sci. 37:478–483.

195. Wymore, L. A., Watson, A. K., and Gotlieb, A. R. 1987. Interaction between *Colletotrichum coccodes* and thidiazuron for control of velvetleaf (*Abutilon theophrasti* L.). Weed Sci. 35:377–382.

196. Yoder, O. C. 1983. Use of pathogen-produced toxins in genetic engineering of plants and pathogens. Pages 335–353, *in*: Genetic Engineering of Plants, T. Kosuge, C. P. Meredith, and A. Hollaender, eds. Plenum Press, New York.

197. Yorinori, J. T. 1984. Biological control of milk weed (*Euphorbia heterophylla*) with pathogenic fungi. Pages 677–681, *in*: Proc. VI Int. Symp. Biol. Control Weeds, E. S. Delfosse, ed. Agric. Canada, Canadian Govt. Publ. Centre, Ottawa.

198. Yorinori, J. T., and Gazziero, D. L. P. 1988. Control of milk weed *Euphorbia heterophylla* with *Helminthosporium* sp. *In*: Abstr. VII Int. Symp. Biol. Control Weeds. Biol. Control Weeds Lab., USDA-ARS, Rome.

3

Nematodes as Biological Control Agents of Weeds

Paul E. Parker

Introduction

The use of nematodes as biological control agents is not a new concept nor is it restricted to the area of weed control. To the contrary, the majority of work and successes with nematode biological control agents has occurred in the area of biological control of insects. Considerable energies have been expended with such diverse targets as the larch sawfly (*Cephalicia lariciphila*) (5), carpenter worms (*Prionoxystus robinae* Peck) of figs (11), and Colorado potato beetle (*Leptinotarsa decemlineata* Say) (13). Many of these systems employing the insect parasitic nematode *Neoaplectana carpocapsae* Weiser show promise as biological control options (12). The utilization of plant parasitic nematodes for the biological control of weeds is newer and less proven than the insect–nematode biological control strategies. All schemes involving the control of weeds by nematode agents center on the application of foliar gall-forming types.

Gall-forming nematodes have been known since 1743, when Needham (18) discovered motionless eelwormlike organisms inside grains of wheat. *Anguina tritici* (Steinbuch) Chitwood, the casual agent for the galling of the wheat kernel, is representative of the family Anguinidae, a grouping of the nematodes noted for their ability to induce malformations (galls) on foliage and floral portions of plants. The type genus *Anguina* almost exclusively infects monocots. Other genera within the family attack monocots and/or dicots (Figure 3.1 [1, 7, 9]).

Gall formation has been termed the growth reaction of the plant against the invasion of the parasite. In the case of gall-forming nematodes it appears that the nematode derives all the benefits from gall development, including shelter, nourishment, and perhaps even dispersal. Others have proposed that the development of galls may be a passive means of defense by the plant in an effort to shield itself from the full impact of the parasite. Gall formation has also been described as a localization of the parasite that over time has forced the parasite into an extreme form of specialization. Foliar galls formed by nematodes are character-ized by a pronounced hypertrophy and cell proliferation of the mesophyll and by presence of a central cavity containing nematodes (14). Gall development may be related to increased auxin levels within infected plant tissues, and there is speculation that the host–cell auxin level may be augmented by the nematode (34).

An important characteristic of foliar gall-forming nematodes of the Anguinidae is their ability to survive dehydration. This phenomenon of anhydrobiosis has been known since the time of Leeuwenhoek (2). A requirement for this state is a period of slow dehydration. In the induction phase of anhydrobiosis the nema-todes coil into tight spirals. Glycerol induction occurs, and the formation of polyalcohols like glycerol may contribute to the ability of the nematode to survive desiccation. The recovery phase is hypothesized to consist of a latent period (metabolism resumes) and a later recovery period during which movement is resumed. Studies with the free-living mycophagous nematode *Aphelenchus avenae* indicate that lipid and glycogen declined rapidly upon dehydration, whereas glycerol and trehalose contents both increased rapidly. Strong correla-tions between survival in dry air and glycerol and trehalose contents were ob-served. The observed increase in glycerol may contribute to the survival of the nematode under dehydrated conditions by replacing the water structure around macromolecules and membrane systems. When rehydration occurs the morpho-logical and metabolic changes that occur during the process of anhydrobiosis are reversed upon rehydration. A long period of slow dehydration is required for the nematodes to reenter another cycle of anhydrobiosis (3,15).

Figure 3.1. Some Foliar Gall-Forming Nematodes of Dicots

Primary Host Plant		Nematode	
Family	Genus Species	Family	Genus Species
Boraginaceae	*Amsinckia intermedia*	Tylenchidae	*Anguina amsinckia**
Compositae	*Achillea millefolium*		*Subanguina millefolii*
	Acroptilon repens		*S. picridis**
	Arctotheca calendula		*S. mobilis*
	Artemisia asiatica		*S. moxae*
	Balsamorrhiza sagittata		*A. balsamophila*
	Centaurea rigida		*A. danthoniae*
	Chartolepis bickersteini		*A. chartolepidis*
	Consinia stenocephala		*A. danthoniae*
	Wyethia amplexicaulis		*A. balsamophila*
Plantaginaceae	*Plantago aristata*		*S. plantaginis**
Polygonaceae	*Polygonium alpestre*		*S. polygoni*
Primulaceae	Primula florindae		*A.* klebahni
Solanaceae	Solanum elaeagnifolium		Orrina phyllobia*

*Mentioned as potential biological control agents. Adopted from Brzesk (1) and Hirschmann (9).

Folliar Gall-Forming Nematodes of Dicots

Goodey (8) investigated the gall-forming nematode *Subanguina millefoli* (Low)
Brzeski. The parasite has a wide plant host distribution, including several species
of *Achillea*. Observations of *S. millefoli* on *millefolium* L. indicated that twisting
and distortion of infected plant parts did not adversely affect the "general health"
of the plant. Leaflets and stems were commonly infected by the nematode. Cells
closest to the gall cavity contained granular protoplasm. Vascular tissues were
more abundant than in healthy plant tissues (8). In studies of *A. balsamophila*
(Thorne) Filipjev infective nematode larvae entered young leaves of developing
crowns of *Balsamorhize sagittata*. Infective larvae entered in mass at various
points. At 3 weeks time, larvae matured into adults and began depositing eggs.
During the normal growth cycle of *B. sagittata* the plants wither and die back in
July or August. At this point the newly hatched larvae become quiescent. With
the onset of spring the larvae escape the galled plant debris and congregate around
developing crowns. Galled tissue of *Wyethia amplexicaulis,* another host of *A.
balsamophila,* tends to exhibit considerable proliferation of leaf tissues, with the
center of the gall being hollow. Cells with dense protoplasmic contents surround
the hollow central chamber. The observations by Goodey with these two nematodes highlight the type of life cycle and development that is typical of foliar gall-
forming nematode (6).

Biological Control Prospects for Anguinidae

Within North America several weed problems lend themselves to the potential
use of plant parasitic nematodes as biological control agents. Where the nematode
and weed have evolved naturally in North America, an augmentative biological
approach is probably most appropriate, but in an introduced exotic weed, a
classical biological control approach is feasible. Both methods have advantages
and disadvantages. For an augmentative approach large numbers of biological
control agents at a relatively low price are required. Introduced exotic agents
require close examination to ensure they are free of pathogens and hyperparasites.
Both approaches require attention to potential impacts of the biological agent
upon plants of agronomic importance and native floras.

Common Fiddleneck

Anguina amsinckia parasitize the common fiddleneck, *Amsinckia intermedia,*
an annual, native weed in California. This weed poses a potential problem to
grazing livestock, since, if ingested in sufficient quantities, it can be toxic (16).
Infection of plants by the nematode tends to be geographically localized. Actual
galling of plant tissues is often limited to the fruits and leaves. Galls are characterized by a central cavity surrounded by dead empty cells. More galling may occur

within fruit and the apical meristems where the nematode is protected from desiccation (33).

Higher humidity levels provided by the stem apex appear to change when the inflorescence elongates. Pantone et al. consider this a factor that curtails nematode movement and subsequent infection. The exposure period during which floral parts are susceptible to nematode infection is short and may limit the utility of this organism as a biological control agent (25). Other factors, such as host range, may also influence use of this agent (23,24).

Silverleaf Nightshade

Silverleaf nightshade, *Solanum elaeagnifolium* Cav., a native to North America, is an economically important perennial weed throughout much of the southwestern United States (22). It has also become a significant pestiferous weed in several countries.

In Australia the weed is considered a serious threat to both crops and pastures. The first infestation in Australia was reported in 1901. Control has been difficult because of the ability of the weed to recover following cultivation. *S. elaeagnifolium* is especially annoying in the states of south Australia and Victoria, where it competes with pasture grasses, reducing yields (4).

The entry of *S. elaeagnifolium* into India is conjectured as being via a contaminant of imported food grains (17). The foliar gall-forming nematode *Orrina phyllobia* (native to North America) also gained entrance to India possibly through infected, dried plant debris (31).

South Africa also recognizes *S. elaeagnifolium* as a serious weed pest. Herbicides have not been successful in controlling the weed, and mechanical control is not practical because of the deep-spreading root system that easily regenerates new plants upon being damaged. In 1973, Neser viewed *S. elaeagnifolium* as a prime target for biological control (19).

In the life cycle of *O. phyllobia,* adults and preinfective juveniles develop only within the moist microhabitat of foliar and stem galls. When galled leaf and plant material abscises and dries, the adults and preinfective juveniles die. The infective stage larvae are able to enter a state of anhydrobiosis in which they can remain viable for several years. This is especially useful in a biological control program, since the inactive form can be applied at a time convenient for the grower. The nematode will revive when conditions are suitable for its development. The dormant stage is able to revive within several hours following rehydration of the gall material. The natural infection cycle commences during periods of extended moisture when the larvae rehydrate, become motile, and exit the galls. In this free state within the upper soil surface, the infective larvae can infest preemergent or emerged shoots. In their dried anhydrobiotic state nematodes are dispersed by wind-blown gall and other infected plant debris. Orr (29) determined that large numbers of infective larvae can be readily introduced into *S. elaeagnifolium* populations by merely spreading dried gall material within the weed stand. Once introduced, the nematodes spread rapidly and significantly reduce top growth of the weed.

Typical gall development is detectable 0–1.5 days following the initial infection period. Larvae can be observed moving among the plant trichomes and become concentrated among apical leaf folds. This concentration of nematodes at the apical leaf folds continues, with actual penetration of the leaf epidermis occurring sometime during the first day (31). Larvae of *O. phyllobia* are often found lodged in wounded areas of stems. The nematode appears to be attracted toward members of the Solanaceae, particularly *Solanum* species. Larvae are not selective of which stems they ascend, virtually ascending any available stem that exhibits proper moisture conditions (30). Penetration usually occurs through the adaxial surface. At 2–4 days, minute swellings less than 2 mm in diameter appear along leaf veins, midribs, petioles, axillary buds, and stems. Within infected tissues, nematodes tend to localize in the palisade and cortical parenchyma. By the third day of development the infected areas contain larger intercellular spaces and the number of nematodes observed increases. At this point nematodes can be observed molting, and distortion and swelling of palisade layer commences. Minute swellings along veins and petioles enlarge into small galls less than 5 mm in diameter. Occasionally, nematodes advance into mid-portions of the leaf. At 3–4 days, gravid females and small numbers of eggs can be found within the developing galls. Seven to 9 days following initial infection, galls consist of loosely packed tissue composed primarily of intercellular spaces with all developmental stages of the nematode present. Larvae begin to migrate into previously uninfected areas. Highly developed galls at volumes up to 5 cm^3 are evident at 15 days. By 30 days, second-generation females mature and begin active egg deposition. At this time actual nematode numbers within individual galls may exceed several hundred thousand (32).

The potential of utilizing *O. phyllobia,* an endemic North American species, as a biological control agent was first contemplated by Notham and Orr (Fig 3.2) (21). *Orrina phyllobia* is commonly found in roadside areas and abandoned fields throughout most of Texas and northeast Mexico (35). Preliminary host range studies indicated that *O. phyllobia* was host specific for *S. elaeagnifolium.* Several crops, including *Lycopersicon esculentum* Mill., *S. tuberosum* L., and *Gossypium hirsutum* L., were tested in the original plant screening. All proved negative for infection by *O. phyllobia.* Later studies indicates that *S. melongena* L. did serve as a limited host plant (20,30). Recent work in South Africa indicates that the host range of *O. phyllobia* may be more extensive then previously known. Neser et al. expressed doubts about widespread utilization of *O. phyllobia* due in part to concern in South Africa over the potential impact of the nematode upon endemic Solanaceae species. Screening of endemic natives showed galling on *S. coccineum* Jacq., *S. Burchelli* Dun., and *S. panduriforme* E. Mey., although

Figure 3.2. *Orrina phyllobia* on *Solanum elaeagnifolium*. (A) Emerging shoot of *S. elaeagnifolium* showing heavy galling. (B, C) Close-up view of galling along mid rib of *S. elaeagnifolium* leaf, typical of gall formation under field conditions. (D) Central hollow cavity of gall. (E) Field production plots for *O. phyllobia* at Mission, TX.

galling on the South African endemic species was less extensive than that experienced with *S. elaeagnifolium* (20).

Mass Production of *Orrina Phyllobia*

An economical mass production scheme is required to utilize *O. phyllobia* in an augmentation biological control program against *S. elaeagnifolium* within the United States. At the U.S. Department of Agriculture, Animal and Plant Health Inspection Service Mission Biological Control Laboratory, Mission, Texas, a field production scheme was developed and employed in the mass production of *O. phyllobia*. The field production scheme requires an adequate dense stand of *S. elaeagnifolium* for optimum production. Most natural stands of the weed are too restricted in size or too scattered to be of use in mass production. Also the availability of irrigation lessens dependency upon rainfall for production purposes and allows multiple cropping during the growing season.

Several different methods were employed in the establishment of the *S. elaeagnifolium*. The seed was planted directly into rows or broadcast onto the prepared field. Germination by both of these methods with and without irrigation was nil. Successful germination was achieved by germinating seed in the greenhouse in seed flats and transplanting to peat pots. Plants were grown from 20 to 30 cm in height prior to transplanting to the field. Seedlings were transplanted 30 cm apart with 76-cm row spacing. Field production plots were irrigated and weeded. Plants grew well under these conditions, with the plant canopy covering the entire plot after 4 months. Total area planted by this method has exceeded 1 ha. Prior to the inoculation of field plots, mature *S. elaeagnifolium* plants were disked under the soil surface. Plots were inoculated by broadcasting nematode inoculum (dried galled leaves containing 30,000+ nematodes per gram) onto the plot at a rate of 3.4–11.2 kg/ha. Light disking of plots helped to incorporate inoculum into the upper soil surface. A modified sprinkler system was employed to maintain moist conditions for a minimum of 16–20 hours following inoculation. Galls resulting from inoculation were evident several weeks following initial infection.

Harvesting of gall material commenced 4–5 weeks following the infection period, depending upon local weather conditions. Moist climatic conditions speeded up gall formation, whereas drier, hotter conditions tended to slow gall development. Galled plants were individually collected so as not to dilute the nematode inoculum with uninfected plant tissue.

Collected infected galls were dried under forced air at a temperature of 37°C for 24–36 hours. Drying time was dependent on air humidity levels. The dryer employed at the Mission Laboratory was a modified tobacco dryer design. Gall material was placed on screened trays, which were placed in racks with forced heated air circulating through the unit. Following drying, stems and berries were stripped off and leaf gall inoculum was double bagged in polyethylene bags. All inoculum was then stored at −17°C.

Unpublished estimates of the cost of this production system ranges from $55–60 per pound of dried galled leaf material. This price was not considered competitive with currently available weed control strategies. Other techniques of mass-producing the nematode may offer opportunity to reduce costs associated with inoculum production and make the biological control option more competitive (26,27).

Russian Knapweed

Acroptilon repens L. DC. (Russian knapweed) is a persistent, introduced perennial weed in North America. The combination of an extensive root system coupled with the plant's allelopathic properties may be partially responsible for its strong competitive ability. In dense infestations it inhibits other plant growth and reduces yields of important agronomic plants. Besides its strong competitive growth, *A. repens* is also considered a toxic weed, being especially poisonous to horses (38). Probable entry into North America occurred about 1900 with importations of Turkestan alfalfa (39). In North America *A. repens* is relatively free of specialized parasites and is not extensively attacked by polyphagous feeders. *A. repens* propagates primarily by vegetative means. The inability of *A. repens* to set seed has lessened the potential effectiveness of introducing various seed head biological control agents, such as *Aceria acroptiloni, Urophora maura,* and *Dasyneura* spp. Emphasis has been directed toward biological control agents that attack vegetative portions of the plants (36). The nematode *Subanguina picridis* (family: Anguinideae) causes extensive damage to *A. repens.* A native to southern portions of the U.S.S.R., it has a worldwide distribution.

Extensive work has been done by A. K. Watson (36,37,39–41) in Canada on the release, establishment, and biology of *Subanguina picridis,* a foliar gall-forming nematode of *Acroptilon repens* L. The organism is currently used within the U.S.S.R. as a biological control agent of Russian knapweed (10). Host races may have developed, although the possibility of up to six different species of the genus has been dismissed. All "races" of the genus are known to attack members of the Cynareae tribe. Morphological differences between various populations exist, depending upon which host plant the nematode is retrieved from, but these differences are not considered distinct enough to justify separate species status (41).

Recent work by Watson (39) indicates that the actual host range of *S. picridis* is more extensive than originally speculated. Gall formation was noted on various members of Centaureinae, including *A. repens, Centurea diffusa, C. maculosa, Carduus nutans, Cirsium flodmanii, Cynara scolymus,* and *Onapordum acanthium.* Representatives from two other families, Echinopinae and Mutisieae, have also exhibited galling infection by *S. picridis.* For the purposes of Watson's studies, host plant response was divided into two categories, (1) resistant plants and (2) susceptible plants. Resistant host plants exhibited good growth with poor parasite reproduction, and susceptible plants showed poor growth and good parasite reproduction. Watson considered only one of the screened species to be susceptible, *C. maculosa*. Of economically important crops screened, *Carthamus tinctorius* (safflower), a close relative to *A. repens,* was resistant with no gall formation. *S. picridis* did induce gall formation and was able to reproduce on *C. scolymus* (Globe artichoke), although the infection level was low (40).

Discussion

Several of the nematodes discussed have been utilized or evaluated for use as biological control agents of native and introduced weeds. The overview presented indicates that they may provide a beneficial role in biological control strategies of weeds.

An important point that needs to be carefully considered is the potential impact of the nematode biological control agents upon plants of agronomic importance and native floras. This is especially true where there is the possibility of potential impact on an endangered plant species. Pantone et al. (23,24) recognized a more extensive host range with *Anguina amsinckiae* than previously recorded, and Neser described potential conflicts of *Orrina phyllobia* with certain South African native floras (20). With increased emphasis being placed on the value of native plants, the impact of potential biological control agents must be fully assessed. Pemberton's work (28) with the native plant issue with regard to insect natural enemies of leafy spurge *(Euphorbia esula* L.) exemplifies a thorough approach to native plant and biological control agent conflicts. The potential role of plant parasitic nematodes in the United States will have to be carefully balanced between its weed control benefits and potential conflict with nontarget plant species.

Literature Cited

1. Brzeski, M. W. 1981. The genera of *Anguinidae* (Nematoda, Tylenchida). Rev. Nematol. 4(1):23–34.

2. Crowe, J. H., and K. A. Madin. 1974. Anhydrobiosis in tardigrades and nematodes. Trans. Amer. Microsc. Soc. 93(4):513–524.

3. Crowe, J. H., and K. A. C. Madin. 1975. Anhydrobiosis in nematodes: evaporative water loss and survival. J. Exp. Zool. 193:323–334.

4. Cuthbertson, E. G., A. R. Leys, and G. McMaster. 1976. Silverleaf nightshade—a potential threat to agriculture. Agric. Gaz. NSW 87:11–13.

5. Georgis, R., and N. G. M. Hague. 1982. Preliminary field trial of the nematode *Neoaplectana carpocapsae* against prepupae of the larch sawfly *Cephalcia lariciphila*. IRCS 10:616.

6. Goodey, J. B. 1948. The galls caused by *Anguillulina balsamophila* (Thorne) Goodey on the leaves of *Wyethia amplexicaulis* Nutt. and *Balsamorhiza sagittata* Nutt. J. Helminthol. 22:109–116.

7. Goodey, J. B. 1965. Ditylenchus and Anguina. Pages 47–58, *in* J. F. Southey, ed., Plant Nematology. Bull. Minist. Agr. Fish. Fd., London.

8. Goodey, T. 1938. Observations on *Anguillina millefolii* (Law, 1874) Goodey, 1932, from galls on the leaves of yarrow, *Achillea milefolium* L. J. Helminthol. 14:93–108.

9. Hirschmann, H. 1977. *Anguina plantaginis* n. sp. parasitic on *Plantago aristata* with a description of its development stages. J. Nematol. 9(3):229–243.

10. Kasimova, G. A. 1978. A record of the knapweed nematode *Paranguina picridis* on Russian knapweed in Azerbaidzhan. Dokl. AKAD. Nauk Azberbaidzhanskoi SSR 34(1):51–53.

11. Lindegren, J. E., and W. W. Barnett. 1982. Applying parasite nematodes to control carpenter-worms in fig orchards. Calif. Agric. November–December, pp. 7–9.

12. Lindegren, J. E., J. E. Dibble, C. E. Curtis, T. T. Yamashita, and E. Romero. 1981. Compatibil-ity of NOW parasite with commercial sprayers. Calif. Agric. March–April, pp. 16–17.

13. MacVean, C. M., J. W. Brewer, and J. L. Capinera. 1982. Field test of antidesiccants to extend the infection period of an entomogenous nematode, *Neoaplectana carpocapsae*, against the colorado potato beetle. J. Econ. Entomol. 75(1):97101.

14. Mani, M. S. 1964. Ecology of plant galls. Junk Publishers, The Hague.

15. Madin, K. A. C., and J. H. Crowe. 1975. Anhydrobiosis in nematodes: carbohydrate and lipid metabolism during dehydration. J. Exp. Zool. 193:335–342.

16. Nagamine, C., and A. R. Maggenti. 1980. "Blinding" of shoots and a leaf gall in *Amsinckia intemedia* induced by the *Anguina amsinckia* (Steiner and Scott, 1934) (Nemata, Tylenchidae), with a note on the absence of a rachis in *A. amsinckia*. J. Nematol. 12(2):129–132.

17. Narayanan, T. R., and D. Meenakshisundaram. 1955. *Solanum elaeagnifolium*, Cav. A new weed of cultivated lands in Madras. Madras Agric. J. 42:482–484.

18. Needham, T. 1744. A Letter from Mr. Turbevil Needham, to the President; concerning certain chalky tubulous concretions, called Malm: with sorme Microscopical Observations on the Farina of the Red Lily, and of Worms discovered in Smutty Corn. *Phil Trans.* R. Soc. Lond. 42:634–641.

19. Neser, S. Biological control of weeds in South Africa. Pretoria, 1973.

20. Nesser, S., H. G. Zimmerman, H. E. Erb, and J. H. Hoffmann. 1988. Progress and prospects for the biological control of two Solanum weeds in South Africa. Proc. VII Int. Symp. Biol. Contr. Weeds, 6–11 March, Rome, Italy.

21. Notham, F. E., and C. C. Orr. 1982. Effects of a nematode on biomass and density of silverleaf nightshade. J. Range Manag. 35(4):536–537.

22. Orr, C. C., J. R. Abernathy, and E. B. Hudspeth. 1975. *Nothanguina phyllobia*, a nematode parasite of silverleaf nightshade. Plant Dis. Rep. 59(5):416–417.

23. Pantone, D. J. 1987. Host range of *Anguina amsinckiae* within the genus *Amsinckia*. Rev. Nematol. 10(1):117–119.

24. Pantone, D. J., and C. Womersley. 1986. The distribution of flower galls caused by *Anguina amsinckiae* on the weed, common fiddleneck, *Amsinckia intermedia*. Rev. Nematol. 9(2):185–189.

25. Pantone, D. J., J. A. Griesbach, and A. R. Maggenti. 1987. Morphometric analysis of *Anguina amsinckiae* from three host species. J. Nematol. 19(2):158–163.

26. Parker, P. E. 1986. Nematode control of silverleaf nightshade, a biological control pilot project. Weed Sci. 34(Suppl 1):33–34.

27. Parker, P. E., and E. Rivas. 1985. Field production of *Orrina phyllobia* for the biological control of silverleaf nightshade. Abstracts of Presentations at the 1985 Annual Meeting of the American Phytopathological Society, August 11–15, Reno, NV.

28. Pemberton, R. W. 1984. Native plant considerations in the biological control of leafy spurge. Proc. VI Int. Symp. Biol. Contr. Weeds, Delfosse, E. S. (ed.). 19–25 August, Vancouver, Canada. Agric. Can., pp. 365–390.

29. Robinson, C. C. Orr, and J. R. Abernathy. 1978. Distribution of *Nothanguina phyllobia* and its potential as a biological control for silverleaf nightshade. J. Nematol. 10(4):362–366.

30. Robinson, A. F., C. C. Orr, and J. R. Abernathy. 1979. Behavioral response of *Nothanguina phyllobia* to selected plant species. J. Nematol. 11(1):73–77.

31. Sivakumar, C. V. 1982. Longevity of *Orrina phyllobia* in leaf galls of *Solanum elaeagnifolium*. Nematologica 28:126–127.

32. Skinner, J. A., C. C. Orr, and A. F. Robinson. 1980. Histopathogensis of the galls induced by *Nothanguina phyllobia* in *Solanum elaeagnifolium*. J. Nematol. 12(2):141–150.

33. Steiner, G., and C. E. Scott. 1934. A nematosis of *Amsinckia* caused by a new variety of *Anguillulina dipsaci*. J. Agric. Res. 49:1087–1092.

34. Viglierchio, D. R. 1971. Nematode and other pathogens in auxin related plant-growth disorders. Bot. Rev. 37:1–21.

35. Wapshere, A. J. 1988. Prospects for the biological control of silver-leaf nightshade, *Solanum elaeagnifolium*, in Australia. Aust. J. Agric. Res. 39:187–197.

36. Watson, A. K., 1976. The biological control of Russian knapweed with a nematode. Proc. IV Int. Symp. on Biol. Contr. Weeds, August 30–September 2, pp. 221–223.

37. Watson, A. K. 1978. Biology and host specificity of *Paranguina picridis*. Abst. 1978 Meeting Weed Science Soc. Am., pp. 10–11.

38. Watson, A. K. 1980. The biology of Canadian weeds. Can. J. Plant Sci. 60(3):993–1004.

39. Watson, A. K. 1986. Biology of *Subanguina picridis*, a potential biological control agent of Russian knapweed. J. Nematol. 18:149–154.

40. Watson, A. K. 1986. Host range of, and plant reaction to, *Subanguina picridis*. J. Nematol. 18:112–120.

41. Watson, A. K. 1986. Morphological and biological parameters of the knapweed nematode, *Subanguina picridis*. J. Nematol. 18:154–158.

4

Options with Plant Pathogens Intended for Classical Control of Range and Pasture Weeds

W. L. Bruckart and S. Hasan

Introduction

Biological pest control requires detailed study of the interaction between each biocontrol agent and the target species in order to achieve maximum results. Good science remains at the heart of every successful biological control research program. Understanding the biology, ecology, and taxonomy of the host (target species or pest) and its natural enemies (plant pathogens and insects = biocontrol agents) is essential for safe and effective implementation of this strategy.

The use of plant pathogens for biological control of weeds has been reviewed several times, most recently by Charudattan and DeLoach (14) and by Hasan (21). There are two main strategies for use of plant pathogens in weed management, differentiated principally by the number of inoculations required for effective weed control. They are known either as the classical or the inundative (=bioherbicide) approach (21,42). Natural enemies used in the classical sense, ideally, require little or no attention following release, whereas agents used as bioherbicides need to be applied on a regular, usually annual, basis. A third strategy, augmentation, requires periodic reestablishment of a biocontrol agent, but to a lesser extent than that required for inundation. Characteristics of each approach and the value of each to society are discussed by Tisdell et al. (43).

The classical approach has been used successfully from the beginnings of biological pest control (18), and success begets tradition; successful cases become models for research on new problems. The concept of the classical approach has been applied most recently to utilization of pathogens for control of rush skeleton weed (*Chondrilla juncea*) by *Puccinia chondrillina* in both Australia (17) and the United States (39), blackberry (*Rubus constrictus*) by *Phragmidium violaceum* in Chile (30), and hamakua pa-makani (*Ageratina riparia*) by a *Cercosporella* sp. in Hawaii (44). The classical approach also serves as a model for research on biological control of weeds at the USDA Foreign Disease–Weed Science Research Unit (8).

Distinguishing between the inundative and classical approaches has been useful when considering the needs for weed control (43). Weeds of ranges and pastures are relatively uneconomical to control because of the low monetary return per hectare from agriculture in these settings, and the classical approach has been regarded as the most appropriate strategy. Natural enemies utilized in such situations are expected to be a low-cost approach over a long period of time, because they should establish themselves, increase, and spread without further augmentation. Control of skeleton weed in Australia and the United States demonstrates the efficacy of this approach (17,39).

However, successful implementation of the bioherbicide concept (5,6,16,24) and renewed interest in genetic modification of fungi (13,40) have stimulated thinking about manipulation of plant pathogens as part of weed control in ranges and pastures. In this chapter the use of plant pathogens for control of range and pasture weeds is reevaluated and alternative possibilities are considered. Situations are described in which manipulation may be justified for plant pathogens intended for control of range and pasture weeds.

Why Manipulate a Good Thing?

Manipulation refers to deliberate, direct or indirect intervention that improves performance or safety of a pesticide. The process of manipulation includes not only augmentation, but also any deliberate change to the host, the parasite, or the environment that increases efficacy of the biocontrol agent. It is an essential element of integrated pest management (IPM) and integrated weed management systems (IWMS). Manipulation plays a necessary part in the successful use of mycoherbicides, including strain selection for improved yield and pathogenicity, formulation for better shelf life and delivery, and irrigation prior to inoculation to create a more favorable microenvironment for infection. Plant pathogens for control of range and pasture weeds also are amenable to such practices.

One theory is that a natural enemy introduced for biological control will reach equilibrium with the target. However, the equilibrium model for successful biological control may not be appropriate, and there are cases for successful pest control in the absence of strict equilibrium (29). The importance of equilibrium in biological control of weeds has been considered by Auld and Tisdell (3), who also identify situations, in theory at least, of unstable equilibrium. This has implications in the manipulation of plant pathogens for control of range and pasture weeds.

Natural enemies for weed control are usually not manipulated in ranges and pastures, except during redistribution. Options to improve weed control by insect natural enemies have been described by Andres (2). Factors that favor establishment and increase of insects for weed control, such as fertilization, herbicide treatments, or pruning, are known also to increase insect feeding damage (2). However, augmentation of insects using field-collected adults or larvae is more commonly practiced than modification of the environment or artificial rearing

(2). Generally, when weed control by one insect is less than desired, another insect(s) has been evaluated to replace or augment it.

A plant pathogen not performing up to expectations would be an obvious candidate for manipulation if constraints (see ref. 42) can be overcome with reasonable effort and the candidate otherwise is worthy on the basis of laboratory, greenhouse, and field research. There may also be reason to manipulate biocontrol agents that are considered successful if the rate of weed control can be increased sufficiently by the extra input or if, by manipulation, extreme fluctuations in weed density can be reduced and pressure from the pathogen on the target population can be stabilized.

The decision to manipulate depends upon economic, environmental, and political considerations. Auld et al. (4), in a discussion of social considerations and weed control, present situations where weed control may be justified for reasons other than economics.

Need

Economical and environmentally safe weed control within the shortest period of time is the object of every weed control program. Annual losses from several weeds are very great, and for certain weeds current control measures are expensive or inadequate, or they pose a risk to the environment. Leafy spurge (*Euphorbia esula-virgata*) is an example of this situation in the Great Plains of the United States. The area infested by leafy spurge in North Dakota doubled to 862,000 acres (329,100 ha) between 1973 and 1982, and costs for losses in hay and cattle production plus the expense of control measures were estimated to be $12.9 million in 1982 (26). Herbicides are the most practical option for control at this time, and herbicides most commonly recommended are picloram; 2,4-D; dicamba; and glyphosate (1). These chemicals are either broad-spectrum, very expensive, or persistent. The use of such measures in the effort to control his weed, considering their cost and environmental impact, indicates the need for weed control.

Natural enemies used in the classical sense are expected to be environmentally safe as a result of evaluations that precede release, but reduction of the target weed density to a manageable level may take several years without manipulation, especially in situations where environment is not optimal for disease development. Use of the resources normally reserved for annual inputs, such as use of herbicides to control leafy spurge described above, may be justified to manipulate a plant pathogen that can control the target without adverse environmental consequences. Resources are being committed on the basis of need, even though economics do not necessarily justify their use.

Key Constraints

A pathogen introduced into a high population of susceptible plants is expected to cause an epidemic, and a high level of disease is expected to result in a reduction in the density of the target plant. However, there may be constraints that limit effectiveness of a pathogen (42). Often these constraints are manifest where environmental conditions are less than optimal for disease. Cullen (17) reported that control of Form "A" of skeleton weed by *P. chondrillina* was better in New South Wales, Australia, where the climate is wetter and the soils less sandy, than in the states of Victoria or South Australia. Trujillo (44) also indicated that control of hamakua pa-makani by *Cercosporella* sp. reached 95% in most areas of its distribution in Hawaii, but where the temperature was less than optimal and rainfall was low, plant density was reduced by only 80% after 8 years.

Auld and Tisdell (3) suggest that in situations of unstable equilibrium or where there is equilibrium at high weed biomass, factors that cause additional stress to the target weed, such as an increase in the biomass of the biocontrol agent, could have a profound effect on weed control. We suggest in this treatment that use of other manipulations may have the same result.

It may be possible to overcome constraints and make effective use of a pathogen by augmentation or other means. We suggest that manipulation of a pathogen by augmentation or other means may have the same result. Templeton et al. (42) identified key constraints in the use of *C. gloeosporioides* for biological control of northern joint vetch (*Aeschynomene virginica*) to be low carryover and poor dissemination of inoculum. Application of more inoculum than occurs in nature when the joint vetch normally flowers results in successful weed control. The introduction of *Puccinia canaliculata* into stands of yellow nut sedge (*Cyperus esculentus*) when the plants were just emerging also resulted in the development of disease in epidemic proportions that caused at least a 50% reduction in tuber density after the first year (9,33).

Potential

Many plant pathogenic fungi lend themselves to augmentation, since they can be readily mass-produced, shipped as spores, and applied using standard agricultural equipment or other simple procedures. These features have been important in the successful use of fungi as bioherbicides, including the development of *P. canaliculata,* an obligate pathogen for biological control of yellow nut sedge (9,33).

There are opportunities also to integrate use of plant pathogens with chemical and cultural control measures, as has been shown for biological control of yellow nut sedge by *P. canaliculaa* in Georgia (12) and for biological control of velvetleaf (*Abutilon theophrasti*) by *Colletotrichum coccodes* (23,45). Integration of plant pathogens can be accomplished by varying the time of inoculation, number of inoculations, inoculum concentration, or method of inoculation.

"Advanced concepts and techniques" in the biological sciences can be applied to the development of plant pathogens for biological control of weeds, according to Hatzios (22), who considers both genetic approaches and process or system engineering (i.e., fermentation, mass production, and formulation). Use of recombinant DNA and genetic engineering has been proposed by Charudattan (13) and TeBeest and Templeton (40) for development of plant pathogens as mycoherbicides. Protoplast fusion has been attempted (7,41), and some reduction in the host range of *Sclerotinia sclerotiorum* using directed mutagenesis was reported by Miller et al. (27,28). The full potential of these approaches has not yet been demonstrated. Other objectives of directed mutagenesis of *S. sclerotiorum* have been suggested, including development of auxotrophic and nonsclerotial isolates (D. C. Sands, personal communication).

Progress toward overcoming the major limitation in field application of mycoherbicides, the need for free moisture during the infection process, has resulted from research in formulation chemistry and development of an oil-in-water invert emulsion. This carrier provides the moisture needed for spores of *Alternaria cassiae* to infect sicklepod (*Cassia obtusifolia*) in the absence of dew (35).

When to Manipulate:
Key Constraints and Classical Weed Control

The decision to manipulate depends upon several factors. Among these are efficacy of the pathogen under field conditions, constraints limiting full expression of efficacy, potential for the pathogen based upon laboratory and greenhouse data, alternatives to weed control (including use of other natural enemies), costs associated with alternative strategies, and options for manipulation based upon the key constraint(s). Key constraints must be identified before justification to manipulate is possible. Several ways to overcome key constraints have been described. Specific ideas are next presented that may also be useful in development of fungi as mycoherbicides.

Plant Competition and Interplanting

Plant disease in combination with plant competition has been found to affect growth and survival of the infected species significantly (10,11,31,32). Harper (20), in reviewing methods to study plant-to-plant interactions, has shown that often one plant species will predominate over another in tests under controlled conditions. Which species predominates may be changed by varying the environment (20). One important environmental factor is plant disease.

Plant species known to be beneficial for forage can be evaluated for ability to compete with the weed. Use of a beneficial competitor along with a plant pathogen may serve (1) to increase the rate of weed control, (2) to fill the niche vacated by the weed with a plant species of value, and (3) to reduce the probability that an undesirable plant species will become established.

Interplanting may be useful in situations where a fungus has an alternate host. *Uromyces striatus,* the cause of a systemic rust disease of leafy spurge in North Dakota, also infects alfalfa (*Medicago sativa*) as the alternate host during part of the life cycle. It has been observed that leafy spurge rarely is found in stands of alfalfa (37), and since alfalfa rust is not regarded as a limiting factor in North Dakota alfalfa production, planting alfalfa into stands of leafy spurge to facilitate control of a very serious weed has been proposed.

Augmentation of Rust Fungi

Rust fungi have not been manipulated in the control of range and pasture weeds. However, this possibility has been demonstrated during development of *P. canaliculata* for biological control of yellow nut sedge. The key constraint with *P. canaliculata* is the limited amount of inoculum available when the plants emerge. Inoculation of yellow nut sedge when it emerges results in a severe epidemic (9,33), because the plants are young and very susceptible and the weather conditions are generally favorable. Once established, the epidemic continues into the summer and causes significant reduction in tuber densities at the end of the season. Fungi in the Uredinales may be very useful for weed control, and their development as mycoherbicides may have application in augmentation for improved control of range and pasture weeds.

Integration with Herbicides and Plant Growth Regulators

Biological control of weeds has been enhanced by combining pathogens with herbicides and plant growth regulators (PRGs). Hodgson et al. (23) demonstrated that *C. coccodes,* for control of velvet leaf, and thidiazuron, which inhibits velvet leaf growth, acted synergistically when combined in field trials. The result was increased control of velvetleaf and increased yield of soybean, *Glycine max*. The interaction was supported in greenhouse studies by Wymore et al. (45). Similar results occurred with *P. canaliculata* and low rates of bentazon herbicide for control of yellow nut sedge in both Maryland (9) and Georgia, where five herbicides were tested (12). Also, *Cochliobolus lunatus* from the Netherlands, which kills seedlings of barnyard grass (*Echinochloa crus-galli*), was effective against older plants when combined with a sublethal dosage of atrazine (36), and imazaquin and *A. cassiae* were found to have a synergistic effect when applied together for the control of sicklepod (34).

Development and Use of Options in Classical Weed Control

Manipulation of biological control agents targeted for weeds of ranges and pastures will require support for implementation. Alternatives to commercial ventures that may be suitable have been discussed by Starler and Ridgway (38) for augmentation in biological control of insects and mites. Among the options for development and implementation are pest management consultants (private practitioners), grower and farm cooperatives, pest control districts, and state and federal agencies.

Pest Management Consultants (Private Practitioners)

Pest management consultants involved with rangeland agriculture would be in an excellent position to enhance the use of plant pathogens for classical weed control. They would be knowledgeable about the agriculture, the needs of the farmer, and the options suitable for pathogens available. Areas where less than optional environmental conditions occur, for example, would be recognized by individuals with understanding and expertise, and options available for manipulation could be identified and used.

Grower and Farmer Cooperatives

There are many grower and farmer cooperatives that serve specialized interest groups, and these organizations are able to support programs, many in biological control (38), that would serve to meet their specific needs in pest control. Cooperatives of this nature have the ability to subcontract work (e.g., research on development of plant pathogens and mass production of rust fungi) to private consultants, and they have ways to bear risks of user losses (38).

Pest Control Districts

Many of the weeds considered for classical biological control infest agriculture on a regional basis. Yellow star thistle (*Centaurea solstitialis*) infests over 3 million ha in the state of California alone, and it is serious in other states as well (25). Leafy spurge is a serious weed in at least five states in the Great Plains (19). Weed control is more effective if it can be implemented on a regional basis. Development and manipulation of plant pathogens would also be much easier if costs and responsibilities were supported by a large number of individuals and organizations, including state and federal agencies. Pest control districts would enable coordination of pest control policies to make full use of options, including plant pathogens and their manipulation. Legal support at the state level also would be available to enforce necessary procedures and policy (38).

State and Federal Agencies

Research and regulatory activities within state and federal organizations have supported both development and implementation of plant pathogens for classical weed control. Ideas to improve efficacy of plant pathogens are being explored within the USDA Agricultural Research Service (ARS) and the Cooperative States Research Service (CSRS). Also, the USDA Animal and Plant Health Inspection Service (APHIS) has a facility at Mission, Texas, the Biological Control Laboratory, to rear insects for weed control. Mass production of plant pathogens for augmentation also could be accomplished at this facility. Considerable effort was made recently to mass-produce *Nothanguina phyllobia,* a nematode, for biological control of silverleaf nightshade (*Solanum elaeagnifiloum*) at this facility (G. Cunningham, personal communication).

There are several regional research organizations that include state experiment stations, including the Southern Regional Research Project S–136 (now S–234), which was instrumental in development of *A. cassiae* for biological control of sicklepod (15). Other symposia have been organized that represent cooperative research at a less formal level. Objectives of these meetings are to determine the current status and options for weed control and to identify new directions for research. Among these are the annual Leafy Spurge symposium and a Knapweed Symposium.

Summary and Conclusions

Plant pathogens are excellent candidates for biological control of range and pasture weeds. Release of a pathogen with no subsequent manipulation, known as the classical approach, is the most economically desirable strategy for weed control in ranges and pastures. The population of a plant pathogen used for weed control in ranges and pastures may or may not reach equilibrium with the target, but it is expected to reduce target weed densities below an economic threshold. Manipulation may be justified in order to enhance efficacy of a pathogen for weed control. If the pathogen and the target weed populations do not reach a stable equilibrium, manipulations may be used to tip the equilibrium in favor of the pathogen and against the target. Options for manipulation include mass production and augmentation of the pathogen (including rust fungi), formulation, timing of inoculation, amount of inoculum, management of plant competition, genetic modification, and integrating pathogens with herbicides. Options to manipulate a pathogen qualified for biological control should be considered during evaluations.

The decision to use the biological control approach and whether to manipulate or find anther natural enemy has to be justified on the basis of cost, benefit, and environmental considerations (which are becoming a larger part of decisions regarding pest control strategy). Development of plant pathogens for use in biological control of weeds, particularly in ranges and pastures, requires the commitment of money and trained individuals to collect and evaluate candidate organisms. In many cases, satisfactory weed control has occurred within a reasonable time with natural enemies used in the classical sense, representing the ideal in terms of cost and permanence. However, farmers, ranchers, weed control personnel, and scientists should be alert to opportunities for manipulation of a promising agent. Optimal use of plant pathogens for biological weed control results only if there is a clear understanding of interactions between host, pathogen, and environment, particularly with regard to key constraints. Ideas generated during development of the mycoherbicides are likely to have application in control of range and pasture weeds, and new ways to manipulate the host–parasite–environment interaction are likely to be discovered.

Manipulation of plant pathogens for improved weed control will require inputs beyond those normally considered for classical biological control. The costs for manipulation of plant pathogens may be borne by interest groups at the commodity, local, regional, state, or federal levels if commercial development is not practical. Development and implementation of plant pathogens for classical weed control is limited only by our imaginations.

Literature Cited

1. Alley, H. P., and Messersmith, C. G. 1985. Chemical control of leafy spurge. Pages 65–78, *in:* Leafy Spurge, A. K. Watson, ed. Weed Sci. Soc. Am. Monogr. 3, Champaign, IL.

2. Andres, L. A. 1982. Integrating weed biological control agents into a pest-management program. Weed Sci. 30 (Suppl. 1):25–30.

3. Audl, B. A., and Tisdell, C. A. 1985. Biological weed control—equilibria models. Agric. Ecosystems Environ. 13:1–8.

4. Auld, B. A., Menz, K. M., and Tisdell, C. A. 1987. Weed Control Economics. Academic Press, Orlando, FL.

5. Bowers, R C. 1982. Commercialization in microbial biological control agents. Pages 157–173, *in:* Biological Control of Weeds with Plant Pathogens, R. Charudattan and H. L. Walker, eds. Wiley, New York.

6. Bowers, R. C. 1986. Commercialization of Collego—an industralist's view. Weed Sci. 34 (Supp. 1):24–25.

7. Brooker, N. L., TeBeest, D. O., and Spiegel, F. L. 1988. Nuclear number and pathogenic variability in *Alternaria crassa* protoplasts. (Abstr.) Phytopathology 78:1519.

8. Bruckart, W. L., and Dowler, W. M. 1986. Evaluation of exotic rust fungi in the United States for classical biological control of weeds. Weed Sci. 34 (Suppl. 1):11–14.

9. Bruckart, W. L., Johnson, D. R., and Frank, J. Ray. 1988. Bentazon reduces rust-induced disease in yellow nutsedge (*Cyperus esculentus*). Weed Technol. 2:299–303.

10. Burdon, J. J., and Chilvers, G. A. 1977. The effect of barley mildew on barley and wheat competition in mixtures. Aust. J. Bot. 25:59–65.

11. Burdon, J. J., Groves, R. H., Kaye, P. E., and Speer, S. S. 1984. Competition in mixtures of susceptible and resistant genotypes of *Chondrilla juncea* differentially infected with rust. Oecologia 64:199–203.

12. Gallaway, M. B., Phatak, S. C., and Wells, H. D. 1987. Interactions of *Puccinia canaliculata* (Schw.) Lagerh, with herbicides on tuber production and growth of *Cyperus esculentus* L. Trop. Pest Manag. 33:22–26.

13. Charudattan, R. 1985. The use of natural and genetically altered strains of pathogens for weed control. Pages 347–372, *in:* Biological Control in Agricultural IPM Systems, M. A. Hoy and D. C. Herzog, eds. Academic Press, Orlando, FL.

14. Charudattan, R., and DeLoach, C. J. 1988. Management of plant pathogens and insects for weed control. Pages 245–263, *in:* Weed Management in Agroecosystems: Ecological Approaches, M. A. Altieri and M. Liebman, eds. CRC Press, Boca Raton, FL.

15. Charudattan, R., Walker, H. L., Boyette, C. D., Ridings, W. H., TeBeest, D. O., VanDyke, C. G., and Worsham, A. D. 1986. Evaluation of *Alternaria cassiae* as a mycoherbicide for sicklepod (*Cassia obtusifolia*) in regional field tests. Southern Coop. Ser. Bull. 317, Ala. Exp. Sta., Auburn University, AL.

16. Churchill, B. W. 1982. Mass production of microorganisms for biological control. Pages 139–156, *in:* Biological Control of Weeds with Plant Pathogens, R. Charudattan and H. L. Walker, eds. Wiley, New York.

17. Cullen, J. M. 1985. Bringing the cost benefit analysis of biological control of *Chondrilla juncea* up to date. Pages 145–152, *in:* Proc. VI Int. Symp. Biol. Contr. Weeds, E. S. Delfosse, ed. Agric. Can., Ottawa.

18. DeBach, P., ed. 1964. Biological Control of Insect Pests and Weeds. Reinhold, New York.

19. Dunn, P. H. 1979. The distribution of leafy spurge (*Euphorbia esula*) and other weedy *Euphorbia* spp. in the United States. Weed Sci. 27:509–516.

20. Harper, J. L. 1977. Population Biology of Plants. Academic Press, New York.

21. Hasan, S. 1988. Biocontrol of weeds with microbes. Pages 129–151 *in:* Biocontrol of Plant Diseases, Vol. 1, K. G. Mukerji and K. L. Garg, eds. CRC Press, Boca Raton, FL.

22. Hatzios, K. K. 1987. Biotechnology applications in weed management: now and in the future. Adv. Agron. 41:325–375.

23. Hodgson, R. H., Wymore, L. A., Watson, A. K., Snyder, R. H., and Collette, A. 1988. Efficacy of *Colletotrichum coccodes* and thidiazuron for velvetleaf (*Abutilon theophrasti*) control in soybean (*Glycine max*). Weed Technol. 2:473–480.

24. Kenney, D. S. 1986. DeVine—the way it was developed—an industralist's view. Weed Sci. 34 (Suppl. 1):15–16.

25. Maddox, D. M., Mayfield, A., and Poritz, N. H. 1985. Distribution of yellow starthistle (*Centaurea solstitialis*) and Russian knapweed (*Centaurea repens*). Weed Sci. 33:315–327.

26. Messersmith, C. G., and Lym, R. G. 1983. Distribution and economic impacts of leafy spurge in North Dakota. N. D. Farm Res. 40:8–13.

27. Miller, R. V., Ford, E. J., and Sands, D. C. 1987. Reduced host-range mutants of *Sclerotinia sclerotiorum*. Phytopathology 77:1695 (Abstr.)

28. Miller, R. V., Ford, E. J., and Sands, D. C. 1987. Induced auxotropic and non-sclerotial isolates of *Sclerotinia sclerotiorum*. (Abstr.) Phytopathology 77:1720.

29. Murdoch, W. W., Chesson, J., and Chesson, P. L. 1985. Biological control in theory and practice. Am. Natur. 125:344–366.

30. Oehrens, E. B., and Gonzales, S. M. 1977. Dispersion, ciclo biologico y danos causdos por *Phragmidium violaceum* (Schulz) Winter en zarzamora (*Rubus constrictus* Lef. et M. y *R. ulmifolius* Schott.) en las zonas centro-sur y sur de Chile. Agro. Sur. 3:87–91.

31. Paul, N. D., and Ayres, P. G. 1986. Interference between healthy and rusted groundsel (*Senecio vulgaris* L.) within mixed populations of different densities and proportions. New Phytol. 104:257–269.

32. Paul, N. D., and Ayres, P. G. 1987. Effects of rust infection of *Senecio vulgaris* on competition with lettuce. Weed Res. 27:431–441.

33. Phatak, S. C., Sumner, D. R., Wells, H. D., Bell, D. K., and Glaze, N. C. 1983. Biological control of yellow nutsedge with the indigenous rust fungus *Puccinia canaliculata*. Science 219:1446–1447.

34. Quimby, P. C., Jr., and Boyette, C. D. 1986. *Alternaria cassiae* can be integrated with selected herbicides. Abstr. Weed Sci. Soc. Am. 26:52.

35. Quimby, P. C., Jr., Fulgham, F. E., Boyette, C. D., and Connick, W. J., Jr. 1988. An invert emulsion replaced dew in biological control of sicklepod—a preliminary study. Pages 264–270 (Vol. 8, ASTM STP 980), *in:* Pesticide Formulations and Application Systems, D. A. Hovde and G. B. Beestman, eds. American Society for Testing and Materials, Philadelphia, PA.

36. Scheepens, P. C. 1987. Joint action of *Cochliobolus lunatus* and atrazine on *Echinochloa crusgalli* (L.) Beauv. Weed Res. 27:43–47.

37. Stack, R. W., and Statler, G. D. 1989. Unexplained nonconcurrent distribution of leafy spurge and alfalfa in noncropped areas of eastern North Dakota. Proc. N. D. Acad. Sci. 43:86.

38. Starler, N. H., and Ridgway, R. L. 1977. Economic and social considerations for the utilization of augmentation of natural enemies. Pages 431–450, *in* Biological Control by Augmentation of Natural Enemies, R. L. Ridgway and S. B. Vinson, eds. Plenum Press, New York.

39. Supkoff, D. M., Joley, D. B., and Marois, J. J. 1988. Effect of introduced biological control organisms on the density of *Chondrilla juncea* in California. J. Appl. Ecol. 25:1089–1095.

40. TeBeest, D. O., and Templeton, G. E. 1985. Mycoherbicides: progress in the biological control of weeds. Plant Dis. 69:6–10.

41. TeBeest, D. O., and Weidemann, G. J. 1985. Preparation of protoplasts from conidia of *Colletotrichum gloeosporioides* f. sp. *aeschynomene*. (Abstr.) Phytopathology 75:1361.

42. Templeton, G. E., TeBeest, D. O., and Smith, R. J., Jr. 1979. Biological weed control with mycoherbicides. Ann. Rev. Phytopathol. 17:301–310.

43. Tisdell, C. A., Auld, B. A., and Menz, K. M. 1984. On assessing the value of biological control of weeds. Protection Ecol. 6:169–179.

44. Trujillo, E. E. 1985. Biological control of hamakua pa-makani with *Cercosporella* sp. in Hawaii. Pages 661–671, *in:* Proc. VI Int. Symp. Biol. Contr. Weeds, E. S. Delfosse, ed. Agric. Can., Ottawa.

45. Wymore, L. A., Watson, A. K., and Gotlieb, A. R. 1987. Interaction between *Colletotrichum coccodes* and thidiazuron for control of velvetleaf (*Abutilon theophrasti*). Weed Sci. 35:377–383.

Host–Parasite Interactions

5

Host-Range Testing:
Safety and Science

G. J. Weidemann

A critical consideration in the development of a biological agent for weed control is the determination of host range. Irrespective of potential benefits, the safety of nontarget cultivated and wild plants must be ensured prior use. Host-range testing has become more important as societal concerns about preservation and protection of rare and endangered plant germ plasm increases. This concern is particularly pronounced with regard to the importation of phytophagous insects and plant pathogens from overseas as classical biocontrol agents. With endemic pathogens used as inundative inoculum, or bioherbicides, increased disease pressure on cultivated plants or potential conflict-of-interest issues must be avoided.

This chapter focuses on the development and current status of host-range safety testing and the scientific basis of host range and host specificity. Aspects of host-range testing have been reviewed previously (54,56,66,68).

Safety

Host-range testing schemes were initially developed for assessing the safety of nontarget hosts in relation to insects imported from regions of host diversification. Although the emphasis remains on classical biological control, host-range tests developed for insect pests have been utilized to demonstrate safety of imported plant pathogens used for classical biological control or for endemic agents used as bioherbicides.

Prior to the work of Harris and Zwolfer (33), host-range testing to satisfy regulatory requirements for importation emphasized determination of the insect's capacity to feed on crop plants in starvation trials or the insect's ability to reproduce on enomomic hosts (37). Although these tests demonstrated relative safety to crop plants, they gave little information on the level of specificity or ability of the insect to attack ecologically or genetically related plant species. Harris and Zwolfer (33,71) attempted to place host-range testing on a rational biological basis by proposing host-range testing on related plant species. They recommended a laboratory host-range test of plants taxonomically related to the weed species, hosts of related insects, plants with occasional host records in the literature, and plants with common morphological or biochemical features. Starvation tests on unrelated economic hosts were used only to confirm the host range. Additional emphasis was placed on studying the biology of the insect and determining the biological or physical basis of host selection. Although several modifications have been proposed since then, this approach continues to serve as the basis of current host-range testing.

Wapshere (63–65) modified the scheme of Harris and Zwolfer (33) and proposed a centrifugal-phylogenetic test. He recommended initially testing a small group of taxonomically related plants with morphological and biochemical similarities to the target weed, and gradually expanding the number of tested species to include more distantly related plants until specificity was established (Figure 5.1). He also proposed testing cultivated plants that were related to the weed, poorly characterized for associated pests, evolved apart from the agent, attacked by related pests, and previously recorded as possible hosts. Despite thorough testing, he considered it possible to fail to determine host range adequately with organisms that attack plants irregularly distributed in several plant families, organisms specific to two alternate hosts in different plant taxa, and organisms attacking several phylogenetically separated plant groups (64). A disadvantage of phylogenetic testing is the heavy reliance on the reliability of taxonomic knowledge concerning the host and pest (63) and the assumption that host specificity is closely related to phylogenetic relationships.

Phylogenetic host-range testing has been used to justify the importation of numerous insects and several plant pathogens for biological control (54). The single example of attack of an economic host, sesame (*Sesamum indicum*), following release of an imported insect, *Teleonema scrupulosa*, in Uganda did not utilize a phylogenetically based host-range test (54,64). Testing of taxonomically related crop plants would have demonstrated the vulnerability of sesame.

Phylogenetic testing is most precise with highly host-specific insects or pathogens that are well characterized in the literature. For instance, Hasan (34,35) demonstrated restriction of the rust *Puccinia chondrillina* to certain biotypes of rush skeleton weed (*Chondrilla juncea*) prior to importation into Australia. The

Figure 5.1. Testing Sequence Used in Centrifugal-Phylogenetic Host-Range Determinations

Plants Tested	Specific to Taxon If Plants Not Attacked
Forms of the target plant	Clone or biotype
Species in the genus	Species
Members of the tribe	Genus
Members of the subfamily	Tribe
Members of the family	Subfamily
Members of the order	Family

Source: Adapted from Ref. 54.

precise delimitation of host range is more questionable with pests that are less host-specific. This is particularly true with facultative parasites that are often most promising as bioherbides. Charudattan (16) has proposed modifying the test requirements based on the level of specificity of plant pathogens. A centrifugal-phylogenetic test would be used with highly host-specific pathogens, whereas pathogen taxa known to be less specific would also include plants ecologically and economically important at the release site and known or reported to be suscepts of the pathogen.

In addition to the selection of appropriate test plants, the development of well-defined test conditions is an important consideration to ensure consistent, accurate host-range testing. Careful attention must be given to parameters such as environmental conditions for infection and disease development; host factors, such as plant age, fertility, and genetic variability; and pathogen factors, such as production of virulent inoculum and genetic stability (59,68).

Previous authors (26,68,71) have suggested that the host range of an insect or pathogen is often wider in laboratory studies than would occur in the field, possibly causing unnecessary rejection of a biocontrol agent. In laboratory studies, optimum environmental conditions for infection and disease development of plant pathogens are typically used that may bear little resemblance to expected field conditions (68). Also, plants grown in growth chambers or greenhouses are often more susceptible to infection than field-grown plants (56,68). For instance, Politis (48) found globe artichoke (*Cynara scolymus*) to be moderately susceptible to *Puccinia carduorum* in host-range tests in the greenhouse. However, the pathogen was not reported on artichoke in locations where the target weed, musk thistle (*Carduus nutans*), was heavily infected in close proximity to the economic host. The use of more realistic experimental conditions and a two-phase testing program of laboratory tests followed by field tests for two to three seasons has been proposed to determine more accurately actual field specificity (68).

Plant age can also profoundly affect infection and disease severity, and disease susceptibility can often change with plant age (49). Infection may only occur during certain stages of plant growth, and inoculation at other times may be unsuccessful. Typically, seedling plants are tested (59) for convenience and because plants are commonly most susceptible at the seedling stage. However, this may not always be true. For instance, TeBeest (unpublished) demonstrated increased susceptibility of morning glory species to the rust *Coleosporium ipomoeae* after flowering. For critical host plants or for pathogens known to have age-related infection periods, testing at more than one developmental stage may be necessary. Often it is difficult to determine whether susceptibility is related to host factors or other parameters such as environment. For example, *Colletotrichum malvarum* rarely causes disease of prickly sida (*Sida spinosa*) after mid-June in Arkansas. Failure to cause infection is related to restrictive temperature conditions rather than plant age (TeBeest, unpublished).

Inoculation technique and the pathogen inoculum used can influence plant disease. Differences in host range have been demonstrated, depending on the inoculation technique used (17,21,50,51). For instance, Ridings (50,51) noted differences in the susceptibility of several hosts to *Phytophthora palmivora* in preemergence and postemergence tests. Charudattan (17) demonstrated differences in the host range of *Fusarium roseum* "Culmorum" between preemergence seed assays and postemergence inoculations. The inoculation technique selected should parallel expected use in the field.

The type of inoculum used, such as conidia or mycelium, physiological state of the inoculum, and amount of inoculum applied can often influence results. Uniform inoculum produced on the identical substrate and of uniform age must be used for all experiments to permit direct comparisons. A uniform suspect, such as the target weed, should be included in all experiments as a check of inoculum effectiveness (58,59).

The number of species required in a phylogenetic host-range test is not fixed. The size of the test depends on the known level of pathogen specificity, the size and relative importance of the plant taxa attacked, and the number of related cultivated and ecologically important plant species that occur in the geographic region of use. Schroeder (54) examined 12 examples of host-range testing of imported insects and found that the size of the test ranged from 21 to 83 species, with 40–50 plant species being most common. For imported pathogens, up to 121 plant species have been tested (18,30). An examination of several host-range tests of endemic pathogens demonstrated a comparable range of tested species from a low of 22 (1) to a high of 82 (70).

Differences exist in the level of specificity considered acceptable between pathogens imported for classical biological control and endemic pathogens used as bioherbicides. It is generally accepted that imported pathogens present a greater potential threat to nontarget plants (42,66) and a thorough host-range test that demonstrates a high level of specificity to the target weed is required prior to importation. Also, any potential risks must be carefully considered. For example, the Australian government rejected a request for the importation of the rust *Phragmidium violacearum* for *Rubus* control, despite its successful use in Chile, because of the objections of commercial blackberry growers and beekeepers in Australia and New Zealand (29).

Strict specificity is not always required for endemic pathogens, because bioherbicides, like chemicals, are site-directed and are present naturally in the region of intended use (16,42,53,66). Potential hosts have been exposed previously in areas where the pathogen is endemic (42,66). Detailed host-range information is still required, however, to avoid potential conflicts of interest, to avoid potential hazards associated with introductions into an area where the pathogen did not occur previously (42), or to accommodate changes in cropping practices. Even relatively nonspecific pathogens may be utilized if they can be applied selectively without endangering nontarget plants. For instance, *Colletotrichum gloeosporiodes* f.sp. *aeschynomene* has been used successfully as the mycoherbicide

COLLEGO (58) for control of the leguminous weed northern joint vetch (*Aeschynomene virginica*) because other susceptible, commercially important legumes are not grown in the areas of use (57,70). *Phytophthora palmivora,* used as the mycoherbicide DeVine (50), also includes several cultivated plants in its host range, but use is limited to citrus orchards. Potential hazards are avoided by using label restrictions to limit use where nontarget suscepts are grown (15). Scheepens and VanZon (53) and DeJong (24) have demonstrated the safety of using *Chondrostereum purpureum* to control wild blackcherry (*Prunus serotina*), despite the susceptibility of cultivated cherry, by using epidemiological models to determine low-risk use areas in the Netherlands. Brosten and Sands (7) have demonstrated the potential use of *Sclerotinia sclerotiorum* for control of Canada thistle (*Cirsium arvense*) despite its wide host range by developing nutritionally dependent strains that are unable to overseason (52). Lack of specificity, comparable to these examples, would most likely be unacceptable for an imported pathogen.

Science

A thorough consideration of the scientific basis of host-range testing must address the biology and genetics of host specificity and genetic diversity in host and pathogen populations. Because of our limited understanding of host specificity, the use of taxonomic relationships to predict genetic specificity is uncertain at best. Phylogenetic relationships are often unclear, making it difficult to predict the complex genetic interactions of host and parasite.

Research on the host range of *Colletotrichum gloeosporioides* f.sp. *aeschynomene* serves to illustrate the lack of a clear relationship between plant phylogeny and pathogen specificity. Daniel et. al. (22) initially tested 30 plant species and 46 cultivars of economic and wild plants and found only the target weed *Aeschynomene virginica* and a related weed, *A. indica*, to be susceptible. Plant selection for testing was based on taxonomic relationships in the Leguminosae and exposure of potential hosts to the pathogen during aerial application or by off-target drift (58). Crop cultivars were selected on the basis of widespread use such that the cultivars tested constituted nearly 90% of the planted crop acreage in the area of expected use (58). No potential hazards were detected and the pathogen has now been used as a commercial bioherbicide since 1982 without incident.

More recently, TeBeest (57) tested 77 plant species and 43 genera in 10 families and found five genera in the subfamily Papilionideae to include susceptible species, including 23 of the 26 pea (*Pisum sativum*) cultivars. However, susceptibility differed markedly and only the target weed was killed. Because several legume genera included suscepts, an additional test of 82 species and 47 genera of legumes was conducted to look for possible phylogenetic relationships (70). Nine genera in six tribes of the subfamily Papilionideae included at least one susceptible species (Figure 5.2). However, susceptible taxa could not be related

Figure 5.2.   Legume Tribes and Genera Containing Species Susceptible to
Colletotrichum gloeosporioides f.sp. *aeschynomene*

Tribe	Genus	No. Species Tested	No. Species Susceptible
Aeschynomeneae	*Aeschynomene*	6	3
Cicereae	*Cicer*	1	1
Genisteae	*Lupinus*	6	5
Indigofereae	*Indigofera*	2	2
Loteae	*Lotus*	2	1
Vicieae	*Lathyrus*	4	1
	Lens	1	1
	Pisum	1(32)*	1(29)*
	Vicia	7	6

Source: From Ref. 70.
*Twenty-nine of 32 cultivars of *Pisum sativum* were susceptible.

to reported phylogentic relationships within the subfamily or to common morphological or biochemical features (47). Studies of rust diseases of grasses (4,28)
have demonstrated a similar lack of a clear association between host phylogenetic
relationships and pathogen specificity.

In a review of biochemical taxonomy of the Leguminosae, Turner (61) attributed the lack of phylogenetic relationships of biochemical characters to possible
errors in taxonomic placement using other criteria such as morphological features,
parallel development of biochemical characters from a distant ancestor, or independent development of similar biochemical characters. All these possible mechanisms proposed by Turner (61) could be equally applied to genetic relationships
in host–parasite interactions. Harris and Zwolfer (33) and Wapshere (67) stressed
understanding the physical or biochemical basis of insect specificity. However,
the understanding of insect specificity was considered inadequate and complicated
by other factors (67). Similarly, despite recent progress (31,40) the understanding
of host specificity in plant pathogens remains limited.

Plant pathogens vary widely, ranging from highly host-specific obligate parasites to facultative necrotrophs with a wide host range (6). However, even
pathogen species considered to have a wide host range may consist of subspecies
populations with more limited host preferences. For instance, *Colletotrichum
gloeosporioides* is considered a group species (55) reported on hundreds of host
species (2). However, numerous subspecies groups exist that have relatively
restricted host ranges (5,22,45,70). Even within subspecies groups host specificity may be variable. For example, Nikandrow and co-workers (unpublished) have
tested a host range of several isolates of *Colletotrichum orbiculare* obtained from
Xanthium spinosum and found them to have differences in host range. Several
isolates included economic hosts, such as safflower (*Carthamus tinctorius*), in
their host range, whereas others did not (Figure 5.3). The ability of pathogenic
fungi to adapt rapidly to changing host populations and to develop reproductive

Figure 5.3. Disease Severity of Selected Isolates of *Colletotrichum orbiculare* Obtained from *Xanthium spinosum* to Potential Host Plant Species

Plant family	Host		Disease Severity (0–3)*		
				Isolate	
		1	001/3	004/15b	005/11
Aesteraceae	*Xanthium spinosum*	3	3	3	3
	X. cavanillesii	1	2	1	1
	X. orientale	0	0	0	0
	X. italicum	1	1	1	1
	X. occidentale	1	1	0	1
	Carthamus tinctorius cv. Gila	2	2	0	1
	Cynara cardinunculus	1	1	0	0
	Silybum marianum	1	2	1	3
	Gerbera jamesonii	0	2	0	0
Cucurbitaceae	*Citrullus lanatus* var caffer cv. Candy red	0	0	1	1
	Cucumis melo cv. Honeydew	0	0	1	0
	Cucumis sativus	0	0	0	0

*Source: From A. Nikandrow (unpublished).
Disease rating: 0 = no disease to 3 = plant death.

barriers suggests that genetically isolated populations differing in host specificity can arise even in close proximity (12).

Plant pathogens have typically been categorized into subspecies groups on the basis of host preference. However, these groups tend to be somewhat artificial, arising as they do from the testing of cultivated plants (39). The man-directed selection of simple genetic resistance in cultivated plants may create an image of extreme specificity in pathogen populations that may not be representative when examined across the entire spectrum of potential hosts (6,39). Limited work on diseases in wild plant populations (25) suggests that host range is often wider and more variable than expected from studies of economic plants.

Studies of the grass rust *Puccinia coronata* (27,28) in Israel have demonstrated that host range is wider than expected from studies of cereal crops. Also, host range was not closely related to taxonomic relationships in the grasses. Geographic proximity and opportunity for adaptation were as important as phylogenetic relationships (4,25,27). *Puccinia coronata* was found to exist as numerous pathotypes with overlapping host ranges such that form species and races were indistinguishable (27). Such work suggests that host specificity should be considered a continuum in a population of a plant pathogen, composed of individuals or subpopulations with a range of specificities (13).

Both host and pathogen populations often exhibit a high degree of genetic variability for disease interactions (23). Plant disease is a strong selective force in both host and pathogen that promotes genetic variability (8,11) and influences genetic diversity (8,11,32) and plant competition (8,19).

Plant species can cope with disease by maintaining genetic resistance in the population or by avoiding the pathogen in time or space (8,43). For instance, northern joint vetch is uniformly susceptible to *Colletotrichum gloeosporioides* f.sp. *aeschynomene* throughout Arkansas, but natural epidemics are limited by the uneven distribution of the weed and poor dissemination capacity of the pathogen (58).

For plant species continually exposed to disease pressure, genetic resistance is the primary means of confronting a pathogen. For instance, studies of powdery mildew on wild barley (62) have demonstrated a high level of disease resistance and complex resistance genotypes in plant populations obtained from geographic areas with high disease pressure. In arid regions with less disease pressure barley populations generally had lower overall levels of disease resistance.

Adaptation to selective pressure can be related in part to the reproductive system in plants (3,9,43). However, even among plants with only asexual reproductive systems genetic variability exists. Rush skeleton weed (*Chondrilla juncea*) is an apomict, producing viable seed without fertilization, yet exists as a collection of biotypes that differ in leaf morphology and in resistance to the rust *Puccinia chondrillina* (35,38).

Plant pathogens also maintain a high level of genetic variability to allow rapid adaptation to changing host populations. Even among asexually reproducing species, genetic diversity is common (36). Hasan (34) demonstrated pathogenic variation in *Puccinia chondrillina* comparable to the variability in its host rush skeleton weed. Numerous isolates obtained throughout the Mediterranean were unable to attack the rush skeleton weed biotypes in Australia, and extensive testing was required before a virulent isolate was found. Webb and Lindow (69) demonstrated that isolates of *Ascochyta pteridis* collected from different geographic locations varied widely in virulence to bracken fern (*Pteridium aquilinum*), sporulation, and tolerance to desiccation.

Host-range tests must be designed to account for the genetic variability that exists in populations of the host and pathogen. Testing of one or a few genotypes of potential hosts or of the pathogen to determine pathogenic potential may be inadequate (42,56). With imported pathogens, the breadth of the host range should be considered (42). If one or more important economic hosts are closely related to the known suspects of the pathogen, this may indicate the possibility of future adaptation (42). With endemic pathogens, knowledge concerning the range of variability in the host population may indicate the potential for selection of resistant biotypes from the target weed population following repeated use. Also, demonstrated pathogenic variability in the biocontrol agent would suggest the potential for selection of more virulent genotypes if host resistance is encountered. For instance, prickly sida (*Sida spinosa*) varies in susceptibility to the fungus *Colletotrichum malvarum* throughout its range (Figure 5.4) (TeBeest, unpublished). A comparable study of *C. malvarum* isolates would indicate the

Figure 5.4. Control of *Sida spinosa* Obtained from Several Locations with *Colletotrichum malvarum*

Sida Spinosa Seed Source	Percent Control
Rohwer, AR	51
Stillwater, OK	63
Griffin, GA	75
Stuttgart, AR	76
Ben Hur, LA	81
Marianna, AR	89
Stoneville, MS	90
Baton Rouge, LA	91
Burbon, MS	93
Urbana, IL	99

Source: From D.O. TeBeest (unpublished).

extent of pathogenic variability in the population and the potential for selecting virulent genotypes in locations where host resistance is high.

Concern has been expressed about the potential of insects and pathogens to adapt to new hosts following release. This concern is more critical with introduced organisms for classical biocontrol that have not coevolved with potential hosts in the geographic area of use and thus present an unknown threat in a new environment. With endemic agents the potential for adaptation seems remote. Pathogens would have had access to potential hosts in close proximity over a period of time and adaptation would have occurred naturally (42). The possibility of a mutational change during commercial development and use is unlikely. Endemic pathogens used as biopesticides are applied as stable, formulated products prepared from the same isolate maintained in a manner to ensure genetic uniformity and stability (20). Adaptation might be possible if the pathogen were used outside the region of endemic occurrence and the new region included a closely related host, or if a new, susceptible economic host were introduced into the area of use.

During the development of *Phytophthora palmivora* as a bioherbicide for strangler vine (*Morrenia odorata*) in citrus (16,50,51) concern was expressed for the potential of *P. palmivora* to hybridize with other *Phytophthora* species that were citrus pathogens or the potential for adaptation to citrus as a host. Studies were conducted to assess genetic variability, the potential for genetic exchange with related species and the potential for host adaptation. Genetic variability was determined by comparing the colony morphology of single zoospore cultures as well as cultures arising from UV-irradiated zoospores to related isolates (50). Matings were conducted with related *Phytophthora* species to assess the potential for genetic exchange (50,51). Oospores were produced in some crosses but could not be germinated (50). The potential for genetic adaptation was tested by successive serial transfers to citrus fruit without demonstrating increased virulence

(50). No change in host range or virulence was detected (16), and the pathogen was deemed sufficiently safe to justify commercial use.

The primary concern with host adaptation is reserved for introduced agents for classical biocontrol. Careful consideration must be given to the range of genetic variability and reproductive potential of plant pathogens. Those pathogens with sexual reproductive systems, related species or subspecies in the introduction area, and common hosts with related pathogens that might support hybridization may constitute an unacceptable risk. For example, several cereal rusts have demonstrated the ability to hybridize on wild grasses and produce progeny with different host ranges (10,28,44). Similar studies by Nelson and Kline (41,46) have demonstrated the potential of related *Cochliobolus* species to cross and produce viable hybrids with changed host specificities. Host adaptation also has been documented with insects (71). For imported pathogens, tests for potential adaptation by serial passage through related hosts (42) or experimental crosses with related pathogens (50) may be warranted.

As genetically modified organisms make their way from the laboratory to the field, public concerns with the potential for genetic exchange will need to be addressed (15,60). Little work has been done to date on the ecology of genetically modified fungal pathogens. Research on hybridization and genetic exchange in plant pathogenic fungi suggests that hybrid progeny often have reduced fitness and are potentially less competitive in the environment (42). Recent work with plasmid-transformed fungi has demonstrated similar results (TeBeest, unpublished). Studies of hybridization and heterokaryosis in *Colletotrichum* (14) have demonstrated the potential for nuclear exchange in *C. gloeosporioides* but have not shown stable hybridization or genetic recombination. Although ecological studies of genetically engineered fungi are in their infancy, the evidence to date suggests that the potential for genetic exchange with related fungi and production of undesirable hybrids is no greater, and is perhaps less of, a threat than that with natural strains.

Summary

Previous experiences with host-range testing of biological control agents suggest that phylogenetically based host-range testing provides a useful assessment of host safety. The lack of a clear relationship of phylogeny to genetic specificity, the limited understanding of the molecular basis of host specificity, and a poor appreciation of the range of genetic diversity that exists in host and pathogen populations limit the development of a completely predictable system. Future host-range tests must be designed to account for genetic variability in the host and pathogen, particularly with imported pathogens. Further improvements in testing protocols must await the development of a better understanding of host specificity, adaptation, and genetic diversity.

Literature Cited

1. Anderson, R. N., and Walker, H. L. 1985. *Colletotrichum coccodes:* a pathogen of eastern black nightshade (*Solanum ptycanthum*). Weed Sci. 33:902–905.

2. Arx, von, J. A. 1970. A revision of the fungi classified as Gloeosporium. Bibl. Myc. 23:1–203.

3. Barrett, S. C. H. 1982. Genetic variation in weeds. Pages 73–98, *in:* Biological Control of Weeds with Plant Pathogens, R. Charudattan and H. L. Walker, eds. Wiley, New York.

4. Baum, B. R., and Savile, D. B. O. 1985. Rusts (Uredinales) of Triticeae: evolution and extent of coevolution, a cladistic analysis. Bot. J. Linn. Soc. 91:367–394.

5. Boyette, C. D., Templeton, G. E., and Smith, R. J., Jr. 1979. Control of winged waterprimrose (*Jussiaea decurrens*) and northern jointvetch (*Aeschynomene virginica*) with fungal pathogens. Weed Sci. 27:497–501.

6. Brian, P. W. 1976. The phenomenon of specificity in plant disease. Pages 15–22, *in:* Specificity in Plant Disease, R. K. S. Wood and A. Graniti, eds. Plenum Press, New York.

7. Brosten, B. S., and Sands, D. C. 1986. Field trials of *Sclerotinia sclerotiorum* to control Canada thristle (*Cirsium arvense*). Weed Sci. 34:377–380.

8. Burdon, J. J. 1982. The effect of fungal pathogens on plant communities. Pages 99–112, *in:* The Plant Community as a Working Mechanism, E. I. Newman, ed. Blackwell, Oxford, England.

9. Burdon, J. J., and Marshall, D. R. 1981. Biological control and the reproductive mode of weeds. J. Appl. Ecol. 18:649–658.

10. Burdon, J. J., Marshall, D. R. and Luig, N. H. 1981. Isozyme analysis indicates that a virulent cereal rust pathogen is a somatic hybrid. Nature 293:565–566.

11. Burdon, J. J., and Shattock, R. C. 1980. Diseases in plant communities. Pages 145–219, *in:* Applied Biology, Vol. V, T. H. Coaker, ed. Academic Press, New York.

12. Burnett, J. H. 1983. Speciation in Fungi. Trans. Br. Mycol. Soc. 81:1–14.

13. Caten, C. E. 1987. The concept of race in plant pathology. Pages 21–37, *in:* Populations of Plant Pathogens, M. S. Wolfe and C. E. Caten, eds. Blackwell, Oxford, England.

14. Chacko, R. J., Weidemann, G. J., and TeBeest, D. O. 1989. Heterokaryosis in *Colletotrichum gloeosporioides* f.sp. *aeschynomene* Mycol. Soc. Am. Newsletter (in press) (Abst.).

15. Charudattan, R. 1985. The use of natural and genetically altered strains of pathogens for weed control. Pages 347–372, *in:* Biological Control in Agricultural IPM Systems. Academic Press, New York.

16. Charudattan, R. 1989. Release of fungi: large-scale use of fungi as biological weed control agents. Proc. Conf. on Risk Assessment in Agric. Biotechnol., J. Marois and A. Browning, eds.

17. Charudattan, R., Freeman, T. E., Cullen, R. E., and Hofmeister, F. M. 1980. Evaluation of *Fusarium roseum* "Culmorum" as a biological control for *Hydrilla verticillata:* Safety. Pages 307–323, *in:* Proc. V Int. Symp. Biol. Control Weeds, E. S. Del Fosse, ed. CSIRO, Melbourne.

18. Charudattan, R., Zettler, F. W., Cordo, H. A., and Christie, R. G. 1980. Partial characterization of a potyvirus infecting the milkweed vine, *Morrenia odorata*. Phytopathology 70:909–913.

19. Chilvers, G. A., and Brittain E. G. 1972. Plant competition mediated by host-specific parasites— a simple model. Aust. J. Biol. Sci. 25:749–756.

20. Churchill, B. W. 1982. Mass production of microorganisms for biological control. Pages 139–156, *in:* Biological Control of Weeds with Plant Pathogens, R. Charudattan and H. L. Walker, eds. Wiley, New York.

21. Cother, E. J. 1975. *Phytophthora drechsleri:* pathogenicity testing and determination of effective host range. Aust. J. Bot 23:87–94.

22. Daniel, J. T., Templeton, G. E., Smith, R. J., and Fox, W. T. 1973. Biological control of northern jointvetch in rice with an endemic fungal disease. Weed Sci. 21:303–307.

23. Day, P. R. 1974. Genetics of Host–Parasite Interaction. Freeman, San Francisco.

24. DeJong, M. D. 1988. Risk to fruit trees and native trees due to control of black cherry. PhD. Thesis. Landbouwuniversiteit te Wageningen, Wageningen, Netherlands.

25. Dinoor, A., and Eshed, N. 1984. The role and importance of pathogens in natural plant communities. Ann. Rev. Phytopathol. 22:443–466.

26. Dinoor, A., and Eshed, N. 1987. The analysis of host and pathogen populations in natural ecosystems. Pages 75–88, *in:* Populations of Plant Pathogens, M. S. Wolfe and C. G. Caten, eds. Blackwell, Oxford, England.

27. Eshed, N., and Dinoor, A. 1980. Genetics of Pathogenicity in *Puccinia coronata:* pathogenic specialization at the host genus level. Phytopathology 70:1042–1046.

28. Eshed, N., and Dinoor, A. 1981. Genetics of pathogenecity in *Puccinia coronata:* the host range among grasses. Phytopathology 71:156–163.

29. Field, R. P., and Bruzzese, E. 1985. Biological control of blackberries: resolving a conflict in Australia. Pages 341–349, *in:* Proc. VI Int. Symp. Biol. Control Weeds, E. S. DelFosse, ed. Agric. Can., Can. Gov. Publ. Ctr., Ottawa.

30. Freeman, T. E., and Charudattan, R. 1985. Conflicts in the use of plant pathogens as biocontrol agents for weeds. Pages 351–357, *in:* Proc. VI Int. Symp. Biol. Control Weeds, E. S. DelFosse, ed. Agric. Can., Can. Gov. Publ. Ctr., Ottawa.

31. Gabriel, D. W. 1986. Specificity and gene function in plant–pathogen interactions. Am. Soc. Microsc. News 52:19–25.

32. Harper, J. L. 1977. Population Biology of Plants. Academic Press, New York.

33. Harris, P., and Zwolfer, H. 1968. Screening of phytophagous insects for biological control of weeds. Can. Entomol. 100:295–303.

34. Hasan, S. 1972. Specificity and host specialization of *Puccinia chondrillina*. Ann. Appl. Biol. 72:257–263.

35. Hasan, S. 1980. Plant pathogens and biological control of weeds. Rev. Plant Pathol. 59:349–356.

36. Hastie, A. C. 1981. The genetics of conidial fungi. Pages 511–547, *in:* Biology of Conidial Fungi, Vol. 2, G. T. Cole and B. Kendrick, eds. Academic Press, New York.

37. Huffaker, C. B. 1962. Some concepts on the ecological basis of biological control of weeds. Can. Entomol. 94:507–514.

38. Hull, V. J. and Groves, R. H. 1973. Variation in *Chondrilla juncea* L. in southeastern Australia. Aust. J. Bot. 21:113–135.

39. Johnson, R. 1976. Genetics of host–parasite interactions. Pages 45–62, *in:* Specificity in Plant Diseases, R. K. S. Wood and A. Graniti, eds. Plenum Press, New York.

40. Keen, N. T., and Staskawicz, B. 1988. Host range determinants in plant pathogens and symbionts. Ann. Rev. Microbiol. 42:421–440.

41. Kline, D. M., and Nelson, R. R. 1971. The inheritance of factors in *Cochliobolus sativus* conditioning lesion induction on gramineous hosts. Phytopathology 61:1052–1054.

42. Leonard, K. J. 1982. The benefits and potential hazards of genetic heterogeneity in plant pathogens. Pages 99–112, *in:* Biological Control of Weeds with Plant Pathogens. R. Charudattan and H. L. Walker, eds. Wiley, New York.

43. Levin, D. A. 1975. Pest pressure and recombination systems in plants. Am. Nat. 109:437–451.

44. Luig, N. H., and Watson, I. A. 1972. The role of wild and cultivated grasses in the hybridization of forma speciales of *Puccinia graminis*. Aust. J. Biol. Sci. 25:335–342.

45. Mortensen, K. 1988. The potential of an endemic fungus, *Colletotrichum gloeosporioides,* for biological control of roundleaved mallow (*Malva pusilla*) and velvetleaf (*Abutilon theophrasti*). Weed Sci. 36:473–478.

46. Nelson, R. R., and Kline, D. M. 1969. Genes for pathogenicity in *Cochliobolus heterostrophus*. Can. J. Bot. 47:1311–1314.

47. Polhill, R. M., and Raven, P. H. 1981. Advances in Legume Systematics. Royal Botanic Gardens, Kew, Surrey, England.

48. Politis, D. J., Watson, A. K., and Bruckart, W. L. 1984. Susceptibility of musk thistle and related composites to *Puccinia carduorum*. Phytopathology 74:687–691.

49. Populer, C. 1978. Changes in host susceptibility with time. Pages 239–262, *in:* Plant Disease: An Advanced Treatise, Vol. II, J. G. Horsfall and E. B. Cowling, eds. Academic Press, New York.

50. Ridings, W. J. 1986. Biological control of stranglervine in citrus—a researcher's view. Weed Sci. 34 (Suppl. 1):31–32.

51. Ridings, W. H., Mitchell, D. J., Schoulties, C. L., and El-Gholl, N. E. 1976. Biological control of milkweed vine in Florida citrus groves with a pathotype of *Phytophthora citrophthora*. Pages 224–240, *in:* Proc. IV Int. Symp. Biol. Control Weeds, T. E. Freeman, ed. University of Florida, Gainesville.

52. Sands, D. C., Ford, E., and Miller, R. V. 1989. Genetic manipulation of fungi for biological control of weeds. WSSA Abst 29:123. (Abst.)

53. Scheepens, P. C., and VanZon, H. C. J. 1982. Microbial herbicides. Pages 623–641, *in:* Microbial and Viral Pesticides, E. Kurstak, ed. Dekker, New York.

54. Schroeder, D. 1983. Biological control of weeds. Pages 41–78, *in:* Recent Advances in Weed Research, W. W. Fletcher, ed. Comm. Agric. Bur., Kew, Surrey, England.

55. Sutton, B. C. 1980. The Coelomycetes. Comm. Mycol. Instit., Kew, Surrey, England.

56. TeBeest, D. O. 1985. Techniques for testing and evaluating plant pathogens for weed control. J. Agric. Entomol. 2:98–134.

57. TeBeest, D. O. 1988. Additions to host range of *Colletotrichum gloeosporiodes* f.sp. *aeschynomene* Plant Dis. 72:16–18.

58. Templeton, G. E., TeBeest, D. O., and Smith, R. J., Jr. 1984. Biological weed control in rice with a strain of *Colletotrichum gloeosporioides* (Penz.) Sacc. used as a mycoherbicide. Crop Protection 3:409–422.

59. Templeton, G. E., Weidemann, G. J., and Smith, R. J. 1986. Biological Weed Control. Pages 99–109, *in:* Research Methods in Weed Science, N. D. Camper, ed. Southern Weed Sci. Soc.

60. Turgeon, G., and Yoder, O. C. 1985. Genetically Engineering Fungi for Weed Control. Pages 221–230, *in:* Biotechnology: Applications and Research, P. N. Cheremisinoff and R. P. Onellette, eds. Technomic, Lancaster, PA.

61. Turner, B. L. 1971. Implications of the biochemical data: a summing up. Pages 549–558, *in:* Chemotaxonomy of the Leguminosae, D. Boulter and B. L. Turner, eds. Academic Press, New York.

62. Wahl, I., Eshed, N., Segal, A., and Sobel, Z. 1978. Significance of wild relatives of small grains and other wild grasses in cereal powdery mildews. Pages 83–100, *in:* The Powdery Mildews, D. M. Spencer, ed. Academic Press, New York.

63. Wapshere, A. J. 1973. A comparison of strategies for screening biological control organisms for weeds. Pages 151–158, *in:* Proc. 2nd Int. Symp. Biol. Control of Weeds, P. H. Dunn, ed. Comm. Agric. Bur., Kew, Surrey, England.

64. Wapshere, A. J. 1974. A strategy for evaluating the safety of organisms for biological weed control. Ann. Appl. Biol. 77:201–211.

65. Wapshere, A. J. 1975. A protocol for programmes for biological control of weeds. PANS 21:295–303.

66. Wapshere, A. J. 1982. Biological control of weeds. Pages 47–56, *in:* Biology and Ecology of Weeds, W. Holzner and N. Numata, eds. W. Junk, The Hague.

67. Wapshere, A. J. 1983. Problems in the use of plant biochemistry for establishing the safety of biological control agents for weeds: the *Chondrilla* and *Echium/Heliotropium* cases. Entomophaga 28:287–294.

68. Watson, A. K. 1985. Host specificity of plant pathogens in biological weed control. Pages 577–586, *in:* Proc. VI Int. Symp. Biol. Control Weeds, E. S. DelFosse, ed. Agric. Can., Can. Gov. Publ. Ctr., Ottawa.

69. Webb, R. R., and Lindow, S. E. Influence of environment and variation in host suspceptibility on a disease of bracken fern caused by *Ascochyta pteridis*. Phytopathology 77:1144–1147.

70. Weidemann, G. J., TeBeest, D. O., and Cartwright, R. D. 1988. Host specificity of *Colletotrichum gloeosporioides* f.sp. *aeschynomene* and *C. truncatum* in the *Leguminosae*. Phytopathology 78:986–990.

71. Zwolfer, H., and Harris, P. 1971. Host specificity determination of insects for biological control of weeds. Ann. Rev. Entomol. 16:159–178.

6

Ecology and Epidemiology of Fungal Plant Pathogens Studied as Biological Control Agents of Weeds

David O. TeBeest

Introduction

The increase in interest in the exploitation of fungal plant pathogens as weed control agents resulted in a rapid increase in the number of plant pathogens discovered and studied as potential biological control agents. These studies accounted, in part, for development of the mycoherbicide and classical strategies discussed in the earlier chapters by Charudattan and Watson. Some of these pathogens were biologically successful, that is, they were shown to control weeds efficiently under field conditions, whereas many others were not. Most certainly, the epidemiological and ecological requirements and characteristics of these plant pathogens played a key role in determining the successful use of plant pathogens as biological control agents under these two strategies.

For the purposes of discussion, I have adopted the definition of epidemiology proposed by Zadoks and Schein (86), as studies in which comparisons are made of the influence of host, pathogen, and environment on disease development and spread, whereas ecology of plant pathogens is in the broader sense a study of the relationship of a pathogen as a member of a community to its surroundings.

Epidemiology of Biological Control Agents

It has been repeatedly stated that two components of the environment that limit the utilization and effectiveness of biological control agents are temperature and moisture as humidity or free water (30,36,70). These components have been studied for several pathogens in controlled environments and in the field in efforts to describe the conditions under which the organisms would be most effective.

Alternaria cassiae, *Alternaria crassa*, and *Alternaria macrospora* have been studied as potential bioherbicides for control of sicklepod, *Cassia obtusifolia*, jimsonweed, *Datura stramonium*, and spurred anoda, *Anoda cristata*, respectively. The optimal environmental conditions for control of sicklepod by *A. cassiae* included at least 8 hours of free moisture at 20°–30°C (76). However,

Walker and Boyette (78) showed that increased control of the weed was obtained
if seedlings inoculated with the fungus received additional dew periods after the
initial dew period. Regionwide field tests with this pathogen illustrated the variation in control caused by differences in environments at the time of inoculation
(22). Van Dyke and Trigiano (75) examined spore germination of *A. cassiae* in
field and growth chamber studies and found that spores germinated within 2–3
hours on inoculated seedlings and that approximately six germ tubes were produced per spore. Appressoria were produced terminally on germ tubes within 20
hours and formed over stomates and directly on the cuticle.

Alternaria crassa has potential as a biological control agent for controlling
jimsonweed. Boyette (7) found that although spores germinated optimally between 20° and 30°C, the fungus grew optimally over a wider range from 25° to
35°C. Somewhat surprisingly, infectivity on *D. stramonium* was reduced at
temperatures below 20°C and above 30°C, a range somewhat different from
growth and germination optima and perhaps reflecting an interaction with the
host.

In sharp contrast to *A. cassiae,* the optimum moisture requirements for infection
and control of *A. cristata* by *A. macrospora* was greater than 24 hours at 20°–
30°C (76). Although a very long requirement, sufficient moisture occurred during
field tests to provide conditions suitable for infection to reduce plant number and
dry weight. Capo (17) found that the optimal range for infection of spurred
anoda by *A. macrospora* was 15°–25°C, with a projected maximum at 21.5°C.
Maximum disease developed with dew periods of 20 hours. Incubation temperatures between 20° and 32°C were best for disease development with maximum
development projected at 26°C. Capo (17) also showed that there was a significant
interaction between dew temperature, duration, and incubation temperature. As
incubation temperatures increased or decreased from the optium, dew requirements increased for disease development. However, at any given dew temperature, the longer the dew period the greater the number of lesions produced on
inoculated leaf tissue. Unless inoculated to very young seedlings, stem lesions
generally grew slowly even under optimal conditions (17,76). Control was reduced if stem lesions did not develop to girdle stems. Higher levels of control of
A. cristata were achieved when *Fusarium lateritium* was applied 5 days after
incoulation with *A. macrospora* (24). It was concluded that *F. lateritium* colo-
nized lesions produced by *A. macrospora* and that the interaction resulted in
increased control of the weed.

The most extensively studied group of pathogens considered as potential bioherbicides includes several species and form species within the genus *Colletotrichum.*
Colletotrichum coccodes has been isolated from several weed hosts (58). An
isolate of *C. coccodes* isolated from black nightshade (3) required 16 hours or
longer of dew for maximum disease development. This requirement was thought
to limit the potential use of this isolate in the field, and it was further suggested
as an explanation for occasional appearance of the disease during extended periods

of wet whether. This isolate of *C. coccodes* from nightshade did not infect velvetleaf, *Abutilon theophrasti*. An isolate of *C. coccodes* which infected *A. theophrasti* was reported in 1988 (83). This isolate incited disease on velevetleaf over a wide range of dew periods and temperatures; however, the most rapid and destructive development of disease was reported to occur following a 24-hour dew period at 24°C. At cooler temperatures and or after shorter dew durations, the pathogen caused only premature defoliation of inoculated leaves. The optimum temperatures for mycelial growth and spore germination of this isolate of *C. coccodes* were 27°C and 24°C, respectively. These conditions compare favorably with the results which indicated that *C. coccodes* caused the most severe damage to velvetleaf following dew periods of 18–24 hours at an air temperature of 24°C or 30°C. Their results also indicated that there were significant effects of air temperature and dew period duration on disease level but that there was no interaction. However, their results also showed that disease development on inoculated seedlings given an optimal dew treatment was slowed or reduced when seedlings were incubated at cool, suboptimal incubation temperatures. *Colletotrichum coccodes* when applied with a growth regulator reduced biomass and increased the control of velvetleaf when compared to either the fungus or growth regulator alone in cool conditions encountered in Quebec, Canada, or warmer temperatures encountered in Maryland (37).

Colletotrichum gloeosporioides f.sp. *aeschynomene* was registered in 1982 as COLLEGO, for control of *Aeschynomene virginica* in soybeans and rice in several Mississippi River delta states (70). The fungus is naturally widely distributed in Arkansas and was found in most of the counties in which rice was grown and could be found on plants throughout much of the growing season but at no time was disease severe enough to kill plants or reduce seed production significantly (64). The disease was considered to be endemic rather than epidemic even though the host, *A. virginica,* was susceptible and could be controlled if spores were artificially disseminated into an infested rice field. Temperature and moisture conditions conductive for rapid disease development were found to be rather broad (69). Anthracnose on northern joint vetch developed rapidly at incubation temperatures ranging from 20° to 32°C, with an optimum near 28°C. Moisture requirements for establishment of infection included dew periods in excess of 12 hours at the optimal temperature of 28°C with longer periods required when dew period temperatures were 20°, 24° or 32°C. At 36°C, disease development was severely limited. It is interesting to note that disease development was also slowed by incubation in growth chambers in which daily temperature regimes were not constant during the day and night. The results of these studies indicated that anthracnose on northern joint vetch should not be limited by normal environmental conditions found during the summers. The limited amount and spread of the disease noted by Smith et al. (54) was interpreted to be the result of other factors as well (66).

Colletotrichum gloeosporioides f.sp. *jussiaeae* has been successfully used as a biological agent for control of winged water primrose, *Jussiaea decurrens,* in rice fields of Arkansas (9). In greenhouse tests water primrose seedlings were killed after inoculation with spores followed by a single 16-hour dew period at 28°C. Brumley (11) found that the minimum time and temperature with respect to the dew period was 16 hours at 28°C. Furthermore, the optimum air temperature following a dew treatment was also 28°C. These conditions fit closely the requirements for spore germination and appressorium formation by this fungus. These requirements are very similar to those required by *C. gloeosporioides* f.sp. *aeschynomene* and tests in which both pathogens were used simultaneously clearly showed that both diseases developed rapidly in the field (9).

Colletotrichum gloeosporioides f.sp. *malvae* was isolated from infected *Malva pusilla* in 1982 and later shown to have potential for biological control of *M. pusilla* (50). Under natural conditions, the disease does not develop into epiphytotic proportions until late in the year. Round-leaved mallow plants inoculated with spore suspensions were killed within 3 weeks in greenhouse tests. Makowski (45) reported that *M. pusilla* was best controlled if inoculated plants received a minimum of 20 hours of dew, or repetitive 16-hour dew periods, at temperatures below 30°C. *Abutilon theophrasti,* also susceptible to infection by this fungus, was best controlled if inoculated seedlings received 48 hours of dew at temperatures below 15°C. The minimum time dew period for development of anthracnose on *M. pusilla* corresponded to the timing of the appressorium formation and the beginning of host cell penetration. *A. theophrasti* outgrew infections at the optimum conditions for anthracnose development. *C. gloeosporioides* f.sp. *malvae* was efficacious in field tests (45,50); however, high moisture conditions or precipitation were required within 48 hours after inoculation for rapid disease development (45).

In 1982 Kirkpatrick et al. (41) reported that *Colletotrichum malvarum* effectively controlled prickly sida, *Sida spinosa,* in greenhouse and field studies in Arkansas. Of 38 plant species tests, *S. spinosa* was the most susceptible to infection by this anthracnose fungus. Disease severity was affected by dew temperature and was most severe at 24°C when given 16 hours of dew. Disease severity was significantly reduced at 20°C and 28°C, and at 16°C and 32°C no lesions were produced. Exposure of seedlings to dew treatments before inoculation did not affect disease levels, whereas two or three consecutive exposures after inoculation increased disease severity. Field studies conducted during favorable periods showed that the fungus could infect and control seedlings of different sizes equally, control ranging from 84% to 95% 3 weeks after inoculation. It is noteworthy that a natural epidemic of a *Colletotrichum* sp. on *S. spinosa* (possibly *C. malvarum* or *C. dematium*) was reported to have occurred in field tests with *S. spinosa* in cotton and that the level of infection of *S. spinosa* reached levels requiring fungicide treatments to control the disease (12).

An attempt to initiate epiphytotics of *Colletotrichum xanthii* on bathurst burr, *Xanthium spinosum,* was reported by Butler (15) in 1951. For three successive seasons the fungus caused widespread destruction of the burr, though previously unrecorded as a pathogen in the area. It appears that the fungus spread rapidly from one site in New South Wales to occupy a total of 50 different districts after 3 years. It was observed that rainfall was the most important factor affecting the natural occurrence of the disease. Additionally, it was established that *C. xanthii* developed on plants between 20° and 28°C, conditions generally found in the area during the time of rapid increase. Artificial inoculation of burr plants was generally successful; however, reliable estimates of control were not available (15). During these tests it was also observed that the destructiveness of the disease on burr was seriously reduced by periods of hot, dry weather. Recent work by McRae (47,48) showed that *C. orbiculare* (= *C. xanthii*) was an effective biological control agent for *X. spinosum*. The optimum dew period temperature for anthracnose development was between 20° and 25°C, whereas the optimum post-dew period incubation temperature was reported to be 30°C (48). Anthracnose severity generally increased with the duration of the dew period with maximum disease development resulting from a 48-hour dew period at 20°C, although significant levels of disease developed after only 8 hours of dew. Disease development was also increased with two consecutive dew periods of at least 12 hours each. Delaying the onset of dew by 4 or more hours, when plants were maintained at 20–40% relative humidity, reduced mean disease ratings significantly (4).

Several other *Colletotrichum* species have been investigated and reported as potential biological control agents for weeds. Cardina et al. (18) have reported that the optimum conditions for infection of Florida beggarweed, *Desmodium tortuosum*, by *Colletotrichum truncatum* were 14–16 hours of 100% relative humidity at 24°–29°C. Infection was significantly less severe when incubated at 18°C. The duration of the incubation period also influenced the rate of disease development. Disease development was most rapid, with at least 14 hours of high relative humidity. Smith et al. (64) have recently reported that the effect of *Colletotrichum gloeosporioides* on Eurasian water milfoil, *Myriophyllum spicatum*, was greatest at 20°C. The fungus was confined to the surface of plants, however, and rarely penetrated cells other than the epidermis. At 20°C, treated plants did not increase in biomass, whereas at 15° and 20°C the fungus did not produce a detectable reduction in plant growth.

Colletotrichum graminicola and *Gloeocercospora sorghi* were studied as potential biological controls for Johnson grass, *Sorghum halepense* (23). Comparison of dew period requirements, as indicated by an index of disease levels for 12 hours of dew compared to a 24-hour dew period, for *C. graminicola* showed that only 10% of maximum disease development was achieved after a 12-hour dew period, whereas for *G. sorghi* only 40% of the disease levels achieved after 24 hours of dew were reached after only a 12-hour dew treatment. Seedlings of several *Sorghum* spp. exhibited moderate to high levels of leaf damage in most compatible reactions with 24 hours of dew.

Other plant pathogens tested as potential biological control agents have exhibited similar requirements for infection to occur. For example, infection of purslane, *Portulaca oleracea*, by *Dichotomophthora portulacae* was influenced by temperature, time, and inoculum density (42). Conidia of the fungus germinated within 1 hour and infected seedlings within 2 hours. Infection was optimum at 27°C, whereas the optimum temperature for spore germination was reported as 33°C, although hyphal growth was best at 30°C. Moisture was required for conidial germination and infection.

Infection of bracken fern, *Pteridium aquilinum*, by *Ascochyta pteridis* occurred as low as 10°C but required at least 18 hours of leaf wetness at this temperature (80). The infection frequency increased with increasing temperature up to 20°C, whereas the length of the leaf wetness period required for infection decreased over the temperature range of 10°–20°C.

Many of the pathogens discussed here are frequently disseminated when suspended in water droplets. However, the rust fungi currently studied as potential classical biological control agents are readily wind disseminated and can spread great distances very rapidly (10). For example, *Puccinia canaliculata* was noted as causing a severe epidemic in 1978 in natural stands of *Cyperus esculentus* (54). *P. canaliculata* infected hosts in greenhouse tests without a moisture treatment following inoculation (16). Similarly, the dew requirement for *P. carduorum* is also very short; the fungus infects *Carduus nutans* over a wide range of dew periods and temperatures, with infection occurring after only 4 hours at 14°C and after 8 hours at 8°C. Optimal conditions were reported as 12 hours of dew between 17° and 24°C (56,57).

Puccinia chondrillina was introduced into the United States and Australia for control of *Chondrilla juncae* (25,43). A strain of *P. chondrilla* released in California in 1975 spread through several *P. chondrilla* populations in California and Oregon within 2 years (43). Later introductions were also successful in Idaho and Washington. The fungus continued to infect the plant as long as temperature and moisture conditions were conducive. The time required to produce a new generation of spores is also temperature dependent (6). Although the largest pustules were produced at 24°C and were produced more quickly than at 8°C and 16°C, the greatest amount of infection occurred at 8°C and 16°C. At 8°C and 16°C, 75% of the leaf area was infected, whereas at 24°C only 25% was infected. The amount of infection, measured by the number of infection sites, appeared to be related to spore germination. At 8°C and 16°C, 80% and 70% of the spores germinated on water agar, whereas at 24°C only 25% of the spores germinated in 18 hours. Hasan and Jenkins (34) have previously reported that uredospores of *P. chondrillina* infected *C. juncea* at temperatures of 10°–33°C and that the minimum time for infection was 2 hours at 20°C. Uredospores of this fungus will germinate at temperatures from 0° to 33°C (35). It is not surprising, then, considering the wide temperature range in which the fungus spores germinate and infect plants, that the fungus spread as rapidly as reported after release in both the United States and Australia (25,43).

Uredo eichorniae and *Uromyces ponteridae* have also been studied for potential use as biological control agents for water hyacinth, *Eichhornia crassipes,* and pickerelweeds, *Pontederia* spp. (21). Although only 20% of freshly harvested uredospores of these rusts were germinable on water agar, increased germination was attained by the addition of several stimulants. Spores were germinable at temperatures ranging from 10° to 30°C, with an optimum of 20°C. Consistent host infection was achieved, however, only when newly infected plants were incubated with gaseous 2-heptanone, 5-methyl–2-hexanone, or 2-hexanone. Uredospores of *U. pontederiae* infected *P. lanceolata* leaves at 15°, 20°, and 25°C, but the fungus resporulated only at 20°C and 25°C. Spores germinated but did not form appressoria at 30°C and failed to germinate altogether at 35°C.

Two soil-borne plant pathogens have been studied as biological control agents. *Fusarium solani* f.sp. *cucurbitae* has effectively controlled Texas gourd, *Cucurbita texana,* in controlled environments and in field tests (8,81,82). In controlled environments, seedlings were killed over a range of air temperatures from 16° to 40°C with an optimum air temperature of 26°–30°C (8). Application of the fungus to soil during the growing season indicated that control declined 50–70% within 12 weeks after soil infestation (79). Soil populations of the fungus rapidly declined after 6 weeks.

Phytophthora citrophthora, reidentified as *Phytophthora palmivora* (29), has been used since 1981 for control of *Morrenia odorata,* strangler vine, in Florida citrus groves. Unlike *F. solani* f.sp. *cucurbitae, P. palmivora* persists for several years after even a single application (40). Ridings (59) has reported that survival of *P. palmivora* was affected by soil conditions, with survival much better in moist soil. Ridings et al. (60,61) reported that the fungus effectively controlled the weed under normal field conditions in two locations in Florida where daily low temperatures ranged from 14° to 22°C and daily high temperatures ranged from 32° to 38°C. Under these conditions 50% control was reached within 4 weeks after inoculation.

From the preceding discussion it is tempting to speculate on the nature of the requirements for infection among the different species being studied as biological control agents. The rust fungi, wind disseminated, appear to have short moisture period requirements for infection. On the other hand, the various species of *Colletotrichum, Ascochyta,* and *Gloeoscercospora,* which produce spores in a slimy matrix or within pycnidia, are water disseminated and seem to require longer moisture periods. In terms of temperature requirements, such relationships are less obvious; nearly all have an optimum near 24°C. In nearly all cases disease is most severe near temperatures conducive to growth and germination of spores. It would also appear to be generally true that the amount of disease increases as the length of the moisture period increases until a maximum length of time is reached after which no further increases are noted, with the provision that the dew period temperature permits germination to occur. Temperatures during the dew period that are nonconductive to germination or infection structure formation restrict the amount of disease that can develop regardless of incubation temperature.

It has been suggested that dew period requirements of 16 hours or longer could be limiting to the practical use of fungi as biological control agents for weeds (76). Although virulence and desiccation survival has been shown to vary among isolated of *A. pteridis* (73), there would appeal to be little available information on the variability of dew period requirements among isolates of weed pathogens.

Ecology of Biological Control Agents

Many of the plant pathogens studied as biological control agents have been considered to be endemic diseases. *Endemicity* has taken on two usages, however. In one, *endemicity* is taken to mean that an organism is native to a particular geographical area (72). In a pathological sense, however, *endemic* refers to the magnitude of disease oscillations over time. In this sense, an endemic disease is one in which the product of iRc, where i is the infectious period and Rc is the corrected basic infection rate, fluctuates about 1 (86). Thus, the amplitude of fluctuations of disease is small, suggesting that the diseases are present at relatively low levels over time, although large increases as noted by Buchanan (12) are possible and expected. The maintenance of disease at low levels has been taken to suggest a coevolution into "natural endemicity" (86). Thus, many pathogens studied as biological control agents are endemic in all senses of the word. An interesting question is then obvious, "Why are these diseases endemic, or found at low levels, rather than epidemic and at explosive levels?" An insight into this question concerning the regulation of endemicity may be gained by examination of data on the ecology and survival of several biological control agents already found in the literature.

The overwintering survival of infectious propagules may play a role in the relatively low levels of the diseases often found early in the year caused by various biocontrol agents. *Alternaria crassa* is reported to be seed-borne on *Datura metel* var. *muricata*. Seed-borne infection was considered to have caused three kinds of damage in *Datura:* preemergence killing, postemergence damping off, and seedling blight (31). No data were given on the proportion of seeds infected by the fungus, although 94–96% of visibly discolored seeds were infected. Similarly, TeBeest and Brumley (67) reported that *Colletotrichum gloeosporioides* f.sp. *aeschynomene* was seed-borne on *A. virginica* and that severely infected seed did not germinate, whereas somewhat lighter infections resulted in infected seedlings that often were killed by anthracnose. Furthermore, a survey of several seed lots collected from field-grown plants indicated that 7.4–41.9% of emergent seedlings were infected by the *C. gloeosporioides* even after storage for several years. It was suggested that repeated use of the fungus as a bioherbicide could contribute to increased incidence of the disease as a result of seed infection and infestation (67).

Although *C. gloeosporioides* f.sp. *aeschynomene* could survive and overwinter on infected seed, data presented by TeBeest (66) clearly indicated that survival of the fungus as spores in soil was very limited. The fungus was not recovered from soil collected from artificially infested field plots after 9 weeks and was only occasionally recovered from debris buried in soil after as little as 2 weeks. However, recovery from infected plant debris was more successful. The fungus was successfully cultured from debris left in the field for 7 months if debris was left unburied. Similarly, the fungus was recovered from steam-sterilized water, but not from rice irrigation water, after 180 days. In laboratory tests a population of the fungus declined to less than 1% of the initial population within 4 weeks after infestation but increased 8–80 times in sterile soil in the same time period. *C. gloeosporioides* f.sp. *aeschynomene* does not produce sclerotia to aid in survival (66).

Weeds may contribute to the survival of an isolate of *C. coccodes* that causes an anthracnose on tomato between crop rotations (58). More significantly, however, this isolate produces sclerotia that can survive in soil for long periods of time (5). Spores of this *C. coccodes* were also short-lived in soil, surviving no longer than 3 weeks (5). The isolate of *Colletotrichum coccodes* isolated from velvetleaf does not damage other plants (83). In controlled environments there was a 75% reduction in viable spores recovered from inoculated leaves within 24 hours after inoculation. Survival of spores of this isolate was improved with the addition of a sorbitol formulation to the inoculum (84). With the sorbitol solution a twentyfold increase in the number of viable spores was recovered in controlled environments. Although similar results were obtained in the field, spore survival time was shorter. The addition of materials to increase survival of spores or infectious propagules is a major goal for research in mycoherbicides as discussed in the chapter by Quimby and Boyette in this volume. *Colletotrichum gloeosporioides* f.sp. *malvae* did not persist in soil when soil was infested with spores (45). It was not reisolated from soil after 1 week and survival of the fungus in buried infected plant refuse declined rapidly within 12 weeks. Like *C. gloeosporioides* f.sp. *aeschynomene*, *C. gloeosporioides* f.sp. *malvae* was reisolated from approximately 10% of samples tested when samples were left on the soil surface.

McRae (47) demonstrated that *C. orbiculare* could also be reisolated after 6 months from infected plant refuse left above the soil line. However, *C. orbiculare* could not be reisolated from infected tissues at all sites. Butler (15) had previously observed and described the potential for overwintering survival of this fungus.

Based on the preceding information, it seems clear that the *Colletotrichum* species under investigation as biological control agents have a limited capacity for survival in the field. Lacking sclerotia and/or sexual stages normally associated with overwintering survival, all these species survive primarily on infected refuse left above the soil surface. The spores of many *Colletotrichum* species are surrounded by and produced within a mucilaginous matrix. Nicholson and Moraes (53) demonstrated that spores produced in a matrix are protected against desiccation. Spore masses maintained viability for several weeks when maintained at low relative humidities (45%), whereas spores from which the matrix was removed lost viability within hours even at high relative humidities (80–90%). In addition, glycoproteins within the matrix selectively bound purified phenolic compounds (i.e., tannin) and could protect spores from toxic metabolites produced by plants during infection (52). The influence of inoculum consisting of spore masses of *Colletotrichum graminicola* left on plant debris was assessed by Lipps (44) in field experiments with corn varieties differing in resistance to anthracnose. Oat kernel inoculum or corn residues buried beneath corn seed at planting did not increase the incidence of disease; however, the incidence of anthracnose was significantly increased when residues were on soil surfaces. In addition, the incidence of corn anthracnose was negatively correlated with distance from the residue. Recently, Cerkauskas (19) reported the significant increase in sporulation of several *Colletotrichum* species on explants of various economic crop and weed species in the laboratory following surface disinfestation and treatment with paraquat. Cerkauskas (19) further cautioned that a significant increase in inoculum could result from growers using desiccants, such as paraquat, as a harvest aid in their agricultural programs. However, simple burial of refuse would also serve to reduce the level of surviving inoculum (44,66).

The soil-borne plant pathogens studied as biological control agents were expected to survive in soil. As stated previously, *P. palmivora* apparently survived in soil for several years after a single infestation (40). In field tests, soil was infested with chlamydospores (59), although the fungus also produces sporangia and oospores (29). Infection of milkweed vine is quantitatively related to inoculum density (49). In controlled environments, 600–900 chlamydospores per kilogram of soil were required to achieve 50% infection of milkweed vines, and the population of *P. palmivora* increased with incubation on susceptible plants. Kenney (40) reported in 1986 that inoculations made during 1978–1980 were still giving 95–100% control without additional treatments. Ridings (59,61) has reasoned that inoculum in soil would decrease with a reduction in the number of host plants in infested soils, especially under dry conditions.

Several studies have been made on the persistence of *Fusarium solani* f.sp. *cucurbitae* in soil after artificial infestation. Boyette et al. (8) found that the fungus persisted in the field for up to 12 months but that after 12 months inoculum levels were insufficient to cause seedling disease even when seedlings were grown under optimum conditions for disease development. Weidemann and Templeton (81,82) showed that the population of *F. solani* declined to approximately 1000 colony-forming units per gram soil within 6–8 weeks after infestation in field plots. Application of inoculum consisting of microconidia or macroconidia or sodium alginate granules infested with the fungus did not affect long-term survival of the fungus. However, populations of the fungus established with sodium alginate granules did increase for the first few weeks after infestation in contrast to the steady decline in the populations established with spores applied directly to soil.

The rust pathogens, promoted as biological control agents in the classical approach, can be expected to overwinter by means of various spore stages on infected plant debris. *Puccinia chondrillina,* for example, appears to survive in the Mediterranean area by its uredospore stage alone. Teliospores, produced in late summer to early winter among the uredospores, do not appear to have a function. Teliospores, when germinated, produced basidiospores that did not infect *Chondrilla juncea* (35). Spermagonial and aecial stages of this fungus were not found. Adams and Line (1) reported that teliospores of *P. chondrillina* were produced in Washington beginning in mid-July and were still capable of germinating in November. Spermagonia and aecia were not observed until the following spring. The fungus appears to survive as uredospores on infected plants in the United States as well (1,27). Whereas *Puccinia chondrillina* is known to infect only *Chondrilla juncea* (35), *Puccinia canaliculata* is a heteroecious, macrocyclic rust of yellow nut sedge, *Cyperus esculentus* (54). The fungus has been demonstrated to infect *Xanthium strumarium* and *Helianthus annuus* as alternate hosts; however, after extensive surveys the aecial stage was not found occurring on *X. strumarium* in Georgia (16). The fungus appears to overwinter on *C. esculentus* as uredospores or teliospores but at levels too low to appear until late summer (54).

Dispersal mechanisms are important to the distribution and ecology of biological control agents. Wind has been implicated as an important factor in the rapid dispersal of the rusts (25,27,54). Additionally, Burnett et al. (14) have suggested that wind dissemination of sporangiophores of *P. palmivora* from infected plants may have accounted for the appearance of the disease on noninoculated milkweed vines in the vicinity. The various species of *Colletotrichum,* however, generally produce sticky masses of spores in acervuli and these spores require suspension in water to be dispersed by wind. Thus, interplant dispersal of *Colletotrichum* species is usually limited to wind-driven rain, whereas intraplant dispersal could also be accomplished by dew. Insects have also been implicated as a possible vector for *C. gloeosporioides* f.sp. *aeschynomene,* since grasshoppers have been observed feeding on the lesions, although no true vector relationships are known (72).

The effects of light on disease development and resporulation on lesions have been studied in corn inoculated with *C. graminicola* (38). Lesion length, lesions per square centimeter, and sporulation per square centimeter decreased with increasing illuminance. Blanchette and Lee (6) reported that germination of the uredospores of *P. chondrillina* was significantly reduced by light at intensities as low as 0.5 klux and that germ tube growth was restricted at somewhat higher intensities.

The interaction of plant pathogens with noncultivated plants has been discussed in greater detail elsewhere and will not be discussed here (2,13,33,62). It should be noted, however, that resistance to *P. chondrillina* has been widely reported (25,35,43). Also, Callaway et al. (16) noted differences in susceptibility of *X. strumarium* to four collections of *P. canaliculata*. Similarly, not all collections of *Carduus nutans* were susceptible to *Puccinia carduorum*, indicating that biotypes of the *Carduus* species may exist in North America (57).

Epidemiology and Ecology of Genetically Manipulated Fungal Plant Pathogens

Transformation of filamentous fungi with exogenous DNA was recently reviewed by Fincham (28). Likewise, the potential for genetically manipulating fungal plant pathogens for weed control has been discussed in several recent review articles (20,30,71,74). Turgeon and Yoder (74) suggested that gene-cloning technology might be used to transfer the genes controlling production of a toxin to a pathogen to enhance virulence toward a specific host. Specific genes that control toxin production have been identified by conventional genetic analysis in plant pathogens (85). Charudattan (20) cited not only toxin production, but also resistance to pesticides as an area in which gene manipulations might be useful in establishing strains for use in integrated pest management (IPM) systems. Greaves et al. (30) and Templeton and Heiny (71) suggested that biotechnology might be used to address problems associated with virulence, host specificity, and environmental capability, although a deeper understanding of the biology of the pathogen at the organismal and ecosystems level will be needed (71).

The survival of genetically transformed organisms in the environment is currently of intense interest and debate, and many questions remain (32,51,55,73). The ability of *P. palmivora* to produce viable sexual structures in both intra-and interspecific crosses with other isolates of *P. palmivora* and *P. citrophora* clearly suggests a potential means of overwintering and an avenue for genetic recombination (29,60). Introduction of novel genes into this species should be considered carefully. Similarly, the presence of a teleomorphic, or *Glomerella*, stage for the *Colletotrichum* species must also be considered as an avenue of gene exchange with compatible wild-type strains. The possibility of a parasexual mechanism in *Colletotrichum* species should also be investigated as a means of gene transfer, since it is generally assumed that parasexuality is a means of generating and storing variation in the imperfect fungi (26), and Stephan (65) demonstrated heterokaryosis in isolates of *Glomerella cingulata*.

The relative fitness of transformed isolates of a fungal plant pathogen is largely unassessed at this time; however, Keller and Yoder (39) have demonstrated that transformed isolates of *Cochliobolus heterstrophus* were less fit than the wild-type parent. Similarly, TeBeest and Dickman (68) have reported that transformed isolates of *Colletotrichum gloeosporioides* f.sp. *aeschynomene* were less aggressive and less fit than a parental culture of the fungus. The reported loss of fitness of transformed isolates of fungal plant pathogens is in itself an interesting question but should not be assumed to be universally true at this time.

Summary

Several important factors apparently limit the development of natural epidemics of plant pathogens of weeds. These factors may include host heterogeneity, a narrow range of temperature or moisture conditions conducive to infection, poor survival of inoculum from season to season, and limiting dispersal mechanisms. Altogether, these factors moderate the amount of disease and decrease the likelihood of eruptive epidemics. These diseases appear to be truly endemic.

A major concern for utilization of mycoherbicides is the dew or moisture requirement needed to establish infection. In most instances 12 hours of moisture appears to be sufficient to establish the infection at an incidence high enough to establish control. Whether the moisture requirement must be satisfied during one or two consecutive intervals appears to depend on the pathogen under study. Controlled-environment studies also indicate that postinoculation incubation temperatures are critical to establishment of the disease by affecting the latent period. The shortest latent periods generally occur at temperatures near optimum for growth of the pathogen. Generally, successful biological control agents require relatively short moisture periods and have a relatively wide temperature latitude for rapid disease development.

The relative survival of the few biological control agents studied so far appears to depend on the amount of inoculum and on the specialized structures each pathogen may or may not possess. Pathogens with specialized fruiting structures such as teliospores, oospores, ascospores, or sclerotia may be expected to survive with relative ease. In this, weed pathogens are very similar if not identical to their more widely studied counterparts of crop plants. The rules governing survival, dissemination, and development of pathogens of noncultivated plants are not very different from the rules established for pathogens of crop plants. A major difference, perhaps, is the genetic homogeneity of the host plant. Noncultivated plants are generally more diverse than their cultivated counterparts.

Much remains to be learned about the pathogens of noncultivated plants and about their use as biological control agents. Their existence as endemic plant pathogens may be radically altered when transported to a new environment on a new host, or when altered to contain a seemingly innocuous piece of new DNA. The fungi being tested now, however, present a wonderful tool for such studies.

Literature Cited

1. Adams, E. B., and R. F. Line. 1984. Biology of *Puccinia chondrillina* in Washington. Phytopathology 74:742–745.

2. Anikster, Y., and I. Wahl. 1979. Coevolution of the rust fungi on Gramineae and Liliaceae and their hosts. Ann. Rev. Phytopathol. 17:367–403.

3. Anderson, R. N., and H. L. Walker. 1985. *Colletotrichum coccodes:* a pathogen of eastern black nightshade *(Solanum ptycanthum)*. Weed Sci. 33:902–905.

4. Auld, B. C., C. F. McRae, and M. M. Say. 1988. Possible control of *Xanthium spinosum* by a fungus. Agric. Ecosyst. and Environ. 21:219–223.

5. Blakeman, J. P., and D. Hornby. 1966. The persistence of *Colletotrichum coccodes* and *Mycosphaerella ligulicola* in soil, with special reference to sclerotia and conidia. Trans. Brit. Mycol. Soc. 48:227–240.

6. Blanchette, B. L., and G. A. Lee. 1981. The influence of environmental factors on infection of rush skeletonweed *Chondrilla juncea* by *Puccinia chondrillina*. Weed Sci. 29:364–367.

7. Boyette, C. D. 1986. Evaluation of *Alternaria crassa* for biological control of jimsonweed: host range and virulence. Plant Sci. 45:223–228.

8. Boyette, C. D., G. E. Templeton, and L. R. Oliver. 1984. Texas gourd *(Cucurbita texana)* control with *Fusarium solani* f.sp. *cucurbitae*. Weed Sci. 32:649–655.

9. Boyette, C. D., G. E. Templeton, and R. J. Smith, Jr. 1979. Control of winged waterprimrose *(Jussiaea decurrens)* and northern jointvetch *(Aeschynomene virginica)* with fungal pathogens. Weed Sci. 27:497–501.

10. Bruckart, W. L., and W. M. Dowler. 1986. Evaluation of exotic rust fungi in the United States for classical biological control of weeds. Weed Sci. 34 (Suppl 1):11–14.

11. Brumley, J. M. 1978. Evnironmental factors affecting infection and development of anthracnose on winged waterprimrose. MS Thesis. University of Arkansas, Fayetteville.

12. Buchanan, G. A., and E. R. Burns. 1977. Competition of prickly sida in cotton. Weed Sci. 25:106–110.

13. Burdon, J. 1987. Diseases and Plant Population Biology. Cambridge University Press, New York.

14. Burnett, H. C., D. P. H. Tucker, M. E. Patterson, and W. H. Ridings. 1973. Biological control of milkweed vine with a race of *Phytophthora citrophthora*. Proc. Fla. State Hort. Soc. 86:111–115.

15. Butler, F. C. 1951. Anthracnose and seedling blight of bathurst burr caused by *Colletotrichum xanthii* Halst. Aust. J. Agric. Res. 2:401–411.

16. Callaway, M. B., S. C. Phatak, and H. D. Wells. Studies on alternate hosts of the rust *Puccinia canaliculata*, a potential biological control agent for nutsedge. Plant Dis. 69:924–926.

17. Capo. B. T. 1979. The effects of temperature on infection of *Anoda cristata* by *Alternaria macrospora*. MS Thesis. University of Arkansas, Fayetteville.

18. Cardina, J., R. H. Littrell, and R. T. Hanlin. 1988. Anthracnose of Florida beggarweed *(Desmodium tortuosum)* caused by *Colletotrichum truncatum*. Weed Sci. 36:329–334.

19. Cerkausas, R. F. 1989. Latent colonization by *Colletotrichum* spp.: epidemiological considerations and implications for mycoherbicides. Can. J. Microbiol. 19:297–310.

20. Charudattan, R. 1985. The use of natural and genetically altered strains of pathogens for weed control: *In:* Biological Control in Agricultural IPM Systems, M. A. Hoy and D. C. Herzog, eds. Academic Press, New York.

21. Charudattan, R., D. E. McKinney, and K. T. Helpting. 1981. Production, storage, germination and infectivity of uredospores of *Uredo eichhorniae* and *Uromyces pontederiae*. Phytopathology 71:1203–1207.

22. Charudattan, R., H. L. Walker, C. D. Boyette, W. H. Ridings, D. O. TeBeest, C. G. VanDyke, and A. D. Worsham. 1986. Evaluation of *Alternaria cassiae* as a mycoherbicide for sicklepod (*Cassia obtusifolia*) in regional field tests. Southern Cooperative Series Bulletin 317, Department of Information, Alabama Agricultural Experiment Station, Auburn University.

23. Chiang, M., C. G. Van Dyke, and K. J. Leonard. 1989. Evaluation of endemic foliar fungi for potential biological control of johnsongrass (*Sorghum halepense*): screening and host range tests. Plant Dis. 73:459–464.

24. Crawley, D. K., H. L. Walker, and J. A. Riley. 1985. Interaction of *Alternaria macrospora* and *Fusarium lateritium* on spurred anoda. Plant Dis. 69:977–979.

25. Cullen, J. M., P. F. Kable, and M. Catt. 1973. Epidemic spread of a rust imported for biological control. Nature 244:462–464.

26. Day, P. 1974. Genetics of host–parasite interaction. Freeman, San Francisco.

27. Emge, R., J. S. Melching, and C. H. Kingsolver. 1981. Epidemiology of *Puccinia chondrillina*, a rust pathogen for the biological control of rush skeleton weed in the United States. Phytopathology 71:839–843.

28. Fincham, J. R. S. 1989. Transformation in fungi. Microbiol. Rev. 53:148–170.

29. Feichtenberger, E., G. A. Zentmyer, and J. A. Menge. 1983. Identity of Phytophora isolated from milkweed vine. Phytopathology 73:50–55.

30. Greaves, M. P., J. A. Bailey, and J. A. Hargreaves. 1989. Mycoherbicides: opportunities for genetic manipulation. Pestic. Sci. 26:93–101.

31. Halfon-Meiri, A. 1973. *Alternaria crassa*, a seedborne fungus of Datura. Plant Dis. Rep. 57:960–963.

32. Halvorson, H. O., D. Pramer, and M. Rougul. 1985. Engineered organisms in the environment: scientific issues. American Society for Microbiology, Washington, D. C.

33. Harlan, J. R. 1976. Diseases as a factor in plant evolution. Ann. Rev. Phytopathol. 14:31–51.

34. Hasan, S., and P. T. Jenkins. 1972. The effect of some climatic factors on infectivity of the skeleton weed rust, *Puccinia chondrillina*. Plant Dis. Rep. 56: 858–860.

35. Hasan, S., and A. J. Wapshere. 1973. The biology of *Puccinia chondrillina:* a potential biological control agent of skeleton weed. Ann. Appl. Biol. 74:325–332.

36. Heale, J. B., J. E. Isaac, and D. Chandler. 1989. Prospects for strain improvement in entomopathogenic fungi. Pestic. Sci. 26:79–92.

37. Hodgson, R. H., L. A. Wymore, A. K. Watson, R. H. Snyder, and A. Collette. 1988. Efficacy of *Colletotrichum coccodes* and thidiazuron for velvetleaf (*Abutilon theophrasti*) control in soybean (Glycine max). Weed Technol. 2:473–480.

38. Jenns, A. E., and K. J. Leonard. 1985. Effect of illuminance on the resistance of inbred lines of corn to isolates of *Colletotrichum graminicola*. Phytopathology 75:281–286.

39. Keller, N. P., and O. C. Yoder. 1988. Evaluation of effects of foreign DNA on fitness of *Cochliobolus heterostrophus*. Phytopathology 78:1588 (Abst.).

40. Kenney, D. S. 1986. De-Vine—the way it was developed—an industralist's view. Weed Sci. 34 (Suppl. 1):15–16.

41. Kirkpatrick, T. L., G. E. Templeton, D. O. TeBeest, and R. J. Smith, Jr. 1982. Potential of *Colletotrichum malvarum* for biological control of prickly sida. Plant Dis. 66:323–325.

42. Klisiewicz, J. M. 1985. Growth and reproduction of *Dichotomophthora portulacae* and its biological activity on purslane. Plant Dis. 69:761–762.

43. Lee, G. A. Integrated control of rush skeletonweed (*Chondrilla juncea*) in the western U.S. Weed Sci. 34 (Suppl. 1):2–6.

44. Lipps, P. E. 1985. Influence of inoculum from buried and surface corn residues on the incidence of corn anthracnose. Phytopathology 75:1212–1216.

45. Makowski, R. M. D. 1987. The evaluation of *Malva pusilla* as a weed and its pathogen *Colletotrichum gloeosporioides* (Penz) Sacc. f.sp. *malvae* as a bioherbicide. PhD Dissertation. University of Saskatchewan.

46. McLean, K. S., and K. W. Roy. 1988. Incidence of *Colletotrichum dematium* on prickly sida, spotted spurge and smooth pigweed and pathogenicity to soybean. Plant Dis. 72:390–393.

47. McRae, C. F. 1989. The feasibility of using *Colletotrichum orbiculare* as a mycoherbicide to control *Xanthium spinosum* (Bathurst burr). PhD Dissertation. University of New England, Armidale, NSW. Australia.

48. McRae, C. F., and B. A. Auld. 1988. The influence of environmental factors on anthracnose of *Xanthium spinosum*. Phytopathology 78:1182–1186.

49. Mitchell, D. J. 1978. Relationships of inoculum levels of several soilborne species of *Phytophthora* and *Phthium* in infection of several hosts. Phytopathology 68:1754–1759.

50. Mortenson, K. 1988. The potential of an endemic fungus, *Colletotrichum gloeosporioides,* for biological control of round-leaved mallow (*Malva pusilla*) and velvetleaf (*Abutilon theophrasti*) Weed Sci. 36:473–478.

51. National Research Council. 1989. Field testing genetically modified organisms: framework for decisions. National Academy Press, Washington, D.C.

52. Nicholson, R. L., L. G. Butler, and T. N. Asquith. 1986. Glycoproteins from *Colletotrichum graminicola* that bind phenols: implications for survival and virulence of phytopathogenic fungi. Phytopathology 76:1315–1318.

53. Nicholson, R. L., and W. B. C. Moraes. 1980. Survival of *Colletotrichum graminicola:* importance of the spore matrix. Phytopathology 70:255–261.

54. Phatak, S. C., D. R. Sumner, H. D. Wells, D. K. Bell, and N. C. Glaze. 1983. Biological control of yellow nutsedge with the indigenous rust fungus *Puccinia canaliculata*. Science 219:1446–1447

55. Pimental, D. 1985. Using genetic engineering for biological control: reducing ecological risks. *In:* Engineered Organisms in the Environment: Scientific Issues, H. O. Halvorson, D. Pramar, and M. Rogol, eds. American Society for Microbiology, Washington, DC.

56. Politis, D. J., and W. L. Bruckart. 1986. Infection of musk thistle by *Puccinia carduorum* influenced by conditions of dew and plant age. Plant Dis. 70:288–290.

57. Politis, D. J., A. K. Watson, and W. L. Bruckart. 1984. Susceptibility of musk thistle and related composites to *Puccinia carduorum*. Phytopathology 74:687–691.

58. Raid, R. N., and S. P. Pennypacker. Weeds as hosts for *Colletotrichum coccodes*. Plant Dis. 71:643–646.

59. Ridings, W. H. 1986. Biological control of stranglervine in citrus—a researcher's view. Weed Sci. 34 (Suppl. 1):31–32.

60. Ridings, W. H., D. J. Mitchell, C. L. Schoulties, and N. E. El-Gholl. 1976. Biological control of milkweed vine in Florida citrus groves with a pathotype of *Phytophthora citrophthora*. Pages 224–240, *in:* Proc. IV Int. Symp. on Biological Control of Weeds, T. E. Freeman, ed. University of Florida, Gainesville.

61. Ridings, W. H., C. L. Schoulties, N. E. El-Gholl, and D. J. Mitchell. 1977. The milkweed pathotype of *Phytophthora citrophthora* as a biological control agent of *Morrenia odorata*. Proc. Int. Soc. Citriculture 3:877–881.

62. Schmidt, R. A. 1978. Diseases in forest ecosystems: the importance of functional diversity. Pages 287–315, *in:* Plant Disease: An Advanced Treatise, J. G. Horsfall and E. B. Cowling, eds. Academic Press, New York.

63. Smith, C. S., S. J. Slade, J. H. Andrews, and R. F. Harris. 1989. Pathogenicity of the fungus, *Colletotrichum gloeosporioides* (Penz) Sacc. to eurasian watermilfoil (*Myriophyllum spicatum* L.). Aquatic Bot. 33:1–12.

64. Smith, R. J. Jr., J. T. Daniel, W. T. Fox, and G. E. Templeton. 1973. Distribution in Arkansas of a fungus disease used for biocontrol of northern jointvetch in rice. Plant Dis. Rep. 57:695–697.

65. Stephan, B. R. 1968. Untersuchungen zum nachweis der heterokaryose bei Colletotrichum gloeosporioides Penzig unter verwendung auxotropher mutanten. Zentr. Bakteriol. Parasitenk. II. 122:420–435.

66. TeBeest, D. O. 1982. Survival of *Colletotrichum gloeosporioides* f.sp. *aeschynomene* in rice irrigation water and soil. Plant Dis. 66:469–472.

67. TeBeest, D. O., and J. M. Brumley. 1978. *Colletotrichum gloeosporioides* borne within the seed of *Aeschynomene virginica*. Plant Dis. Rep. 62:675–678.

68. TeBeest, D. O., and M. B. Dickman. 1989. Transformation of *Colletotrichum gloeosporioides* f.sp. *aeschynomene*. Phytopathology 79:1173 (Abstr.).

69. TeBeest, D. O., and G. E. Templeton. 1978. Temperature and moisture requirements for development of anthracnose on northern jointvetch. Phytopathology 68:389–393.

70. TeBeest, D. O., and G. E. Templeton. 1985. Mycoherbicides: progress in biological control of weeds. Plant Dis. 69:6–10.

71. Templeton, G. E., and D. I. Heiny. 1989. Improvement of fungi to enhance mycoherbicide potential. Pages 127–152, *in:* Biotechnology of Fungi for Improving Plant Growth, J. M. Whipps and R. D. Lumsden, eds. Cambridge University Press, Cambridge, England.

72. Templeton, G. E., D. O. TeBeest, and R. J. Smith, Jr. 1979. Biological weed control with mycoherbicides. Ann. Rev. Phytopathol. 17:301–310.

73. Tiedje, J. M., R. K. Colwell, Y. L. Grossman, R. E. Hodson, R. E. Lenski, R. N. Mack, and P. J. Regal. 1989. The planned introduction of genetically engineered organisms: ecological considerations and recommendations. Ecology 70:298–315.

74. Turgeon, G., and O. C. Yoder. 1985. Genetically engineered fungi for weed control. *In:* Biotechnology: Applications and Research, P. N. Cheremisinoff and R. P. Ouelette, eds. Technomic, Lancaster, PA.

75. Van Dyke, G. C., and R. N. Trigiano. 1987. Light and scanning electron microscopy of the interaction of the biocontrol fungus *Alternaria cassiae* with sicklepod (*Cassia obtusifolia*). Can. J. Plant Pathol. 9:230–235.

76. Walker, H. L. 1981. Factors affecting biological control of spurred anoda (*Anoda cristata*) with *Alternaria macrospora*. Weed Sci. 29:505–507.

77. Walker, H. L. 1985. *Colletotrichum coccodes:* a pathogen of eastern black nightshade (*Solanum ptycanthum*). Weed Sci. 33:902–905.

78. Walker, H. L., and C. D. Boyette. 1986. Influence of sequential dew periods on biocontrol of sicklepod (*Cassia obtusifolia*) by *Alternaria cassiae*. Plant Dis. 70:962–963.

.

79. Walker, H. L., and J. A. Riley. 1982. Evaluation of *Alternaria cassiae* for biocontrol of sicklepod (*Cassia obtusifolia*). Weed Sci. 30:651–654.

80. Webb, R. R., and S. E. Lindow. 1987. Influence of environmental and variation in host susceptibility on a disease of bracken fern caused by *Aschochyta pteridis*. Phytopathology 77:1144–1147.

81. Weidemann, G. J., and G. E. Templeton. 1988. Efficacy and soil persistence of *Fusarium solani* f.sp. *cucurbitae* for control of Texas gourd (*Cucurbitae texana*). Plant Dis. 72:36–38.

82. Weidemann, G. J., and G. E. Templeton. 1988. Control of Texas gourd, *Cucurbita texana*, with *Fusarium solani* f.sp. *cucurbitae*. Weed Technol. 2:271–274.

83. Wymore, L. A., C. Poirier, A. K. Watson, and A. R. Gotlieb. 1988. *Colletotrichum coccodes*, a potential bioherbicide for control of velvetleaf (*Abutilon theophrasti*). Plant Dis. 72:534–538.

84. Wymore, L. A., and A. K. Watson. 1986. An adjuvant increases survival and efficacy of *Colletotrichum coccodes*, a mycoherbicide for velvetleaf (*Abutilon theophrasti*). Phytopathology 76:1115 (Abstr.).

85. Yoder, O. C. 1981. Toxins in pathogenesis. Ann. Rev. Phytopathol. 18:103–129.

86. Zadoks, J. C., and R. D. Schein. 1979. Epidemiology and Plant Disease Management. Oxford University Press, New York.

7

Parasitism, Host Species Specificity, and Gene-Specific Host Cell Death

Dean W. Gabriel

Introduction

The development of biological control agents for the control of weeds is only in its infancy, and its success as an applied science will parallel and be limited by progress made in our understanding of the basic genetic and pathological sciences. The better our understanding of the structure, function, and expression of genes affecting plant–microbe interactions, the better we will be able first to select, then to recombine, then finally to engineer biological control agents. The reductionistic description of the mechanics of living organisms in terms of their genetic codes is already becoming reality. The potential for a truly engineered, and not merely recombined, living organism that forms an entirely novel species awaits only a more detailed understanding of pathogenicity gene function. It is clear that the evaluation and selection of biocontrol agents can be successful and has encouraged a fair number of research projects with similar strategies (71). Another potentially effective strategy is to recombine genes from different organisms using recombinant DNA techniques to achieve a more potent or efficacious biocontrol agent. Regardless of whether the biocontrol strategy is to establish a resident and persistent pathogen population, to use large amounts of short-lived inocula, or to use industrial-scale fermentation to produce a microbial herbicide, specificity for the target plant is a primary concern.

There are at least two types of specificity, biological and biochemical. The biochemical specificity of unique biological products such as broad-spectrum antibiotics, on the one hand, and narrow-spectrum, host- or gene-specific toxins, on the other, may allow their isolation by fermentation processes. An additional advantage may be conferred by using a living pathogen as a biotic delivery system. The biological agent might produce the herbicidal compound only when growing, and then, it is hoped, only on or in the target. Either way, knowledge of the principles of plant–microbe specificity is the key to an improved biological control agent using a deliberate engineering strategy.

Much progress in the area of plant–microbe specificity has been made at the organismal level, and more recently at the cellular and molecular levels. However, increasing attention is now turning to developing concepts at the population level, an area critical to the success of practical control measures. For example, it is logical to assume that one could more easily achieve biocontrol of an introduced weed species that clonally reproduces than a native plant species with a high level of heterogeneity (5,45). A genetically uniform target population offers less opportunity for resistant variants to have evolved (3). This seems intuitive and rational enough, but what is the experimental basis for statements involving evolutionary arguments? Do we really understand why most plants are resistant to most pathogens? Could it be they are merely not attacked properly? Could it be active defense? Both?

Because the molecular tools used to answer such questions were most easily developed for use with prokaryotes, most of the recent conceptual advances in plant–microbe specificity have been in the area of bacterial parasitism and pathology, particularly involving members of the Rhizobeaceae (for recent reviews, see refs. 14 and 21). Most of the information bearing on the specificity of plant defense has been derived from classical breeding studies involving biotrophic fungi (for reviews, see refs 12 and 17) or comparative biochemical studies of necrotrophic fungi on nonhost plants (for review, see [6]). It should therefore not be surprising that concepts derived from microbial molecular genetics, plant breeding, and comparative biochemistry may seem somewhat disconnected, although attempts have been made to synthesize the information relating to plant defense into comprehensive models with specific molecular predictions (for example, see refs 19, 25, and 37). Not surprisingly, reviews and models dealing with the subject of plant–microbe specificity as it relates to pathogens (not including *Rhizobium*) have emphasized plant defense and strain-specific ("gene-for-gene") incompatibility. The topic of host-species-specific parasitism is rarely broached by pathologists, although many papers have been written on the subject by researchers working on *Rhizobium*. Fortunately, the molecular analysis of cloned genes affecting parasitism and pathogenicity has brought microbiologists and plant pathologists into closer associations, resulting in conceptual advances in both fields.

The purpose of this paper is to present a minireview of the genetics and molecular biology of plant–microbe specificity from an engineering perspective. Recent advances in the area of plant–microbe interactions are now of more than basic science interest; the potential for engineering biological control agents now appears commercially feasible.

Taxonomic Assumptions Relevant to Host Specificity and Ecological Fate

Microbes are ideally classified to the species level on the basis of clusters of relatively stable morphological and biochemical characteristics. For example, a typical range of DNA homology values for strains within a biochemical species is 60–100% (32). Nevertheless, the concept of a species is a human device, and pathogenic bacteria tend to be classified with enormous—and often, inappropriate—weight placed on pathogenic characteristics. Pathogenic microbes may therefore be classified not only on the basis of their intrinsic properties alone, but also on the basis of a host-response phenotype. Very highly related pathogens may be separated into different species solely on the basis of a pathogenic phenotype that appears to depend on the presence or absence of a single gene. For example, strains of *Bordatella pertussis* (causing whooping cough) and *B. parapertussis* (causing "minor" bronchial infection) exhibit cophenetic correlations of 96% (33). Evidently, the only significant difference between these two species is that *B. parapertussis* carries a transcriptionally silent version of the pertussis toxin gene that is active in *B. pertussis* (31). The cloned pertussis toxin gene from *B. pertussis*, when moved by marker exchange (single gene replacement) into *B. parapertussis*, converts the latter species into the former (1).

Conversely, relatively unrelated pathogens may be "lumped" together based on a host–response phenotype. For example, within-species variation of *Rhizobium* is quite high, with DNA homology values as low as 24% (8) and electrophoretic similarity indices (relatedness) as low as 11% (57). Similarly, DNA homology values within *X. campestris* have been reported as low as 4% (68). Even worse, some of *Rhizobium* and *Xanthomonas* strains are classified on the basis of the "host from which first isolated" (40,64). Such arbitrary classifications are not merely of academic interest. Recently, a *Xanthomonas* strain that causes a minor leaf spot disease on citrus was classified as pv. *citri* primarily because it was isolated from citrus, even though the symptoms were quite different (60). This misclassification directly led to a misdiagnosis of "citrus canker" disease (23). Since the organism that causes citrus canker is subject to federal quarantine and eradication rules, this misdiagnosis cost the state of Florida well over $25 million in 4 years (60).

The fact that some relatively unrelated strains may be classified as one species, and that highly related strains may be classified as different species, makes the evaluation and selection of appropriate strains for biological control purposes more difficult. Automatic assumptions of pathogenicity (or the lack thereof) and host range of specific strains of a species cannot be made if a primary basis of classification involves a host-response phenotype. All classifications based on host response phenotype should be viewed with suspicion, as pathogenicity and host range must be tested on a strain-by-strain basis. Fortunately, with the advent of effective microbial taxonomic tools based on genetically stable characteristics of the microbe itself (such as DNA homology, RFLP analyses, multilocus enzyme analyses), the taxonomies of many pathogenic microbes are becoming more appropriate, stabilized, and therefore reliable.

Reliable knowledge of the appropriate taxonomy of any strain used as the basis for engineering a biological control agent is important. If the organism is genetically modified, such knowledge is biologically critical to success, and probably is also required before any regulatory agency approval of a field release experiment is likely. Reliable taxonomic data are also required in any patent applications on such strains. Ultimately, strain-by-strain characterizations of plant pathogens may become necessary because many factors directly affecting virulence and host range are strain specific, and even gene specific (discussed later).

Some generalizations regarding pathogenic potential may safely be made using taxonomies based upon reliable microbial classification techniques. Perhaps surprisingly, plant pathogenic species classified on the basis of their own intrinsic properties exhibit little variation in certain key pathogenic properties, such as host species specificity or host range. To the extent that pathogenic symptoms affect reproductive fitness, the symptoms will also remain characteristic of the pathology elicited by the microbe on its host(s). If the genes affecting parasitism are of high selective value to the population, the genes will be conserved, and if recombination rates are low enough in the population, other genes will remain in fixed associations by linkage disequilibrium (for review, refer to ref. 21). If the selective values of certain combinations are sufficiently high to be preserved over time, they may form a parasite species. Certain parasitism genes of high selective value will therefore be found conserved at the species level or higher. Other parasitism genes, of perhaps more limited selective value on a more limited range of hosts, may therefore be conserved at the pathovar, biovar, or *forma speciales* level. Certain other important pathogenic properties, such as gene-specific avirulence, or gene-specific toxin production, are of no known selective value and are not conserved at the microbial species level (21). The genes involved in conditioning basic parasitism, host species specificity, and gene-specific avirulence or toxin production (host cell death) are considered later and are illustrated in Figure 7.1. Some of the genetic terms used to describe plant–microbe interactions are also illustrated in Figure 7.1.

Parasitism, Virulence, or Basic Plant–Microbe Compatibility

Not all plants support the growth of all microbes; obviously, microbial genes exist that are necessary to condition parasitism. Parasitism is here defined as the ability to achieve reproductive success of the population as a whole at the expense of nutrients that would otherwise be utilized for maintenance, growth, and reproduction of the host. Among necrotrophic pathogens, which kill host tissue in advance of colonization, parasitism always involves pathological symptoms. Yet parasitism does not have to involve overt pathology; among biotrophic parasites, which colonize host tissue prior to any host cell death, pathological damage may be unnecessary for ecological fitness. The vesicular–arbuscular (VA) mycorrhizal

Figure 7.1. Genes Affecting Plant–Microbe Interactions, and Terminology Used

Basic parasitism: Genes conserved at the species level or higher and essential for parasitic reproductive fitness of the population. Genes may or may not induce gross pathogenic symptoms on host. *Examples:* common *nod (Rhizobium), hrp (P. syringae),* indole-acetic acid production *(P. savastanoi, A. tumefaciens).*

Host species specificity: Genes usually conserved below the species level (ie., at the biovar, pathovar, etc. level) and essential for parasitic reproductive fitness primarily on certain host species. *Examples:* Host-specific nodulation, or *hsn (Rhizobium),* host species specificity, or *hss (X. campestris),* tartaric acid catabolism and some *vir* genes *(A. tumefaciens),* pisatin demethylase production *(Nectria haematococca).*

Gene-specific host cell death (hcd): Strain-specific genes conditioning host-gene-specific host cell death (hcd). In necrotrophs, which kill host tissue in advance of colonization, gene function results in production of a host-gene-specific *toxin.* For hcd, the host must carry a specific gene encoding a receptor. In biotrophs, which must colonize living host tissue, gene function results in *avirulence.* For hcd, the host must carry a specific gene encoding *resistance.* With necrotrophs, hcd leads to greater pathology on hosts with the receptor; with biotrophs, hcd leads to less pathology on hosts with the resistance gene. *Examples:* host-specific toxins (eg., victorin, T-toxin) and avirulence *(avr)* gene function.

	Microbe		Classical Terms with Reference to:		
Type	Function	Host Gene	Parasite	Host	Interaction
Necrotroph	Toxin	Receptor	Virulent	Susceptible	Compatible
Biotroph	Avirulence	Resistance	Avirulent	Resistant	Incompatible

The microbe genes are said to be *specific* for the host genes; the net result in either case is hcd.

fungi and *Rhizobium* are examples. Other genera, normally considered pathogenic, are also illustrative. For example, an entirely asymptomatic infection of apple fruit and stems by *X. campestris* occurs naturally, without evident pathology to the host (50). These apple strains are parasites, but not pathogens. Yet with the possible exceptions of *Agrobacterium* and *Rhizobium* species, the physiological mechanisms involved in conditioning basic parasitism are unknown.

With both *Agrobacterium* and *Rhizobium,* parasitic reproductive success is probably due to their ability to drain host energy in the form of unique metabolites—opines—that only the inducing strain population can utilize (38,54). The tumor-inducing genes of *A. tumefaciens, Bradyrhizobium japonicum,* and *P. savastanoi* are evidently of high selective value to these organisms as species. The genes for indole acetic acid production appear so highly conserved as to suggest a common ancestor (61). Such conservation indicates that the pathogenicity factor is critical for parasitic fitness of the pathogen population.

Unfortunately, the genes known to affect parasitic growth thus far identified and cloned from *Pseudomonas syringae* and *X. campestris* (10,47,62) have not been biochemically characterized. Some of the mutants would not cause water-soaking of host tissues, even when inoculated at high levels, indicating that the affected genes are involved in pathogenic as well as parasitic capacity. Those genes cloned from these pathogens that have been biochemically characterized (protease and endoglucanase) may not be necessary for basic parasitic ability of these pathogens (27,70). It appears that these plant degradative enzymes are of more use to these pathogens for saprophytic growth and survival. It is of interest to note that these two bacterial pathogenic species are mainly biotrophic; there might be a strong selective disadvantage to biotrophic parasites to express genes that would result in rapid host cell death.

Host Species Specificity

Host species specificity, also known as host range, refers to the plant species that are parasitized by the microbe. If even one cultivar or plant variety within a species is attacked, that host species is considered to be within the range of the parasite. Several very recent reviews have highlighted the rapid progress being made in elucidating the number and kinds of genes conditioning host species specificity, principally from members of the Rhizobeaceae (14,21). Basically, host species specificity (Hss) genes are parasitism genes that are primarily effective on certain hosts but not on others within the range of the parasite. When such genes are mutagenized, host range is greatly reduced on one or more host species, but parasitism on other hosts is still possible. For example, the Ti plasmid of *A. tumefaciens* was "saturated" with transposon Tn5 in an attempt to identify all genes affecting pathogenicity and host range (36). Some mutations severely affect pathogenicity on all hosts, whereas other mutations affect pathogenicity primarily on one or several hosts but had must less effect on other hosts. Mutations of host species specificity genes have been reported in *Rhizobium spp* (14), *X. campestris* pv. *translucens* (52), *X. campestris* pv. *citrumelo* (41), *P. solanacearum* (49), and *P. syringae* (59).

Some genes for host species specificity have been biochemically characterized. For example, a gene for tartaric acid catabolism from *A. tumefaciens* appears to play a significant role in extending the pathogen's host range to grapes (26). The primary available sugar in grapevine is tartaric acid. A recent similar example involving fungal pathogens involves the gene encoding pisatin demethylase (PDA). When PDA is cloned from *Nectria haematococca* (a pathogen of pea) into *Cochliobolus heterostrophus* (a pathogen of maize), the host species specificity (range) of *C. heterostrophus* is increased from maize alone to maize and pea both (18). Mutations in host species specificity genes would be expected to limit an otherwise wider range of hosts.

Because mutations of genes affecting host species specificity may also have some effect on parasitism in general, the importance of a given gene for parasitism on a specific host may be questioned. For these genes to be considered host species specific, as opposed to being generally essential for parasitism of the specifies on any host, the genes must be conserved only among strains parasitic on the specific host. In other words, a mutation of a host species specific gene from a wide host range *X. campestris* pathovar will not be genetically conserved or complemented by genes from most other *X. campestris* pathovars. As mentioned earlier, such genes from *Rhizobium* have been well characterized and are called host-specific nodulation *(hsn)* genes to distinguish them from common nodulation *(nod)* genes. Common *nod* genes (e.g., *nodD, A, B,* and *C*) are highly conserved at the genus level (42) and mutants of such genes are complemented by *nod* gene clones from other biovars and species. By contrast, *hsn* genes are not highly conserved, are not known to be complemented by genes from other biovars and species, and appear to function to condition nodulation on specific additional host species.

The number of genes suspected or known to be involved in plant pathogenesis is growing quickly; as mentioned, most have not been biochemically characterized. In Figure 7.2 is a summary of a number of genes cloned to date, gene functions characterized, and their known or suspected role in plant pathogenesis.

Gene-Specific Host Cell Death

Genetics

Most described plant parasites are pathogens, but our perception of the importance of pathogenicity per se to a parasitic population may be biased. The genes that contribute most heavily to the host–response phenotype tend to attract man's

Figure 7.2.   Genes Plant–Microbe Interactions

Microbial Genes	Host Genes	Reaction Phenotype
Saprophytic opportunism: Housekeeping + Proteolytic, pectolytic	Passive	Decay of damaged, dead or dying tissue.
Parasitism: Above + Path clones, *hrp (Xanthomonas, Pseudomonas),* common *nod (Rhizobium),* or some *vir* + IAA (Agrobacterium)	unknown	Asymptomatic, water-soaking, nodulation, or galls
Host species specificity: Above + *hss* genes *(Xanthomonas),*	unknown	virulent
tartaric acid catabolism *(A. tumefaciens),* or	tartaric acid	virulent
pisatin demethylase *(Nectria)*	pisatin	virulent
Gene-specific host cell death: Above + toxins *(Helminthosporium)*	receptor genes (e.g., *Vb, urf*13)	Host cell death, severe symptoms
Above + *avr* (rusts, smuts, mildown, apple scab, *Xanthomonas, Pseudomonas, Rhizoblum)*	resistance *(R)* genes	Host cell death reduced symptoms

attention (or at least a grower's attention), but the value of any gene (and its conservation) in a parasite population depends strictly upon its contribution to reproductive fitness. The perhaps unexpected relative unimportance of endoglucanases and proteases to the overall pathogenicity of *X. campestris* pv. *campestris* has already been mentioned. Even the most biotrophic of parasites can cause disease; for example, a very few *R. japonicum* strains may induce chlorosis by producing rhizobitoxin (15), and VA mycorrhizae may cause stunting (53). The limited ecological success of biotrophic pathogens that produce toxins or VA mycorrhizae that cause stunting is evidence that these are not useful traits to the reproductive success of these organisms as parasites. Similarly, avirulence genes (abbreviated *A* or *P* in fungi, *avr* in bacteria), which have a dramatic effect on host–response phenotype, have been described in the literature since 1946 (20) in a wide variety of biotrophic plant pathogens (12). They were first observed as a result of strain-specific variation in virulence on certain plant lines. Such variation occurs at very high frequency (mutation rates up to 4×10^{-4} per cell per division at a single lcous [9]), leading to the conclusion that the final, observed phenotype on a given plant genotype may be almost incidental. Despite efforts to explain away their existence on the basis of some pleiotropic function (cf. "stabilizing selection" [72,73]), there is no evidence whatsoever for any selective value of these genes (4,7,28,46,56). To a biotroph, avirulence genes, like toxins, appear to be incidental and gratuitous (21). What possible function could they serve?

The now classical studies by Flor, published in 1946–1947, revealed that strain-specific variation in virulence/avirulence appeared to be oligogenically inherited (24), and several reports have subsequently confirmed oligogenic inheritance ([30,34,43], reviewed in ref. 21). More important, avirulence was also host gene specific. The specific host genes were called resistance *(R)* genes, and the phenotype of resistance was revealed only when the host carried an *R* gene and the parasite carried a specific *avr* gene (29). This led to the so-called gene-for-gene hypothesis (see Figure 7.3). The genetics of strain-specific resistance and gene-specific avirulence has been well-reviewed (17,21,25). The inheritance of strain-specific host resistance to microbes appears to be monogenic, and the data base for this statement is truly vast. The inheritance of gene-specific avirulence appears to be oligogenic, but the genetic data on this phenotype are not as complete. The reason for this is simple: The most clear-cut differences in strain-specific phenotypes occur in the obligate or near-obligate parasites (see Table 4.3, pp. 96–97 in ref. 12), and genetic analyses of rusts and mildews are difficult to perform. However, since the genetics of avirulence appears to be oligogenic, whereas that of resistance is monogenic, a more accurate description of the genetic relationship would be *gene-for-genes*.

Mechanism

In terms of physiological interpretations of the gene-for-genes hypothesis, several different models have been proposed. The two most important of these

Figure 7.3. The genetic and physiologic patterns of gene-specific host cell death

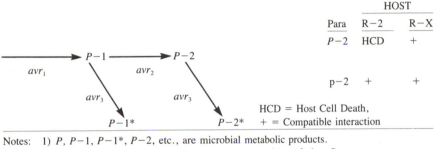

The gene-for-genes hypothesis:

Parasite	Host R_1R_1	r_1r_1
A_1 (fungi) or avr_1 (bacteria)	−	+
a_1 (fungi) or avr_1^- (bacteria)	+	+

Typical "quadratic check"

Parasite	Host R_2R_2	r_2r_2
avr_1, avr_2	−	+
avr_1, avr_2^- or avr_1, avr_2 or avr_1^-, avr_2^-	+	+

More realistic quadratic check

Proposed (25) physiological pattern interpretation of gene-for-genes:

Para	HOST R−2	R−X
P−2	HCD	+
p−2	+	+

HCD = Host Cell Death,
+ = Compatible interaction

Notes:
1) P, $P-1$, $P-1*$, $P-2$, etc., are microbial metabolic products.
2) $R-1$, $R-2$, $R-X$, etc., are the translational products of plant R genes.
3) Only genotype avr_1, avr_2 gives microbial product P−2.
4) avr_3 would be a dominant suppressor of avr_1 or avr_2.
5) Either avr_1 or avr_2 could be considered recessive suppressors of the other.
6) Only product P−2 reacts with resistance gene protein R−2.

have been the "dimer" hypothesis (19) and some variant of the "elicitor-sensor" model (2,12,37). In the dimer hypothesis, the primary protein products of R and *avr* genes interact directly and are themselves a sufficient cause of a unique incompatibility (19). This model is attractive because it neatly explains the wide range of variation observed in host resistance (from immunity to slower pathogen reproduction). However, there has been no direct evidence to support the model, and no evidence that the primary protein products of three sequenced *avr* genes (55,69) can interact in the manner required by the model.

In the most physiologically specific of the elictor–sensor models, plant R genes are proposed as structural loci for membrane-bound protein sensors, and microbial *avr* genes are proposed as enzymes, such as glycosyl transferases, that modify microbial surface components (37). The microbial surface components then interact, not necessarily directly, with the plant sensors to elicit a generalized plant defense response. The signal transduction mechanism(s) is not specified. To be consistent with the nearly universal monogenic inheritance of resistance, these models make the assumption that variation in the defense response is lethal to the plant, and therefore no variation in the structural genes determining the general defense are seen. There are two main difficulties with the elicitor–sensor models. First, the idea of a common defense (the hypersensitive response) is hard to rationalize with the observed variation in host defense (16,22,65); this was one of the primary arguments used to support the dimer hypothesis (19). The term *elicitor* has been used to refer to any general trigger of host hypersensitivity and implies a common defense mechanism. The hypersensitive response is also a wound response. No one would argue that a wound response does not also occur in some gene-for-genes interactions. The real question is whether or not the wound response is the cause or the consequence of resistance. Second, the idea that *avr* genes modify surface components is unnecessarily limited. At least one *avr* gene, *Lr20* of *Puccinia recondita,* appears to be involved in the production of a diffusible, small-molecular-weight, toxinlike molecule (35).

The ion-channel defense model (25) circumvents the difficulties of both the dimer hypothesis and the elicitor–sensor models. The ion-channel defense model is similar to the elicitor–sensor model in predicting that *avr* genes are modification enzymes that produce a product that interacts with an *R* gene protein, which is membrane bound. In terms of plant defense against nonpathogens, opportunistic pathogens, and necrotrophs, the only major difference is that a mechanism for signal transduction is proposed in the ion-channel model. Also, in the ion-channel model, the modification enzymes are predicted to be enzymes involved in secondary metabolism (defined in the general sense as those not critically essentially to the survival of the organism). However, all similarity between these models ends when defense against biotrophs is considered.

In the ion-channel model, the *R* gene proteins are not mere sensors; they *are* the defense. According to the model, when the *R* gene is bound, it creates an ion channel through the membrane in a manner directly analogous to a toxin receptor binding a toxin; resistance to biotrophs is primarily the result of host cell death (see Figure 7.3). Biotrophic organisms cannot afford to kill the cells they must live in close association with. (This is probably why the genes for protease and endoglucanases, discussed earlier, are not useful for the parasitic phase of a biotroph's life cycle.) The ion-channel defense model thus offers the advantage of the dimer hypothesis over elicitor–sensor models in explaining the unique, *R*-gene-specific defense responses observed with biotrophs. The different *R*-gene phenotypes would be a result of the rapidity of the induced Ca^{+2} flux through the ion-channel; phenotypic immunity to the microbe might be the result of nearly instantaneous host cell death; and slow sporulation might be the result of slow collapse. The induction of host defense compounds, such as phenolics and phytoalexins, which generally occurs in both resistant and susceptible host tissue, need not be involved in the defense. The sensitivity of the pathogen to phytoalexins is likely to be a general characteristic of the microbial species and to be dependent on enzymes of high selective value, such as pisatin demethylase, discussed

earlier. One would predict that biotrophs would be highly resistant to all compounds produced by their hosts, perhaps in a host-species-specific manner.

The ion-channel defense model predicts that the development of the hypersensitive response is a phenomenon necessary mainly for defense against saprophytes,
opportunistic pathogens, and necrotrophs. Of course, host cell death is essential
for saprophytic degradation of plant tissue and not a problem for necrotrophs;
these microbes must be stopped by an active host defense. Necrotrophs may have
resistance to some of the defense compounds produced by their hosts, especially
those that persist in dying tissue. There would be strong selection pressure for
such resistance at the microbial species level. Interestingly, most of the research
on elicitor-sensor models, which invoke a common defense mechanism for all
types of pathogen (6,44), involves work with necrotrophic pathogens or nonhost
defenses.

The ion-channel defense model explains the incidental occurrence and gratuitous nature of host-gene-specific avirulence as well as host-gene-specific toxin
production; both avirulence and toxin production are mechanistically similar,
but the results are very different. Host cell death is the common denominator.
Furthermore, the production of at least some *R* genes is likely to be similar to
the production of at least some toxin receptors—incidental. Therefore, neither
avirulence nor toxin production is likely to exhibit any selective disadvantage or
advantage to the microbial population. Any effects observed due to gene-specific
host cell death would be transient and reflective of the extent to which any
particular crop plant genotype is widely planted. Of course, genetic uniformity
in agricultural crops (11) is an unfortunate reality that has resulted in a few
disasters, such as those described later.

Microbes produce a wide variety of secondary metabolites that may serve useful
purposes in particular ecological niches (74). The genes involved in secondary
metabolism mutate readily (51), and if they are not of great selective value, they
would not likely be highly conserved within a species. In fact, most appear to be
strain specific. Several host-specific toxins are really host-gene-specific secondary
metabolites (in the general sense), and available evidence strongly suggests that
avirulence genes encode similar functions (21,25). Certain toxins can add greatly
to the pathogenic phenotype of necrotrophic pathogens, but only if a specific
receptor is available in the host. For example, T-toxin produced by some strains
of *H. maydis* (syn. *Cochliobolus heterostrophus*) is specific for the maize gene
urf13-T (13,39). The maize gene *urf13-T* is a spontaneously generated, rearranged
hybrid of the 3' end of the 26S rRNA gene and the 5' end of the *atp6* gene (13).
Incredibly, the toxin is a gratuitously produced secondary metabolite, and the
receptor is a spontaneously generated corruption of at least two genes, but the
resulting phenotype to a plant pathologist (and maize growers) is nothing short
of spectacular. These two freaks of nature together wiped out over 70% of the
corn planted in 1971 (12).

Pryor has drawn attention to the fact that many resistance genes may have been generated as a result of spontaneous genetic rearrangements similar to that described for *urf13-T* (58). Disease resistance and toxin receptor genes may be functionally analogous, and specific for avirulence and toxin gene products, respectively. Both host and parasite functions may be gratuitous (recently reviewed in refs. 21 and 25). There are strong parallels between host-gene-specific avirulence of biotrophic pathogens and host-gene-specific toxins of necrotrophs. The genetics of both appears the same. Both are dispensable. Both require specific single-gene receptors. Considerable genetic, biochemical, and molecular genetic evidence has been accumulating that supports the idea that the avirulence genes *(avr)* first described in *Melampsora lini* (20), and cloned from *P. syringae* (66) and *X. campestris* (22,67), are genes involved in secondary metabolism or some dispensable function. All such genes and their functions are strain specific, and incidental to, any parasitic capacity the microbial species has on plants within its host range.

Potential Engineering Considerations

It may now be possible to engineer a microbial weed control agent that (1) produces a potent toxin or herbicide and (2) is host species specific. For reasons already discussed, engineering toxin production that confers a selective advantage would be much more likely to succeed in a necrotrophic pathogen than in a biotrophic one. Unfortunately, it is the biotrophs that have the greatest degree of host species specificity. Engineering toxin production in a biotroph may be tantamount to engineering avirulence into the organism. Therefore, such an engineering feat would likely result either in a biotroph that would be ecologically unfit or in a necrotroph that would require strict quarantine.

For certain applications, such as when it is desired that a herbicide or toxin gene(s) be switched on only when the microbe lands on the target plant, this situation may be ideal. Several microbial pathogenicity genes have been demonstrated to be transcriptionally silent until activated by positive acting, environmentally sensitive genes that respond specific compounds from their hosts (e.g., *nod* genes of *Rhizobium* [48], and *vir* genes of *A. tumefaciens* [63,75]). Even though the biotrophic pathogen might be unable to reproduce well, individual parasite units that became metabolically active in response to a host signal would then express the toxin or herbicide gene. Vast quantities of completely biodegradable microbes could be sprayed over a target field. Those microbes that failed to land on the inducing host would be metabolically inactive, and those that were metabolically active would inactivate themselves by destroying their host tissue. From the standpoint of biosafety regulations, a genetically engineered microbe that was both host specific and ecologically unfit might stand a better chance for regulatory agency approval of field release experiments than one that had only one or none of these safety features.

The attempt to engineer a general herbicide or general (not host gene specific) toxin into a necrotrophic pathogen might be unwise. Such a pathogen might gain considerable reproductive fitness attributes that might be unexpected, such as a much wider host range. A more destructive virulence may also evolve from such an organism. It is fortunate, for example, that T-toxin was host gene, and not host species, specific, It may be important, therefore, to characterize carefully the pathogenic potential of both the engineered trait and the recipient microbe. In general, if the preceding analyses of specificity are correct, the broader the spectrum of herbicidal action, and the more necrotrophic the recipient, the more potentially dangerous the experiment would be.

There are a number of good examples of severe pathogenicity gene functions involving the production of secondary metabolites (toxins) that are superimposed—and therefore superimposable—on a basic parasitic ability. Recently, *Bordatella parapertussis,* a mammalian parasite with minor pathogenic capacity, was converted to a much more potent pathogen by transferring a single gene for pertussis toxin from *B. pertussis* (1). Toxin transfer was successful only because *B. parapertussis* already carried the genes essential for parasitic success on the host. The idea that genes involved in secondary metabolism are superimposed on a basic parasitic capacity appears to be generally true of microbial parasites, whether attacking animal or plant hosts. As increasing numbers of microbial virulence factors are reported, the likelihood of finding combinations of such factors that might prove effective in making effective, biodegradable pesticides increases. As our knowledge of pathogenicity gene structure, function, and regulation increases, so does the potential for successfully engineering specific combinations of factors that would make effective, host-species specific, nonpersistent biological control agents.

Literature Cited

1. 1989 From wimp to pathogen. ASM News 55:10.

2. Albersheim, P., and Anderson-Prouty, A. J. 1975. Carbohydrates, proteins, cell-surfaces, and the biochemistry of pathogenesis. Ann. Rev. Plant Physiol. 26:31–52.

3. Barrett, S. C. H. 1982. Genetic variation in weeds. Pages 73–98 *in:* Biological control of weeds with plant pathogens. R. Charudattan and H. L. Walker, eds., Wiley, New York.

4. Bronson, C. R., and Ellingboe, A. H. 1986. The influence of four unnecessary genes for virulence on the fitness of *Erysiphe graminis* f. sp. *tritici.* Phytopathology 76:154–158.

5. Burdon, J. J., and Marshall, D. R. 1981. Biological control and the reproductive mode of weeds. J. Appl. Ecol. 18:649–658.

6. Collinge, D. B., and Slusarenko, A. J. 1987. Plant gene expression in response to pathogens. 9:389–410.

7. Crill, P. 1977. An assessment of stabilizing selection in crop variety development. Ann. Rev. Phytopathol. 15:185–202.

8. Crow, V. L., Jarvis, B. D. W., and Greenwood, R. M. 1981. Deoxyribonucleic acid homologies among acid-producing strains of *Rhizobium.* Int. J. Syst. Bacteriol. 31:152–172.

9. Dahlbeck, D., and Stall, R. E. 1979. Mutation for change of race in cultures of *Xanthomonas vesicatoria*. Phytopathology 69:634–636.

10. Daniels, M. J., Barber, C. E., Turner, P. C., Sawczyc, M. K., Byrde, R. J. W., and Fielding, A. H. 1985. Cloning of genes involved in pathogenicity of *Xanthomonas campestris* pv. *campestris* using the broad host range cosmid pLAFR1. EMBO J. 13:3323–3328.

11. Day, P. R. 1973. Genetic variability of crops. Ann. Rev. Phytopathol. 11:293–312.

12. Day, P. R. 1974. Genetics of Host–Parasite Interaction. Freeman, San Francisco, Pp. 1–238.

13. Dewey, R. E., Levings III, C. S., and Timothy, D. H. 1986. Novel recombinations in the maize mitochondrial genome produce a unique transcriptional unit in the Texas Male Sterile cytoplasm. Cell 44:439–449.

14. Djordjevic, M. A., Gabriel, D. W., and Rolfe, B. G. 1987. *Rhizobium* the refined parasite of legumes. Ann. Rev. Phytopath. 25:145–168.

15. Durbin, R. D. 1982. Toxins and pathogenesis. Pages 423–442, *in:* Phytopathogenic Prokaryotes, Vol. 1. M. S. Mount and G. H. Lacy, eds., Academic Press, New York.

16. Ellingboe, A. H. 1972. Genetics and physiology of primary infection by *Erysiphe graminis*. Phytopathology 62:401–406.

17. Ellingboe, A. H. 1976. Genetics of host–parasite interactions. Pages 761–778, *in:* Encyclopedia of Plant Pathology, New Series, Vol. 4: Physiological Plant Pathology. R. Heitefuss and P. H. Williams, eds., Springer-Verlag, Heidelberg.

18. Ellingboe, A. H. 1981. Changing concepts in host-pathogen genetics. Ann. Rev. Phytopathol. 19:125–143.

19. Ellingboe, A. H. 1982. Genetical aspects of active defense. Pages 143–192, *in:* Active Defense Mechanisms in Plants, Proceedings of a NATO Conference, Cape Sounion, Greece. R. K. S. Wood, ed., Plenum Press, New York.

20. Flor, H. H. 1946. Genetics of pathogenicity in *Melampsora lini*. J. Agr. Res. 73:335–357.

21. Gabriel, D. W. 1989. The genetics of plant pathogen population structure and host–parasite specificity. in: Plant–Microbe Interactions: Molecular and Genetic Perspectives, Vol. 3. T. Kosuge and E. W. Nester, eds., Macmillan, New York.

22. Gabriel, D. W., Burges, A., and Lazo, G. R. 1986. Gene-for-gene recognition of five cloned avirulence genes from *Xanthomonas campestris* pv. *malvacearum* by specific resistance genes in cotton. Proc. Natl. Acad. Sci. USA 83:6415–6419.

23. Gabriel, D. W., Kingsley, M. T., Hunter, J. E., and Gottwald, T. R. 1989. Reinstatement of *Xanthomonas citri* (*ex* Hasse) and *X. phaseoli* (*ex* Smith) and reclassification of all *X. campestris* pv. *citri* strains. Int. J. Syst. Bacteriol. 39:14–22.

24. Gabriel, D. W., Lisker, N., and Ellingboe, A. H. 1982. The induction and analysis of two classes of mutations affecting pathogenicity in an obligate parasite. Phytopathology 72:1026–1028.

25. Gabriel, D. W., Loschke, D. C., and Rolfe, B. G. 1988. Gene-for-gene recognition: the ion channel defense model. Pages 3–14, *in:* Molecular Plant–Microbe Interactions, Proceedings of the 4th International Symposium. D. P. Verma and R. Palacios, eds., APS Press, St. Paul.

26. Gallie, D. R., Zaitlin, D., Perry, K. L. and Kado, C. I. 1984. Characterization of the replication and stability regions of *Agrobacterium tumefaciens* plasmid pTAR. J. Bacteriol. 157:739–745.

27. Gough, C. L., Dow, J. M., Barber, C. E., and Daniels, M. J. 1988. Cloning of two endoglucanase genes of *Xanthomonas campestris* pv. *campestris:* analysis of the role of the major endoglucanase in pathogenesis. Mol. Plant Microbe Interact. 1:275–281.

28. Grant, M. W., and Archer, S. A. 1983. Calculation of selection coefficients against unnecessary genes for virulence from field data. Phytopathology 73:547–551.

29. Green, G. J. 1971. Hybridization between *Puccinia graminis tritici* and *Puccinia graminis secalis* and its evolutionary implications. Can. J. Bot. 49:2089–2095.

30. Green, G. J., and McKenzie, R. I. H. 1967. Mendelian and extrachromosomal inheritance of virulence in *Puccinia graminis* f. sp. *avenae*. 9:785–793.

31. Gross, R., and Rappouli, R. 1988. Positive regulation of pertussis toxin expression. Proc. Natl. Acad. Sci. USA 85:3913–3917.

32. Johnson, J. L. 1984. Nucleic acids in bacterial classification. Pages 8–11, *in:* Bergey's Manual of Systematic Bacteriology, N. R. Krieg and J. G. Holt, eds., Williams and Wilkins, Baltimore.

33. Johnson, R. and Sneath, P. H. A. 1973. Taxonomy of *Bordetella* and related organisms of the families Achromobacteraceae, Brucellaceae, and Neisseriaceae. Int. J. Syst. Bacteriol. 23:381–404.

34. Jones, D. A. 1988. Genetic properties of inhibitor genes in flax rust that alter avirulence to virulence on flax. Phytopathology 78:342–344.

35. Jones, D. R., and Deverall, B. J. 1978. The use of leaf transplants to study the cause of hypersensitivity to leaf rust, *Puccinia recondita,* in wheat carrying the *Lr20* gene. Physiol. Plant Pathol. 12:311–319.

36. Kao, J. C., Perry, K. L., and Kado, C. I. 1982. Indoleacetic acid complementation and its relation to host range specifying genes on the Ti plasmid of *Agrobacterium tumefaciens*. Mol. Gen. Genet. 188:425–432.

37. Keen, N. T. 1982. Specific recognition in gene-for-gene host-parasite systems. Pages 35–82, *in:* Advances in Plant Pathology, Vol. 1. D. S. Ingram and P. H. Williams, eds., Academic Press, New York.

38. Kempt, J. D. 1982. Plant pathogens that engineer their hosts. Pages 443–457, *in:* Phytopathogenic Prokaryotes. M. S. Mount and G. H. Lacy, eds., Academic Press, New York.

39. Kennell, J. C., Wise, R. P., and Pring, D. R. 1987. Influence of nuclear background on transcription of a maize mitochondrial region associated with Texas male sterile cytoplasm. Mol. Gen. Genet. 210:399–406.

40. Keyser, H. H., van Berkum, P., and Weber, D. F. 1982. A comparative study of the physiology of symbioses formed by *Rhizobium japonicum* with *Glycine max, Vigna unguiculata,* and *Macroptilium atropurpurem*. Plant Physiol. 70:1626–1630.

41. Kingsley, M. T., and Gabriel, D. W. 1988. Tn5-induced mutations of *Xanthomonas campestris* pv. *citri* affecting pathogenicity on citrus, common bean and alfalfa. Pages lb–11 *in:* Abstracts of the 4th International Symposium on Molecular Plant–Microbe Interactions. D. P. Verma and R. Palacios, eds., Acapulco.

42. Kondorosi, A., Horvath, B., Rostas, K., Gottfert, M., Putnoky, P., Rodriguez-Quinones. F., Banfalvi, Z., and Kondorosi, E. 1986. Common and host-specific nodulation genes of *Rhizobium meliloti*. Page 88, *in:* Third International Symposium on the molecular genetics of plant–microbe interactions, July 27–31. McGill University, Montreal.

43. Lawrence, G. J., Mayo, G. M. E., and Shepherd, K. W. 1981. Interactions between genes controlling pathogenicity in the flax rust fungus. Phytopathology 71:12–19.

44. Lawton, M. A., and Lamb, C. J. 1987. Transcriptional activation of plant defense genes by fungal elicitor, wounding and infection. Mole. Cell. Biol. 7:335–341.

45. Leonard, K. J. 1982. The benefits and potential hazards of genetic heterogeneity in plant pathogens., Pages 99–112, *in:* Biological Control of Weeds with Plant Pathogens. R. Charudattan and H. L. Walker, eds., Wiley, New York.

46. Leonard, K. J., and Czochor, R. J. 1980. Theory of genetic interactions among populations of plants and their pathogens. Ann. Rev. Phytopathol. 18:237–258.

47. Lindgren, P. B., Peet, R. C., and Panopoulos, N. J. 1986. Gene cluster of *Psuedomonas syringae* pv. *"phaseolicola"* controls pathogenicity of bean plants and hypersensitivity on nonhost plants. J. Bacteriol. 168:512–522.

48. Litt, M., and White, R. L. 1985. A highly polymorphic locus in human DNA revealed by cosmid-derived probes. Proc. Natl. Acad. Sci. USA 82:6206–6210.

49. Ma, Q. S., Chang, M. F., Tang, J. L., Feng, J. X., Fan, M. J., Han, B., and Liu, T. 1988. Identification of DNA sequences involved in host specificity in the pathogenesis of *Pseudomonas solanacearum* strain T2005. 1:169–174.

50. Maas, J. L., Finney, M. M., Civerolo, E. L., and Sasser, M. 1985. Association of an unusual strain of *Xanthomonas campestris* with apple. Phytopathology 75:438–445.

51. Martin, J. F., and Demain, A. L. 1980. Control of antibiotic biosynthesis. Microb. Rev. 44:230–251.

52. Mellano, V. J., and Cooksey, D. A. 1988. Development of host range mutants of *Xanthomonas campestris* pv. *translucens*. Appl. Environ. Microbiol. 54:884–889.

53. Modjo, H. S., and Hendrix, J. W. 1986. The mycorrhizal fungus *Glomus macrocarpum* as a cause of tobacco stunt disease. Phytopathology 76:688–691.

54. Murphy, P. J., Heycke, N., Trenz, S. P., Ratet, P., DeBruijn, F. J., and Schell, J. 1988. Synthesis of an opine-like compound, a rhizopine, in alfalfa nodules is symbiotically regulated. Proc. Natl. Acad. Sci. USA 85:9133–9137.

55. Napoli, C., and Staskawicz, B. 1987. Molecular characterization and nucleic acid sequence of an avirulence gene from race 6 of *Pseudomonas syringae* pv. *glycinea*. J. Bacteriol. 169:572–578.

56. Parlevliet, J. E. 1981. Stabilizing selection in crop pathosystems: an empty concept or reality? Euphytica 30:259–269.

57. Pinero, D., Martinez, E., and Selander, R. K. 1988. Genetic diversity and relationships among isolates of *Rhizobium leguminosarum* biovar *phaseoli*. Appl. Environ. Microbiol. 54:2825–2832.

58. Pryor, T. 1987. The origin and structure of fungal disease resistance genes in plants. Trends Genet. 3:157–161.

59. Salch, Y. P., and Shaw, P. D. 1988. Isolation and characterization of pathogenicity genes of *Pseudomonas syringae* pv. *tabaci*. 170:2584–2591.

60. Schoulties, C. L., Civerolo, E. L., Miller, J. W., Stall, R. E., Krass, C. J., Poe, S. R., and DuCharme, E. P. 1987. Citrus Canker in Florida. Plant Dis. 71:388–394.

61. Sekine, M., Watanabe, K., and Syono, K. 1989. Molecular cloning of a gene for indole–3-acetamide hydrolase from *Bradyrhizobium japonicum*. J. Bacteriol. 171:1718–1724.

62. Somlyai, G., Hevesi, M., Banfalvi, Z., Klement, Z., and Kondorosi, A. 1986. Isolation and characterization of non-pathogenic and reduced virulence mutants of *Pseudomonas syringae* pv. *phaseolicola* induced by Tn5 transposon insertions. Physiol. Molec. Plant Pathol. 29:369–380.

63. Stachel, S. E., Nester, E. W., and Zambryski, C. 1986. A plant cell factor induces *Agrobacterium tumefaciens vir* gene expression. Proc. Natl. Acad. Sci. USA 83:379–383.

64. Starr, M. P. 1983. The genus *Xanthomonas*. Pages 742–763, *in:* The Prokaryotes: a Handbook on Habitats, Isolation, and Identification of Bacteria. M. P. Starr, ed., Springer-Verlag, New York.

65. Staskawicz, B., Dahlbeck, D., Keen, N., and Napoli, C. 1987. Molecular characterization of cloned avirulence genes from race 0 and race 1 of *Pseudomonas syringae* pv. *glycinea*. J. Bacteriol. 169:5789–5794.

66. Staskawicz, B. J., Dahlbeck, D., and Keen, N. T. 1984. Cloned avirulence gene of *Pseudomonas syringae* pv. *glycinea* determines race-specific incompatibility on *Glycine max* (L.) Merr. Proc. Natl. Acad. Sci. USA 81:6024–6028.

67. Swanson, J., Kearney, B., Dahlbeck, D., and Staskawicz, B. 1988. Cloned avirulence gene of *Xanthomonas campestris* pv. *vesicatoria* complements spontaneous race change mutants. Mol. Plant Microbe Interact. 1:5–9.

68. Swings, J., De Vos, P., Van Den Mooter, M., and De Ley, J. 1983. Transfer of *Pseudomonas maltophilia* Hugh 1981 to the genus *Xanthomonas* as *Xanthomonas maltophilia* (Hugh 1981) comb. nov. Int. J. Syst. Bacteriol. 33:409–413.

69. Tamaki, S., Dahlbeck, D., Staskawicz, B., and Keen, N. T. 1988. Characterization and expression of two avirulence genes cloned from *Pseudomonas syringae* pv. *glycinea*. J. Bacteriol. 170:4846–4854.

70. Tang, J. L., Gough, C. L., Barber, C. E., Dow, J. M., and Daniels, M. J. 1987. Molecular cloning of protease gene(s) from *Xanthomonas campestris* pv. *campestris:* expression in *Escherichia coli* and role in pathogenicity. Mol. Gen. Genet. 210:443–448.

71. Templeton, G. E. 1982. Status of weed control with plant pathogens. Pages 29–44, *in:* Biological Control of Weeds with Plant Pathogens. R. Charudattan and H. L. Walker, eds., Wiley, New York.

72. Vanderplank, J. E. 1963. Plant Diseases: Epidemics and Control. Academic Press, New York. Pp. 1–349.

73. Vanderplank, J. E. 1968. Disease Resistance in Plants. Academic Press, New York, Pp. 1–206.

74. Vidaver, A. K. 1983. Bacteriocins: the lure and the reality. Plant Dis. 67:471–475.

75. Winans, S. C., Ebert, P. R., Stachel, S. E., Gordon, M. E., and Nester, E. W. 1986. A gene essential for *Agrobacterium* virulence is homologous to a family of positive regulatory loci. Proc. Natl. Acad. Sci. USA 83:8278–8282.

Genetic Manipulation of
Plant Pathogens

8

Perspectives for Biological Engineering of Prokaryotes for Biological Control of Weeds

George H. Lacy

Perspectives on Biological Control

This is a speculative chapter because very little is known of biological control of weeds using prokaryotes and much less is known concerning the molecular biology of the many possible interactions among combinations of weeds and prokaryotes. Biological control using prokaryotes is viewed differently by many groups. Each viewpoint requires specific responses from those developing biological control agents. In the following paragraphs, I seek to provide some insight on the challenges faced by scientists developing biological control concepts and systems at the molecular level.

From the point of regulatory agencies, biological control of weeds is frightening because many biological control agents are pathogens. Pathogens, despite long records of safe environmental uses in biological control and in breeding resistant crops, strike a particular terror into regulators' hearts that may be disproportionate to any actual threat to ecological stability. Therefore, they may be slow to approve environmental applications. The challenge to the biological control scientist is, therefore, to anticipate environmental-impact-related questions with well-constructed experiments and good-quality data.

Environmentalists are on the horns of a dilemma. On the one hand, they endorse using natural systems to control pests and weeds to reduce nontarget chemical pesticide effects. On the other hand, they question the environmental impacts that may accrue from releases of biological pesticides. The special concern with biological systems is that living organisms, once released, may multiply, in addition to persisting or accumulating in the environment. Here we are challenged to go beyond being just responsible scientists, careful of our responsibilities to agriculture and the environment, and avoid, as well, the public perception of intellectual arrogance and scientific irresponsibility that accompanied those much-publicized nonregulated and nonauthorized releases of "genetically engineered" microbes by American scientists in industry and academia in California and Montana.

Other agricultural scientists, especially pathologists and entomologists, who were trained during the chemical era of pest control, may view biological control with less than enthusiasm. The evolution of their disciplines went through similar eras. The first was the *descriptive era;* during this period the organisms, their biology, and the conditions leading to past situations were described. The *agricultural practices era* followed; controls based on the information discovered in the descriptive era were used to formulate good agricultural practices to avoid pest problems (6,13). The *chemical era* precluded the need to study the biology of the organisms involved—the simple answer was to design a chemical to kill the pest (17). Unfortunately, one by-product of this era may been to downgrade applied agricultural training to "squirt gun chemistry." Ecological damage predictably occurred through lack of experience and, perhaps, an excess of enthusiasm for chemical solutions, and ignorance of ecological consequences. We are entering an *ecological responsibility era,* brought about because we are losing some of the most effective chemicals to Rebuttable Presumption Against Registration (RPAR), the development of resistance to chemicals, and to the overwhelming cost of chemical development and regulation. We are being forced to consider the biology of the interactants in pest situations again. In the end, this will be good for our science; however, in transition, some agricultural scientists will seek, through nostalgia, lack of ecological understanding, and lack of training, the simple answer to turn back the clock to the chemical era. However, most of our colleagues will spur development of biological control by challenging us to demonstrate using good scientific techniques that biological pesticides work as well as chemical pesticides.

Holders of research purse strings have not realized that American agriculture is in trouble without replacements for chemical pesticides that are critical for our current agricultural productivity. No federal- or state-funded programs exist to support research to replace chemical pesticides. The purse string holders see local surpluses and fail to see worldwide shortages. They cut research funding but fail to calculate the tremendous expenses that will accure should we ever fail to have those surpluses. This short-sightedness is due to ignorance and should challenge every scientist involved in biological control to help educate our representatives and funding agencies.

Potential and Problems

This discussion of biological control of weeds is developed on the premise that the controlling agents will be plant pathogenic bacteria. Prokaryotes include bacteria, mollicutes, and blue-green algae. In this discussion I shall write about bacteria unless otherwise noted. Prokaryotes with potential to control weeds occur in all groups, since each contains pathogens of plants (19,38).

Potential for Biological Control with Prokaryotes

Bacteria are very simple organisms biologically and genetically. Genetic engineering of these organisms is relatively easy compared to most eukaryotes (parasitic plants, insects, and fungi). We have a great deal of experience with the molecular biology of plant pathogenic bacteria compared with plant pathogenic fungi. The growth requirements of most bacteria are simple, and they are easily cultivated in standard fermentation equipment.

Problems with Biological Control with Prokaryotes

In general, plant pathogenic bacteria lack resistant structures such as fungi have in spores and sclerotia. They are susceptible to inactivation and death related to environmental stress (37). Packaging of bacteria used as biological control agents is more difficult, since the product will have a short shelf, will be difficult to ship, and may require special storage facilities, such as refrigeration. Application may stress bacteria. In spray applications, chemicals and fungi may remain relatively unchanged; however, shear forces, hydraulic pressure changes, evaporative cooling, and drying all affect bacterial viability. To be effective, biological control agents must persist on the plant surface until conditions permit plant pathogenesis to occur. Bacteria are susceptible to predation by mites and protozoa on roots; to antagonism by other microbes, including bacteria; to drying; and to exposure to infrared and ultraviolet radiation. For pathogenesis to develop, conditions adequate for bacterial multiplication must be present. For successful infection, nutrients, a portal of entry (wound), and free moisture must be available under conditions of adequate temperature. In contrast, some fungal spores may germinate on dry plant surfaces or in high humidity without the requirement for free moisture. Further, bacteria as a class lack mechanisms for direct penetration of plant tissues. The apparent exception to this rule is the production of chitinases by fungi-form actinomycetes such as *Streptomyces scabies* and *S. ipomea*.

Tools of Genetic Engineering for Prokaryotes

The tools for engineering bacteria include the genetics of prokaryotes and techniques of molecular biology.

Bacterial Genetics

Bacteria have four methods for exchanging genetic material. These efficient mechanisms probably account for the adaptive ability and evolutionary success of bacteria. Since neither the genetics nor the molecular biology of many plant pathogenic species is known, a review of known genetic mechanisms follows.

Conjugation: Plasmid and Chromosomal Transfer

In conjugation, cell-to-cell contact must occur, and plasmid- or chromosomally encoded genes are passed from the donor to the recipient bacterium. In all cases examined, plasmids are required for chromosomal exchange (31).

Plasmid Transfer. Plasmids are conjugative or nonconjugative. Small conjugative plasmids are about 30 kiloDaltons (kD), whereas nonconjugative plasmids are usually less than 10 kD. Nonconjugative plasmids may be transferred, or mobilized, by a coresident "helper" plasmid (24). The helper plasmid provides mobility proteins that interact with mobilization nucleotide sequences on the nonconjugative plasmids (45). Recombinant DNA is usually manipulated on nonconjugative plasmids; however, it is possible to mobilize these plasmids with helper plasmids (41).

Transfer of Chromosomal Genes. Conjugative transfer of chromosomal genes is plasmid-mediated and occurs in three ways (22,44):

1. *Plasmid integration in the chromosome*. Conjugative plasmids may integrate chromosomally, probably mediated by insertion sequences shared with the chromosome, and transfers DNA during conjugative plasmid transfer. This type of chromosomal transfer occurs at the lowest frequency.

2. *Prime plasmid integration in the chromosome*. Prime plasmids contain sequences from chromosomal DNA which allow more frequent association and subsequent integration in the chromosome. This type of chromosomal transfer occurs at an intermediate frequency and is analogous to the F' system in *Escherichia coli*. The prime condition indicates that plasmid insertion in the chromosome and chromosomal gene insertion in plasmids is reversible. This is confirmation that bacterial and plasmid genomes are plastic.

3. *High frequency of transfer*. In high frequency of recombination (Hfr) systems, a donor strain has a plasmid stably inserted in its genome which mediates unidirectional chromosomal transfer from the point of insertion during conjugal plasmid transfer. The chromosome is transferred before the plasmid. This type of transfer occurs at the highest frequency.

Transformation

Bacterial transformation involves the uptake of cell-free or "naked" DNA by competent bacteria. Factors determining competency of recipient cells to undergo transformation are not fully understood. The process of transformation differs between Gram-positive and Gram-negative bacteria. For the Gram-positive bacteria, competence is initiated in the recipient cell through the production of a protein competence factor. Double-stranded linear DNA adheres to a recipient cell, and a single strand of this molecule is taken up by the recipient cell. This strand is coated with proteins protecting it from nucleases, forming an eclipse complex, and is integrated by recombination into the recipient chromosome. Gram-negative bacteria apparently do not produce competence factors and require special treatments, notably by $CaCl_2$, to induce competence, and only take up circular DNA molecules (31).

Transfection is the uptake of naked bacteriophage (bacterial virus) DNA by competent bacteria and occurs by a process similar to that of transformation; however, there is no need for recombination, since bacteriophages replicate autonomously.

Electroporation

Bacterial genes cloned on plasmids may be inserted into bacterial cells that have been disrupted by electrical current. Electroporation is more efficient if an osmotic support is used to prevent cellular lysis. Because cell preparation is so simple and the procedure so rapid, many laboratories have abandoned transformation and transfection in favor of electroporation.

Transduction

Transduction results from bacteriophage infection of a recipient bacterial cell. A defective bacteriophage injects bacterial DNA along with or instead of bacteriophage genetic material from a lysed donor bacterium into the recipient bacterium. Transduction is limited by the bacteriophage host range.

Two categories of transduction are recognized: generalized and specialized. In generalized transduction, any fragment of the infected donor's chromosome can be incorporated into a bacteriophage's protein capsule or head, transferred to a recipient bacterium, and incorporated into its genome. Generalized transducing bacteriophages cause the donor bacterium to lyse.

In specialized transduction, the bacteriophage genome becomes inserted in a donor bacterium's genome; this insertion is referred to as a "prophage." Only those segments of the donor genome adjacent to the prophage DNA are transferred. Specialized transducing bacteriophages replicate for generations with the host genome before causing lysis.

Cell Fusion

The least-well-known method for bacterial genetic transfer mechanisms is cell or protoplast fusion. In this process, cell walls are removed from two bacterial cells and the resulting protoplasts fuse. Gentle surfactants and osmotic protectants are required to achieve fusion. The chromosomes of the two cells form a heteroduplex, and recombination may occur. Protoplast fusion is a useful laboratory genetic technique for the transfer of genetic material among organisms as diverse as *Staphylococcus aureus, B. subtilis,* and several other *Bacillus* spp. (9).

Recombinant DNA Techniques

Recombinant DNA techniques that will be useful for engineering prokaryotes for biological control of weeds are described next.

Transposon Mutagenesis

Transposon mutagenesis is a powerful method for locating genes involved in bacterial–plant interactions (3,28). Transposons are biological entities consisting of a gene-coding region, including a transposase endonuclease, located between two inverted nucleotide base sequences. These repeats often resemble insertion sequences. By electron microscopy, inverted repeats form palidromes (or lolli-pops) protruding at right angles to melted and renatured strands of DNA. The unique ability of transposons is that they may insert almost anywhere in DNA. They depend upon the replication of bacterial, plasmid, or bacteriophage DNA to replicate. If the transposon carries a gene for antibiotic resistance, it may be detected on antibiotic-containing media.

Transposons may be introduced to bacteria on plasmid or bacteriophage vectors. In efficient systems, the vectors are incapable of replication in the bacterium so that only transposons that "jump" from the suicide vectors survive to express their antibiotic resistance. Should insertion occur in a gene or regulatory region related to bacterial–plant interactions in pathogenesis, a polar mutation often occurs that eliminates the activity of that gene. Nonpathogenic mutants may be discovered in this manner. A library of transposon mutants may be prepared in this manner. Since bacterial chromosomes only have about 3000 genes, testing a relatively small number of mutants (1200–2000 fragments) will provide a library of mutants.

Cloning: Restriction and Ligation

Cloning DNA fragments is the process of digesting or restricting DNA into fragments that may be recombined in new configurations using ligase, an enzyme that restores the covalent bonds between the ends of DNA fragments (4,26). Restriction endonucleases are very specific in which base sequences they recog-nize. Some endonucleases form blunt-ended DNA fragments with the restriction site between adjacent bases in both strands of double-stranded DNA molecules; others cut the two strands at different points, creating overhanging ends. Hydro-gen bonding between complementary overhanging (or sticky) ends stabilizes the ends, so that ligation occurs more efficiently than between blunt ends. DNA must be cloned into molecular vectors capable of replication; plasmids, cosmids, and bacteriophages serve as vectors.

Genomic Library Construction and Complementation

Genomic libraries consist of fragments of bacterial genome cloned into gene vectors, so that, on a statistical basis, all the fragments of the genome are represented. Incomplete (or partial) endonuclease digestions are used so that every gene will be represented in its complete form in the library. Either cosmids or bacteriophages are used as vectors (4,24). In either case, the vectors are constructed so that the cloned genome fragments are packaged in bacteriophage heads. Upon transduction of bacteria, the cosmids replicate as plasmids and the bacteriophages replicate as viruses. An entire bacterial genome may be represented in as few as 100–200 fragments, depending upon the size of the inserts.

Complementation is accomplished by transferring a cosmid library into transposon mutants. Transfer depends on the conjugal ability of plasmids. If a cosmid clone restores the activity lost by the action of the transposon, complementation has occurred (1). The clone may be characterized further to determine the exact location and structure of the gene.

Molecular Characterization

Cloned DNA may be characterized by subcloning, deletion mapping, restriction mapping, transposon mapping, and nucleotide sequencing. Further, regulatory regions—including repressor, activator, and ribosome binding sites; transcription and translation start and stop points; and signal peptidase cleavage points—may be located by cDNA cloning, protein analyses, and other procedures outside the scope of this chapter. (2).

Site-Specific Mutation

Engineering bacterial biological control agents will require removing genes, adding genes, or modifying genes in the bacterium. Site-specific mutation is a powerful method for engineering bacterial genomes (16,35). This is accomplished by selecting for rare double cross-over events between plasmid or bacteriophage genomes and the bacterial genome. To discover these events, either suicide vectors (bacteriophages or plasmids) or selection against plasmid vectors may be used. Modified DNA must escape, via double-cross-over events, from the vector before it is lost by dilution among daughter cells during replication. Key to this process are areas of homology flanking the modified DNA with the bacterial genome; these areas of homology facilitate cross-over events.

Bacterial Pathogenesis

To manipulate bacterial pathogenesis for biological control, we must understand what pathogenesis is and the principles causing plant damage. In the following paragraphs, I shall describe the disease process and what bacterial factors cause the plant damage we see as disease.

Pathogenesis

Pathogenesis is the process of disease development and occurs among biological entities—a host and a pathogen. These entities exist together before and after disease occurs in a situation in which the pathogen is dependent upon a single host for its nutrition and survival. These definitions and those that follow, although generally used, are not accepted by all pathologists. One important alternative viewpoint is that abiotic agents—such as plant stress, air pollutants, and mechanical devices; grazing animals; and even genetic disorders—are pathogens. A second alternative viewpoint is that even if one cell of a plant is physiologically "abnormal" compared with the "optimum" physiology of that cell, pathogenesis occurs. This viewpoint is not supportable, since "abnormal physiology" cannot be defined in any meaningful way. The physiology of individual cells, tissues, organs, and plants varies with age and responds in a predictable or "normal" manner to stress, light, damage, and even pathogens. In our discussion, I shall disregard these alternative viewpoints, since we are limited to considering prokaryotes as pathogens of weeds.

Infection Court

An infection court is the localization of a susceptible host plant, the weed, with the pathogen. Environmental conditions, such as light, temperature, and moisture; the presence or absence of pathogen predators or antagonists; and the presence or absence of wounds determine the outcome of the interaction at this point. In engineering effective biological control agents, infection court factors such as survival on plant surfaces should be considered. The quantity or quality of extrapolysaccharides may affect survival on leaf surfaces. The discovery that some epiphytic bacteria replicate on leaf tissue even in the absence of free moisture from dew, rain, or guttation (15) suggests that bacteria may conserve water in their extracellular polysaccharides.

Penetration

Prokaryotes, with the exception of *Streptomyces* spp., do not penetrate their host directly, but require wounds from insects, mechanical damage, or pseudo-wounds caused by the emergence of lateral roots or the abscission of leaves or petioles. Binding to wounds seems important for some pathogens such as *Agrobacterium tumefaciens*. Engineering for biological control may consider the type and quantity of lipopolysaccharides that are involved with wound binding (30).

Infection

Infection is that moment that a pathogen becomes established within the host. This may be equated with the establishment of a nutrient association by the pathogen. For prokaryotes it may be measured as the transition point between the lag and exponential growth phases in plant tissues. Infection does not mean that pathogenesis will occur, since host resistance, environmental conditions, and therapy may interfere with continued development. Engineering for biological control may alter the nutrients that may be taken up by the pathogen during infection. By narrowing the range of nutrients that a pathogen may take up, the host range of a pathogen may be limited to a weed species (8).

Colonization

Colonization is the multiplication and spread of the pathogen from the point of infection through plant tissues. During infection and the first part of colonization, no visible evidence of pathogenesis may be apparent. This is the latent period. At some critical point the biological mass of the pathogen will be reached at which plant tissue is damaged. Again, colonization of specific host may be altered by manipulating nutrients affecting multiplication rate. Removal of certain pectate lyases from *Erwinia chrysanthemi* reduces the rate of spread of the pathogen in plant tissues (21).

Disease

Disease is plant damage resulting from interaction of the plant with the pathogen. The severity of disease may be altered by genetic manipulation. This subject is considered in the following section.

Principles of Disease

The principle of disease is the biochemical method by which the pathogen damages plant tissues. Principles of disease may or may not be necessary for the pathogen to infect, colonize, and cause disease in a plant. However, principles of disease affect the extent of plant damage. Principles of disease have as their basic purpose the release of nutrients from plant tissues for the nutritional benefit of the pathogen. Confusion exists in defining the extent of disease; the terms *more damaging* or *less damaging* may be the most clear. Virulence is less useful because pathologists have confused the literature by giving it two meanings: (1) Virulent pathogens are more damaging to the host; less virulent and avirulent pathogens cause less or no damage, respectively. (2) A virulent pathogen is a pathogen that causes disease only on those cultivars of species without specific genetic resistance to that pathogen; an avirulent pathogen does not cause disease on cultivars of the same species that have specific resistance.

Enzymes

Plant cell–degrading enzymes release nutrients from the various components of plant cells by catalytic activity. Soft-rot pathogens are the best-known plant pathogens utilizing this mechanism for plant damage. Extracellular cell-degrading enzymes that have been implicated in plant damage are generally extracellular, endo-cutting proteins. These molecules include pectate lyases, polygalacturonase, proteases, cellulases, phospholipases, and nucleases (7,21). Removal of pectate lyases by site-specific mutation has demonstrated that pathogenesis may be modulated to be less damaging. Adding more genes for pectate lyase may permit the pathogen to cause more damage. However, the organism remained pathogenic and caused at least limited damage (34). Pectate lyase isozyme profiles of *E. chrysanthemi* are correlated with host range (33). This discovery may suggest a strategy for construction of pathogens with greater host specificity for certain weeds.

Polysaccharides

Extracellular polysaccharides have at least two roles in pathogenesis; they protect pathogens from detection by the resistance mechanisms of plants and they plug xylem vessels in wilting diseases (20,42,43). The pathogens causing fire blight of pome fruit trees *(Erwinia amylovora)* and Granville wilt of solanaceous plants *(Pseudomonas solanacearum)* are examples of disease in which pathogenicity, the ability to cause disease, is related to extracellular polysaccharide production. Molecular studies of extracellular polysaccharide biosynthesis indicate that some elements may be interchangeable among bacteria in different genera (40). This may indicate that the production of extracellular polysaccharides in putative biological control agents may be manipulated genetically to increase plant damage or to avoid host defenses.

Growth Regulators

Plant growth regulators such as auxins and cytokinins are involved in the production of hypertrophy and hyperplasia leading to tumor formation. Cells affected by growth regulators often lead nutrients that are beneficial to the pathogen. The best-known example of these diseases is crown gall caused by *Agrobac-*

terium tumefaciens. Both auxin and cytokinin biosynthesis pathways and genes are related by function and DNA homology in different bacterial species (36,46,47). These results indicate that genes for growth regulators may be introduced into or switched with genes in biological control agents to affect pathogenicity.

Toxins

Plant pathogenic bacteria synthesize a battery of toxins, including coronatine, syringomycin, tagetitoxin, rhizobitoxin, phaseolotoxin, and tabtoxin. Interestingly, bacteria that under conditions in which phaseolotoxin is not produced or mutants that cannot produce tabtoxin remain pathogenic but produce leaf spots without additional damage caused by toxins (11,29). This illustrates the point that plant damage may be separated from pathogenesis and provides a strategy for engineering greater levels of pathogenicity into biological control agents for weeds.

Selecting a System to Engineer

Selecting a bacterium to engineer as a biological control agent is a critical step. Since so little is known of bacterial pathogens of weeds, a good deal of speculation must be used. Two areas that may be important for consideration are outlined next.

How the Pathogen Affects the Host

Good candidates for biological control must damage the weed enough to restrict its reproductive ability through reduced growth or death. For instance, leaf spotters and gall-forming pathogens often do not damage the host significantly. These organisms may be engineered to girdle stems through engineering larger galls or spots on petioles and stems that coalesce to kill plants. Soft-rotting or wilting bacteria often have a more profound effect on individual hosts, but the pathogen does not spread readily to adjacent plants. Engineering may adopt the strategy of developing more effective dispersal of these pathogens by mechanical means or by altering relations with vectors such as insects and nematodes.

Environmental Niche

Since bacteria are more susceptible to environmental stresses than fungal biological control agents, special consideration must be given to the target habitats for biological control. Ideally, free moisture should be present.

Phylloplane

The phyllosphere, or the surfaces of leaves and stems, is the harshest environment that plant pathogenic bacteria have to deal with. They are exposed to fluctuations in temperature, humidity, free moisture, radiant energy, and low nutritional condition (5). Some bacteria are successful in colonizing leaf surfaces. Pathogens must reach critical population levels before pathogenesis is initiated (14). However, during pathogenesis, the population levels inside the leaf or stem are much higher than in the phyllosphere. This difference in levels emphasizes the harshness of the environment on the surface. Genetic engineers may consider several strategies for introducing bacteria onto the surface of plants: Improve the water-holding capacity of extrapolysaccharides, improve the ultraviolet resistance of the bacterium, or devise some tactic to make bacteria adhere to fungal hyphaspheres. This last suggestion may provide a mechanism for biological control with two agents—fungal and bacterial.

Soil

The soil habitat is very densely populated (12). Antagonism, predation, survival, and dispersal of bacteria form barriers to successful biological control. Generally, water is not limiting, since it adheres with bacteria on the surfaces of mineral and organic particles in the soil. Few plant pathogenic bacteria are actually soil bacteria. Most persist on plant debris and replicate most rapidly in contact with living host plants. For control of fungi-using bacteria, an effective means of dispersal of the biological control agent is to inoculate seed directly or place the bacterium on granular carriers directly into the furrow with seed. This tactic does not seem useful for weed control. Strategies that allow slow release of encapsulated bacteria incorporated into the soil seem more advisable. Encapsulation would protect bacteria from antagonism and predation and would, with the proper carriers, extend bacterial survival.

Aquatic and Marine Environments

Aquatic and marine environments seem perfect for bacteria used for biological control. In water, fluctuations of temperature and radiation are dampered and there is no shortage of free moisture. The problem here is that almost nothing is known about pathogenesis caused by bacteria of plants, algal and flowering, in marine or aquatic environments. This may suggest that there are no plant pathogenic bacteria in these habitats, but this may not be true, because no systematic studies have been made by pathologists. Certainly we know that bacterial pathogens are distributed by water and that soft-rot organisms are found in both aquatic and marine environments (18,27). Discoveries in the field could suggest controls for water weeds.

Specificity

To protect against nontarget effects of biological pesticides, host specificity must be narrow. This may be accomplished using two strategies: using or engineering a narrow host range pathogen or using a narrow host range vector system.

Currently, a number of avirulence *(avr)* genes are being intensively studied among leaf-spotting pseudomonads and xanthomonads (39). These genes in bacteria trigger plant resistance responses, including hypersensitivity, and effectively determine the host range of these pathogens. Necessary to the function of *avr* are a series of clustered loci, termed *hrp* genes, controlling the expression of pathogenicity and hypersensitivity (25). Since these genes are conserved and interchanged among pathogens, and even nonpathogens, such as *Escherichia coli*, they may be utilized to affect host ranges of pathogens selected as biological control agents. For this to be successful the target weed must not contain resistance genes specific for the recognition of *avr* genes. Therefore, careful surveys of the susceptibilities of the target must be made before engineering is begun. Probably, biological pesticides of this type will be effective in areas in which certain genetic disease susceptibility patterns exist. We would expect that those patterns would shift in response to the selective pressure exerted by the engineered pathogens.

Nonspecific Systems

Our focus in biological control may be too narrow. We tend to think in terms of one-pathogen–one-weed combinations. In this situation we must require that the pathogen have an extremely narrow host range. This may be rare, or it may only be accomplished by extensive genetic manipulation. Simpler systems might be found. Among plant pathogens, many pathogens depend on insect hosts as vectors for dispersal to susceptible plants. These relationships vary from casual (the fire blight causal agent *[Erwinia amylovora]* is transmitted by honey bees *[Apis mellifera]* to other apple *[Malus sylvestris]* blossoms) to completely dependent (the pear decline mycoplasmalike agent is transmitted only by *Psylla pyricola* to other pear *[Pyrus communis]* trees) (32).

Special relationships have developed among wide-host-range pathogens and some insects. The soft-rot organism *Erwinia carotovora*, with possibly the widest host range of any plant pathogen attacking succulent tissues of all plants but conifers, has evolved symbiotic relationships with the onion maggot *(Hylemya antiqua)*, the black onion fly *(Tritoxa flexa)*, the seed corn maggot *(Hylemya platura)*, the iris borer *(Macronoctura onusta)*, and the cactus borer *(Cactoblastis cactorum)*, which, through the narrow host range of the vector, deliver the pathogen to specific target plants. In the case of the onion fly and maggot, normal development of the larva does not occur without *E. carotovora*. Evidently, the bacterium provides nutrients otherwise not available to the insect. The corn maggot and iris borer, and most likely the cactus borer, maintain the pathogen in their digestive tracts. The corn maggot eggs are contaminated with the bacterium and the iris borer maintains the bacterium in the puparia. Presumably, the cactus borer contaminates its offspring in a similar manner.

The remarkable thing about these pathogen-vector systems is that they deliver the pathogen directly to the host. The combination of *C. cactorum* and *E. carotovora* provides us one of the most successful examples of biological control of a weed (prickly pear cactus; *Opuntia* spp.), with a bacterial plant pathogen with no apparent nontarget effects (10). The caterpillars of the cactus borer tunnel into the pads of prickly pear and they feed mostly on cactus tissue rotted by the pathogen. The pathogen, not the caterpillars, was responsible for the death of the weeds. The lesson in this story for us is to consider using vectors to disperse bacteria to the target weeds. In this system we might make the bacterium dependent on the vector for survival and overwintering. Perhaps this could be done by deleting genes for survival in soil, water, and plant debris or by making the bacterium nutritionally dependent upon an amino acid or other nutrient present in the intestine and feces of the caterpillars. In this way the wide-host-range pathogen would be targeted to a narrow range of plants yet not be able to survive apart from the vector to threaten nontarget hosts.

In the same speculative vein, the vector system of delivery for pathogens would be well suited for control of water weeds. Insects either browsing upon or colonizing such water weeds as water hyacinth *(Eichhornia crassipes)*, Eurasian water milfoil *(Myrophyllum spicatum)*, and hydrilla *(Hydrilla verticillata)* would make idea vector systems to investigate using *E. carotovora*, since this pathogen evidently attacks tissues of submerged terrestrial plants (Lacy et al., 1981).

Summary

The information presented in this chapter ranges from the philosophy of approach to biological control using bacteria through systems for genetically manipulating bacteria and systems for pathogenesis to selecting bacteria for genetic engineering. The academic exercise of preparing this manuscript has convinced me that there is potential for biological control with this class of pathogens, should biological control scientists team with molecular biologists.

Literature Cited

1. Allen, C., V. K. Stromberg, F. D. Smith, G. H. Lacy, and M. S. Mount. 1986. Complementation of an *Erwinia carotovora* subsp. *carotovora* protease mutant with a protease-encoding cosmid. Molec. Gen. Genet. 202:276–279.

2. Ausubel, F. M., R. Brent, R. E. Kingston, D. D. Moore, J. G. Seidman, J. A. Smith, and K. Struhl, eds. 1987. Current Protocols in Molecular Biology. Wiley, New York.

3. Berg, D. E., and C. M. Berg. 1983. The prokaryotic transposable element Tn5. Bio/Technology July:417–435.

4. Berger, S. L., and A. R. Kimmel, eds. 1987. Guide to Molecular Cloning Techniques. Academic Press, San Diego.

5. Blakeman, J. P. 1982. Phylloplane interactions. Pages 307–333, *in:* M. S. Mount and G. H. Lacy, eds., Phytopathogenic Prokaryotes, Vol. 1. Academic Press, New York.

6. Civerolo, E. L., 1982. Disease management by cultural practices and environmental control. Pages 343–360, *in:* M. S. Mount and G. H. Lacy, eds., Phytopathogenic Prokaryotes, Vol. 2. Academic Press, New York.

7. Collmer, A., and N. T. Keen. 1986. The role of pectic enzymes in plant pathogenesis. Annu. Rev. Phytopathol. 24:383–409.

8. Coplin, D. L., L. Sequeira, and R. S. Hansen. 1974. *Pseudomonas solanacearum:* virulence of biochemical mutants. Can. J. Microbiol. 20:519–529.

9. Dancer, B. N. 1980. Transfer of plasmids among bacilli. J. Gen. Microbiol. 121:263–266.

10. Dodd, A. P. 1940. The biological campaign against prickly-pear. A. H. Tucker, Brisbane, Australia.

11. Durbin, R. D. 1982. Toxins and pathogenesis. Pages 423–441, *in:* M. S. Mount and G. H. Lacy, eds., Phytopathogenic Prokaryotes, Vol. 1. Academic Press, New York.

12. Foster, R. C., and G. D. Bowen. 1982. Plant surfaces and bacterial growth: the rhizosphere and rhizoplane. Pages 159–185, *in:* M. S. Mount and G. H. Lacy, eds., Phytopathogenic Prokaryotes, Vol. 1. Academic Press, New York.

13. Hagedorn, D. J. 1982. Control of prokaryotes by host breeding. Pages 361–385, *in:* M. S. Mount and G. H. Lacy, eds., Phytopathogenic Prokaryotes, Vol. 2. Academic Press, New York.

14. Hirano, S. S., and C. D. Upper. 1983. Ecology and epidemiology of foliar bacterial plant pathogens. Annu. Rev. Phytopathol. 21:243–269.

15. Hirano, S. S., and C. D. Upper. 1986. Temporal, spatial, and genetic variability of leaf-associated bacterial populations. Pages 235–251, *in:* N. J. Fokkema and J. Van den Heuvel, eds., Microbiology of the Phyllosphere. Cambridge University Press, Cambridge, England.

16. Horsch, R. B., R. T. Fraley, S. G. Rogers, P. R. Sanders, A. Lloyd, and N. Hoffman. 1984. Inheritance of functional foreign genes in plants. Science 223:496–498.

17. Jones, A. L. 1982. Chemical control of phytopathogenic prokaryotes. Pages 339–414, *in:* M. S. Mount and G. H. Lacy, eds., Phytopathogenic Prokaryotes, Vol. 2. Academic Press, New York.

18. Jorge, P. E., and M. D. Harrison. 1986. The association of *Erwinia carotovora* with surface water in Northeastern Colorado. 1. The presence and population of the bacterium in relation to location, season, and water temperature. Am. Potato J. 63:517–531.

19. Kennedy, R. W., and G. H. Lacy. 1982. Phytopathogenic prokaryotes: an overview. Pages 3–17, *in:* M. S. Mount and G. H. Lacy, eds., Phytopathogenic Prokaryotes, Vol. 1. Academic Press, New York.

20. Klement, Z. 1982. Hypersensitivity. Pages 149–177, *in:* M. S. Mount and G. H. Lacy, eds., Phytopathogenic Prokaryotes, Vol. 2. Academic Press, New York.

21. Kotoujansky, A. 1987. Molecular genetics of pathogenesis by soft-rot erwinias. Annu. Rev. Phytopathol. 25:405–430.

22. Lacy, G. H., R. C. Lambe, and C. M. Berg. 1981. Iris soft rot caused by *Erwinia chrysanthemi* associated with overhead irrigation and its control by chlorination. Int. Plant Propagat. Soc. 31:624–634.

23. Lacy, G. H., and J. V. Leary. 1979. Genetic systems in phytopathogenic bacteria. Annu. Rev. Phytopathol. 17:181–202.

24. Leemans, J., D. Inze, R. Villarroel, J. P. Heralsteens, M. DeBlock, and M. Van Montagu. 1981. Plasmid mobilization as a tool for *in vivo* genetic engineering. Pages 401–426, *in:* G. B. Levy, R. C. Clowes, and E. L. Koenig, eds., Molecular Biology, Pathogenicity, and Ecology of Plasmids. Plenum Press, New York.

25. Lindgren, P. B., N. J. Panopoulos, B. J. Staskawicz, and D. Dahlbeck. 1988. Genes required for pathogenicity and hypersensitivity are conserved and interchangeable among pathovars of *Pseudomonas syringae*. Molec. Gen. Genet. 211:499–506.

26. Maniatis, T., E. F. Fritsch, and J. Sambrook. 1983. Molecular cloning: a laboratory manual. Cold Spring Harbor Laboratory, Cold Spring Harbor, NY.

27. McCarter-Zorner, N. J., G. D. Franc, M. D. Harrison, J. E. Michaud, C. E. Quinn, I. A. Sells, and D. C. Graham. 1984. Soft rot *Erwinia* bacteria in surface and underground water in southern Scotland and Colorado, United States. J. Appl. Bacteriol. 57:95–105.

28. Mills, D. 1985. Transposon mutagenesis and its potential for studying virulence genes in plant pathogens. Annu. Rev. Phytopathol. 23:297–320.

29. Mitchell, R. E. 1984. The relevance of non-host-specific toxins in the expression of virulence by pathogens. Annu. Rev. Phytopathol. 22:215–245.

30. New, P. B., J. J. Scott, C. R. Ireland, S. K. Farrand, B. B. Lippincott, and J. A. Lippincott. 1983. Plasmid pSa causes loss of LPS-mediated adherence in *Agrobacterium*. J. Gen. Microbiol. 129:3657–3660.

31. O'Connell, M. P. 1984. Genetic transfer in prokaryotes: transformation, transduction, and conjugation. Pages 2–13, *in:* A. Puehler and K. N. Timmis, eds., Advanced Molecular Genetics. Springer-Verlag, New York.

32. Purcell, A. H. 1982. Evolution of the insect vector relationship. Pages 121–156, *in:* M. S. Mount and G. H. Lacy, eds., Phytopathogenic Prokaryotes, Vol. 1. Academic Press, New York.

33. Ried, J. L., and A. Collmer. 1986. Comparison of pectic enzymes produced by *Erwinia chrysanthemi, Erwinia carotovora* subsp. *carotovora,* and *Erwinia carotovora* subsp. *atroseptica.* Appl. Environ. Microbiol. 52:305–310.

34. Ried, J. L., and A. Collmer. 1988. Construction and characterization of an *Erwinia chrysanthemi* mutant with directed deletions in all of the pectate lyase structural genes. Molec. Plant–Microbe Interact. 1:32–38.

35. Roeder, D. L., and A. Collmer. 1985. Marker-exchange mutagenesis of a pectate lyase isozyme gene in *Erwinia chrysanthemi.* J. Bacteriol. 164:51–56.

36. Schröder, J. 1987. Plant hormones in plant–microbe interactions. Pages 40–63, *in:* T. Kosuge and E. W. Nester, eds., Plant–Microbe Interactions, Vol. 2. Macmillan, New York.

37. Sleesman, J. P., 1982. Preservation of phytopathogenic prokaryotes. Pages 447–484, *in:* M. S. Mount and G. H. Lacy, eds., Phytopathogenic Prokaryotes, Vol. 2. Academic Press, New York.

38. Starr, M. P., ed., 1983. Prokaryotes as plant pathogens. 1–9, *in:* M. P. Starr, H. Stolp, H. G. Truper, A. Balows, and H. G. Schlegel, eds., Phytopathogenic Bacteria: Selections from Prokaryotes. Springer-Verlag, New York.

39. Staskawicz, B. J., D. Dahlbeck, N. Keen, and C. Napoli. 1987. Molecular characterization of cloned avirulence genes from race 0 and race 1 of *Pseudomonas syringae* pv *glycinea.* J. Bacteriol. 169:5789–5794.

40. Torres-Cabassa, A., S. Gottesman, R. D. Frederick, P. J. Dolph, and D. L. Coplin. 1987. Control of extracellular polysaccharide synthesis in *Erwinia stewartii* and *Escherichia coli* K–12: a common regulatory function. J. Bacteriol. 169:4525–4531.

41. Twigg, A. J., and D. Sherratt. 1980. Transcomplementable copy number mutants of plasmids ColE1. Nature 283:216–218.

42. Van Alfen, N. K., B. D. McMillan, and P. Dryden. 1987a. The multi-component extracellular polysaccharide of *Clavibacter michiganense* subsp. *insidiosum*. Phytopathology 77:496–501.

43. Van Alfen, N. K., B. D. McMillan, and Y. Wang. 1987b. Properties of extracellular polysaccharides of *Clavibacter michiganense* subsp. *insidiosum* that may affect pathogenesis. Phytopathology 77:501–505.

44. Watson, J. D., N. H. Hopkins, J. W. Roberts, J. A. Steitz, and A. M. Weiner. 1987. Molecular Biology of the Gene, 4th ed., Benjamin/Cummings, Menlo Park, CA.

45. Willets, N. S., and B. Wilkins. 1984. Processing of plasmid DNA during bacterial conjugation. Microbiol. Rev. 48:24–41.

46. Yamada, T., C. J. Palm, B. Brooks, and T. Kosuge. 1985. Nucleotide sequence of the *Pseudomonas savastanoi* indoleacetic acid genes and homology with *Agrobacterium tumefaciens* T-DNA. Proc. Natl. Acad. Sci. USA 82:6522–6526.

47. Ziegler, S. F., F. F. White, and E. W. Nester. 1987. Genes involved in indole acetic acid production in plant pathogenic bacteria. Pages 18–25 *in:* E. Civerolo, A. Collmer, R. E. Davis, and A. G. Gillaspie, eds., Plant Pathogenic Bacteria. Martinius Nijhoff, Dordrecht, The Netherlands.

9

Genetic Manipulation of Plant Pathogenic Fungi

H. Corby Kistler

Introduction

Fungi are the most common causal agents of plant disease. As such their potential for control of weed plants is great, and in many cases efficient biological weed control may depend upon the use of fungal pathogens. Strategies for developing more effective biological weed control will involve the genetic manipulation of these pathogens. The purpose of this chapter is to discuss potential strategies that may be followed to construct fungal pathogens improved in their ability to act as biological control agents of weeds. Although a comprehensive review of the literature will not be attempted, I hope to illustrate many of these approaches with published examples, or to point out the lack of published information in certain areas. Most examples are from the plant pathology literature, which emphasizes diseases of agriculturally important plants with a goal (if often un-stated) of disease prevention. Nonetheless, basic information concerning the genetic control of pathogenesis and disease epidemics may be applied to both the prevention *and* maintenance of plant disease. Certainly some allowances must be made for differences among the studied systems, most notably the genetic diversity of weed populations compared with agronomic plants. These differences will be mentioned where appropriate.

Techniques for the genetic manipulation of fungal pathogens or fungi in general will not be discussed at length. These topics have been the subject of recent reviews (33,38,45) and books (3,52). The reader is referred to these sources for more details of the techniques enumerated later.

The author wishes to thank H. D. VanEtten and L. Ciuffetti for discussion and for use of pUCH1-PDA. F. Martin is gratefully acknowledged for his editorial comments. Work in the author's laboratory was supported by USDA grant 86-CRCR–1–2195 and the University of Florida Interdisciplinary Center for Biotechnology Research.

Methods

Several methods are available to modify the genotype of potentially useful biological control fungi. However, before modification is undertaken it would be wise to investigate the genetic diversity in the pathogen population already present in nature. Arguments are often made for the environmental release of genetically altered micoorganisms on the basis that processes occurring in the laboratory surely also must be occurring in nature. If this is truly the case, then strains of a particularly useful phenotype may be found in nature without genetic modification. Genetic and ultimately biochemical and physiological variation within a species of a pathogen may be considerable, and assessment of this variation in the native population might be an appropriate first measure in pathogen development.

Among the methods for genetic modification available for fungi are mutation, recombination, and direct gene transfer. Of these methods, mutation, whether chemical or physical (i.e., UV irradiation), is the least satisfactory, since changes in genotype are largely uncontrolled and cannot be directly characterized. This is a problem from both a biorational and a regulatory standpoint. Mutagenized fungi may be screened for improved characteristics or by "brute force" or improved weed control, but because of the nature of the change the mechanism involved may not be demonstrated unequivocally. Therefore, the results would not necessarily be repeatable, nor would the changes, so that no adverse effect could be said to be unlikely upon release.

Genetic recombination may occur in several ways in fungi. Reassortment of traits during mitosis and meiosis has been well established in the filamentous fungi *Aspergillus* and *Neurospora,* which provide textbook examples of the ease and power of these processes when used for studying the transmission of heritable traits. In addition to mitotic and meiotic recombination, fungi can undergo parasexual recombination whereby strains of unlike genotype fuse (plasmogamy), unlike nuclei fuse (karyogamy), recombination occurs, and nuclei are returned to the ploidy of the original strains. The mechanisms by which these processes occur are poorly characterized. Readers interested in techniques for meiotic, mitotic, or parasexual recombinational analysis for particular pathogenic fungi are referred to other sources (viz. ref. 52). Parasexual genetic recombination can be artificially induced by protoplast fusion, which is the topic of Chapter 10 in this volume. Normal meiotic and mitotic recombination have often been deemed to be a more "natural" process than forced parasexuality, and thus fewer objections are raised at the suggestion of the former for breeding pathogen strains with enhanced biocontrol phenotypes. This "natural" process, however, may in many instances be as forced as protoplast fusion, since, for example, some fungi form sexual structures readily in culture but rarely in nature. Potentially fertile strains of other fungi may be separated geographically or by host range, forming sympatric and allopatric barriers that may be breeched in the laboratory.

The more precise and easily characterized method for genetic manipulation of fungi is by direct gene transfer. Although this apparently can be accomplished by microinjection of foreign DNA into recipient fungi (59), as demonstrated by complementing auxotrophies of *Verticillium* spp., clearly the method of choice has become genetic transformation. Fungal cells are made competent by removing the cell walls in an osmotically stabilized medium with commercially available enzymes, and treatment of protoplast with polyethylene glycol and calcium salts. Transformation of over a dozen species of fungal plant pathogens has been demonstrated in this manner. Selectable markers that have been used to transform these fungi are a gene *(hph)* conferring resistance to the antibiotic hygromycin B (18,27,40,47,49,58,65), a gene (*amd*S) that allows the utilization of acetamide as a nitrogen source (47,57), and genes (*tub*–2) that confer resistance to the fungicide benomyl (24,41). Parsons et al. (42) additionally have shown transformation by complementing an arginine auxotroph with a cloned gene (*argB*). Selectable markers for oligomycin (66) and geneticin (46) resistance have been used to transform nonpathogenic fungi and may also be used for fungal pathogens. Thus, methods for transformation are readily available and applicable to a wide range of pathogenic fungi.

Transformation allows for simple, quick, and specific genetic modification. Since transformation of pathogens occurs by integration of the introduced DNA into fungal chromosomes, the genetic alterations are generally stable (however, see the following discussion). These integrations may be site directed through homologous recombination (42,65) after linearization of vectors. Site-directed integration allows for control of position–effect regulation of introduced genes by assuring insertion into a defined chromosomal milieu. Site-directed integration also allows for gene eviction and gene replacement manipulations (31,33,39). Expression of introduced genes may also be modified by selecting variants having single or multiple copies of the introduced genes.

A potential problem for using integrative transformants is that occasionally they may be mitotically unstable. For example, Rodriguez and Yoder (47) found that *Glomerella cingulata* did not retain the transferred *amd*S gene upon repeated subculturing on nonselective media. DNA transferred with the *hph* gene in *G. cingulata* was rearranged during growth on nonselective media. We have found rearrangement of transferred DNA in *Fusarium oxysporum* under both selective and nonselective conditions (27; Kistler, unpublished). Some hygromycin-resistant transformants were unstable phenotypically (Kistler and Benny, 1988), and introduced sequences in these individuals showed altered DNA restriction patterns associated with instability. However, repeated single-spore isolations of one such transformant led to a culture that was stable for both drug resistance and DNA restriction pattern of introduced DNA for 80 mitotic divisions (Figure 9.1). However, even mitotic stability in culture may be misleading. Keller et al. (25) have found that transformants of *Cochliobolus heterostrophus* that are mitotically stable in culture may be unstable when passaged on a host. The authors suggest

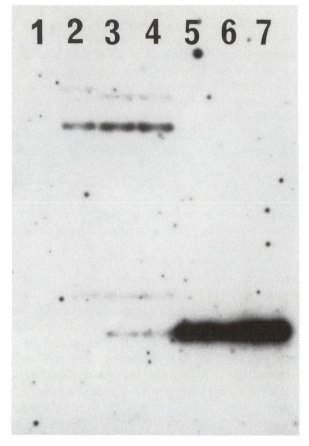

Figure 9.1. Stability of introduced DNA in *Fusarium oxysporum*. *F. oxysporum* strain ATCC 58110 was transformed to hygromycin B resistance with pDH25 as previously described (27). A resulting transformant (Tr1) was grown on media containing 200 μg/ml hygromycin B or in a complete liquid medium without selection. Two independent, faster-growing sectors on plates containing hygromycin B were isolated (S1 and S2). Single spore isolations of Tr1 grown without selection were made after 34 (Tr1A), 89 (Tr1A3) or 114 (Tr1A4) generations. DNA was isolated, digested with *Sal*1 and fractionated on a 0.7% agarose gel. DNA transferred to solid support was hybridized with a radiolabelled *Cla1−Bam*H1 fragment from pDH25 that contains *hph*. Lanes 1–7 contain DNA from (1) untransformed ATCC 58110, (2) Tr1, (3) S1, (4) S2, (5) Tr1A, (6) Tr1A3, and (7) Tr1A4.

that the stability of the introduced DNA may be dependent upon environmental pressure, such as the stress resulting from the process of infecting plants. DNA stability during infection should be addressed for each pathogen to be manipulated in this manner.

Another potentially useful manipulation in pathogenic fungi is cotransformation. Cotransformation is the concomitant uptake and integration of noncovalently attached, nonselected DNA along with transforming DNA containing a selectable marker. Cotransformation frequencies may be very high, ranging from 40% to >90% of characterized fungal transformants (10,26,55,68).

In addition to integrative transformation vectors, we have recently developed a plasmid that replicates free of the chromosome in *Fusarium oxysporum*. This vector confers hygromycin B resistance, and apparently replicates in the fungal nucleus. The plasmid copy number is high enough to be visualized on ethidium-bromide-stained agarose gels of total fungal DNA. This plasmid increases transformation frequency approximately 1000-fold (to >1000 transformants per μg DNA per 10^7 cells) when compared with integrative vectors in *Fusarium oxysporum*, and when coupled with rounds of cotransformation and curing of the autonomous plasmid, may allow for the sequential introduction of a number of genes into a recipient strain using a single vector. A similar plasmid recently has been constructed for *Ustilago maydis* (56).

Genetic Strategies for Strain Improvement

There are several potential strategies for developing fungal pathogens for improved efficiency in weed control: (1) improvement in the ability to be pathogenic to target plants, (2) improvement in the ability to disseminate, (3) improvement in competitive ability, and (4) improvement in safety. Although these goals are sound, the exact manner by which they are achieved is largely untested. Much is unknown about what will contribute to improved efficiency, so presently no screening program can proceed with the assurance that any one characteristic or phenotype will result in the desired goal. Yet with precise genetic manipulation of these pathogens by modification of one gene at a time, the basis for achieving enhanced disease progress may be understood and the principles applied to other biological control problems. These strategies for improved efficiency will now be illustrated with examples from the literature that show both the variety of characteristics that may be modified for beneficial purposes and the variety in methods for genetic manipulation. These examples have been taken from the plant pathology literature and thus discuss exclusively diseases of crop plants. The principles for genetic control of disease progress should apply to weed plants as well.

Improvement in the Ability to Be Pathogenic to Target Plants

Improvement in pathogenic characteristics can be approached in two ways: modifying host range and modifying aggressiveness. Aggressiveness is meant here as the relative ability to cause disease in a quantitative sense, either in individuals or in populations. Therefore, strains that cause more lesions per leaf, more wilt per unit infective propagule, or a greater number of diseased plants per unit time are more aggressive. Although in some instances it may be advantageous to decrease host range or decrease aggressiveness to nontarget plants (see "Improved Safety," which follows), it will more likely be the goal to broaden the host of an existing pathogen in a specific and directed manner and/or to increase the aggressiveness of that pathogen.

The genetic basis for host range in fungal pathogens has been of intense interest to plant pathologists for decades. The host specialization of a pathogen may be to the level of plant genus (or species) or the level of a specific host genotype. Genetic control of specialization to either level may be simple. For example, in the blast fungus *Magnaporthe grisea,* a single gene may condition pathogenicity to the host species goosegrass *(Eleusine indica),* and a second unlinked gene conditions pathogenicity to weeping lovegrass *(Eragrostis curvula)* (36,60). Likewise, monogenic control of ability to cause disease on particular cultivars of rice *(Oryza sativa)* has been demonstrated (8,35). Therefore, genetic modification for expanding host range by sexual crosses may be straightforward.

Screening available strains of a pathogen population for host specialization may yield unexpected and potentially useful results. For example, host range screening for isolates of the fungal species *Nectria haematococca,* which includes the well-known pea root rot pathogen, has established that it is pathogenic to at least nine additional plant species (61). Genetic control of pathogenicity to pea *(Pisum sativum)* and chickpea. *(Cicer arietinum)* is independent as indicated by the fact that field isolates were obtained with aggressiveness toward one, the other, both, or neither plant. Presumably, by screening pathogenicity further, isolates could be found with a unique host range exhibiting aggressiveness to a particular combination plant species for which it is a pathogen. In principle, by proper screening it may be possible to find a field isolate of a pathogen with a particularly useful host without genetic manipulation. The existing genetic variants in a pathogen population are a most important resource, not only for strictly pathogenic considerations but also because they have been proved to persist in nature.

Mutation is also a process by which the host range of a pathogen may be modified. This can be illustrated by the work of Défago and co-workers, also using the fungal pea pathogen *Nectria haematococca* (12,13). This pathogen is not ordinarily considered a pathogen of tomato (61), yet when it is introduced to a wound of ripened tomato fruit it produces a severe necrosis (12). This could be interpreted as essentially a saprophytic reaction, since wounded green tomato fruit are unaffected by these fungi. Green tomato fruit, however, also differ from ripened fruit by their higher level of the inhibitory steroidal glycoalkaloid tomatine. Tomatine, inhibitory to many nonhost pathogens of tomato, including *N. haematococca,* is suggested as a mechanism for resistance to these opportunistic fungi. Indeed, mutants of *N. haematococca* highly tolerant to tomatine were capable of causing a more extensive rot on green tomato fruit (12). A genetic analysis of mutants with respect to tomatine insensitivity and enhanced aggres-

siveness to green tomato indicated complete linkage and monogenic inheritance
(13). This example illustrates several points: the use of a mutation to enhance
aggressiveness to a particular plant (or plant tissue), the use of genetic analysis
to confirm the nature of change, and the use of knowledge concerning a possible
mechanism for disease resistance for overcoming resistance. The role of tomatine
as a preformed inhibitor of fungal invasion may be similar to the role of other
preformed inhibitors suggested to play a role in latent infection, such as the dienes
and 5-substituted resorcinols of avocado and mango, respectively (15,44). If these
compounds are in some instances the sole mechanism for restricting colonization,
mutation of pathogens to overcome inhibition by these compounds should allow
premature infection of unripe fruit. Similar strategies could be envisioned for
other proposed preformed barriers to infection.

Direct gene transfer also has potential for modifying the host range of a fungal
pathogen. An example is the transfer of the gene for pisatin demethylase to other
fungi. Pea root rot pathogens of the fungal species *Nectria haematococca* are
highly tolerant to inhibition by the major, microbe-induced, low-molecular-
weight antibiotic (phytoalexin) produced by pea (62). Tolerance of *N. haematococca* to this isoflavanoid phytoalexin, called pisatin, results from the ability to
detoxify the compound metabolically to a less toxic state. Pisatin detoxification
involves a demethylation reaction mediated by the enzyme pisatin demethylase,
a cytochrome P–450 monoxygenase. The ability to infect pea consegregates
during meiotic recombination with genes coding for high (but not low) pisatin
demethylase activity (62). A gene for pisatin demethylase from *N. haematococca*
has been cloned by expression in *Aspergillus nidulans* (67). This has made
possible a proof for the role of pisatin detoxification in aggressiveness toward
pea. The pisatin demethylase gene *(PDA)*, along with a gene for hygromin B
resistance, was used to transform a variant strain of *Nectria haematococca* that
lacked both pisatin demethylase and pathogenicity to pea. Three transformants
were able to demethylate pisatin in vitro, and two of the three were pathogenic
to pea (9). The reason for the failure of the third of three transformants to be
pathogenic toward pea is unclear, but presumably it could be due to the ability
to create sufficient enzyme levels or to regulate the expression of the gene in the
proper manner. Nevertheless the cotransfer of pisatin demethylase and aggressiveness to pea is a direct demonstration of the role the enzyme plays in pathogenesis
toward pea.

An important question for those interested in manipulating the host range of
fungal pathogens is, "Can the gene for pisatin demethylase be used to extend the
host range of other pathogens to include pea?" The answer to this question seems
to be yes, in the proper genetic background, and given the proper expression of
the gene in question. We have introduced *PDA* into a heterologous fungus,
namely, the wilt pathogen *Fusarium oxysporum*. The pea wilting form of the
fungus *F. oxysporum* f.sp. *pisi* is, like *N. haematococca*, extremely tolerant to
the phytoalexin pisatin and is capable of demethylating it to the same 3-hydroxy

product (19). Other strains of *F. oxysporum* do not cause wilt in pea and lack pisatin demethylatin ability (19, Kistler, unpublished). A vector (pUCH1-*PDA*) containing the gene for pisatin demethylase on a 3.2-kb *Xho*1 fragment cloned in a plasmid having a selectable gene for hygromycin B resistance was introduced into a strain of *F. oxysporum* f.sp. *conglutinans* (ATCC 9990). This fungus is a virulent pathogen of cabbage but is normally sensitive to inhibition by pisatin and has the ability to cause only very small lesions and very slight wilt symptoms in pea (Figure 9.2). Under the assayed conditions an authentic pea wilt pathogen (*F. oxysporum* f.sp. *pisi*) causes either severe cortical rot similar to *N. haematococca* or death through wilting, depending on the assay. Six independent hygromycin B–resistant ATCC 9990 transformants were obtained by transformation with pUCH1-*PDA*. Four of six transformants expressed pisatin demethylase activity as shown by enhanced tolerance to inhibition by 0.5 m*M* pisatin. Two transformants were more tolerant of pisatin than the authentic pea wilt pathogen. However, none of the tested transformants were any more aggressive toward pea than the original recipient strain, either in the ability to wilt pea or in the ability to cause necrosis on wounded pea stems.

These results differ from those obtained by Schäfer et al. (50). Again using pUCH1-*PDA*, they introduced the gene for pisatin demethylase into a heterologous plant pathogen, this time *Cochliobolus heterostrophus,* the cause of southern

Figure 9.2. Effect of Transfer of the Gene for Pisatin Demethylase on a Nonhost Pathogen

Strain	Percent Inhibition by 0.5 m*M*	Pea Lesion Length (mm)*	Wilt Symptons†
Nectria haematococca (T-9)	12	7.7 ± 2.8	ND
Fusarium oxysporum f.sp. *pisi*	66	4.0 ± 1.9	5.0 ± 0.0
Fusarium oxysporum f.sp. *conglutinans* (ATCC 9990)	83	1.3 ± 0.5	1.3 ± 0.5
ATCC 9990 [Pda1]A	37	1.3 ± 0.3	1.0 ± 1.4
ATCC 9990 [Pda1]B	83	2.1 ± 0.7	1.0 ± 0.0
ATCC 9990 [Pda1]C	85	1.7 ± 0.5	0.8 ± 0.5
ATCC 9990 [Pda1]D	47	1.6 ± 0.8	0.8 ± 1.0
ATCC 9990 [Pda1]F	65	1.7 ± 0.5	ND
ATCC 9990 [Pda1]H	60	1.6 ± 0.7	0.4 ± 0.6

*End point percent inhibition and pea lesion assay described by Kistler and VanEtten (28).
†Wilt symptoms scored on intact 3-week-old pea plants and based on a 0–5 scale, with 0 being uninoculated plants, and 5 being dead plants.
ND = not determined.

corn leaf blight. However, Schäfer et al. attempted to maximize gene expression by using very high concentration of hygromycin B for selection, reasoning that increased levels of pisatin demethylase activity might be present in transformants with increased hygromycin tolerance. Enzyme assays indicated that one such transformant had pisatin demethylase activity similar to wild-type *N. haemato-cocca* strains resulting from 18 tandemly repeated copies of the incorporated plasmid (VanEtten et al., personal communication). This strain was able to cause significantly larger lesions on pea stems and leaves than wild-type *C. heterostrophus*. The leaf spotting ability of the altered *C. heterostrophus* on pea (normally a foliar pathogen on corn) was greater than that of *N. haematococca* (normally a root-rotting pathogen). This indicates that in the proper genetic background a single gene properly expressed may be able to expand the host range of a pathogen. The failure of transgenic *F. oxysporum* f.sp. *conglutinans* to cause a greater amount of rot or wilt in pea could be explained by the inability to produce sufficient pisatin demethylase activity during infection. However, even wild-type strains of *F. oxysporum* f.sp. *conglutinans* are capable of colonizing pea epicotyl cortex tissue and may be isolated more than 1 cm away from sites of inoculation without the appearance of any more than a slight amount of disease (Kistler, unpublished). Strains of *F. oxysporum* are often efficient root colonizers on nonhost plants (2,48), so addition of pisatin demethylase in itself may not be expected to increase the pathogen's ability to colonize host tissue and cause necrosis and wilt. Pisatin demethylase is not directly related to cellular necrosis and wilt, but only to the ability of a fungus to survive in potentially inhibitory, pisatin-containing tissue. Since *F. oxysporum* f.sp. *conglutinans* is already capable of colonizing pea cortex, no additional pathogenic advantage may be gained by addition of pisatin demethylase. *C. heterostrophus,* on the other hand, may be an efficient leaf-blighting organism lacking only the ability to overcome the phytoalexin barrier of pea in order to be able to cause more extensive necrosis and cellular death in this new host. The genetic and pathogenic context must thus be taken into consideration when attempting to broaden the host range of a pathogen by the addition of a single gene.

In addition to modifying host range, modifying the aggressiveness of a pathogen may be an effective strategy for improvement of biological control agents. The value of screening pathogen populations for the most pathogenic strain is obvious; this is the most common step in current biocontrol programs. Another strategy that may be effective is to select the most aggressive progeny in sexual crosses of compatible fungi.

Where aggressiveness has been studied in such crosses, the distribution of aggressiveness in progeny is suggestive of a polygenic system (5–7). Naturally occurring strains appear to have unique combinations of genes that allow for high aggressiveness; crossing field isolates often leads to progeny even less aggressive than either parent or reduced in aggressiveness compared with parental means. It should be noted, however, that even by outcrossing between aggressive strains, selecting the most aggressive progeny, and backcrossing these to an aggressive, recurring parent, the result is little or no increase in the ability to cause disease. Aggressiveness may be nearly optimal in wild-type strains. Such fungal breeding programs may be beneficial more to combine other desirable traits, such as extended host range or sexual fertility, than to attempt to create more aggressive pathogens per se. A particularly successful program of this type is illustrated by the improved fertility and recombination of host ranges in *M. grisea* described by Valent et al. (60). Another possible complication is the indication that less aggressive genotypes in some instances may be more fit (7).

Although increasing the aggressiveness of a pathogen by conventional genetic means may be difficult and/or time-consuming, the potential for modifying aggressiveness by molecular genetic methodology presently seems unlimited. The most obvious type of gene to transfer to weed control fungi is a gene for a herbicide. The specificity of effect on the target plant could be assured by either the selectivity of the introduced herbicidal molecule or the host range of the pathogen, or both. Molecules with extreme host-selective toxic effect are known for many plants (51) and in some instances have been shown to be under the control of a single gene (51,70) or perhaps two or more closely linked genes (63). Progress in the isolation of enzymes involved in phytotoxin production (64,69) will make possible the identification and isolation of genes involved in the biosynthesis of these molecules. Once the genes have been isolated, they may be transferred to other fungi to test if pathogenic specialization and/or increased aggressiveness to a target plant is a result. The effectiveness of these strategies will rely on the ability to identify herbicidal compounds host selective for weeds that must be controlled. Alternatively, herbicidal compounds with a broad range of activity may also be transferred to potential weed control pathogens relying on the host range of the pathogen to target its effect. This would open up a wide range of compounds, from both bacteria and fungi, that exhibit herbicidal but limited fungicidal activity. Genes for the biosynthesis of these compounds could be overexpressed by modifying with a strong fungal promoter or by screening for strains having a greater copy number of introduced genes. The problem with this strategy is the potential effect of toxin production on nontarget organisms. Virtually nothing is known about the ecological consequences of this type of modification.

The addition of certain enzymes may also assist in the creation of more aggressive fungal pathogens. An example is the work of Dickman et al. (14) on the fungal cutin degrading enzyme known as cutinase. Cutin is the fatty acid polymer that is the major structural component of the outer, cuticular layer of plants. Some plant pathogenic fungi require cutinase in order to penetrate this barrier at the plant surface (30). However, certain pathogens, such as the *Mycosphaerella* spp. that infect papaya, require wounding in order for successful infection to take place. Dickman et al. (14) have shown that the infection efficiency on unwounded papaya can be greatly increased by the addition of exogenous cutinase. The structure of cutinase genes from a few fungal pathogens is now known (17). It is reasonable to expect that introduction and expression of cutinase in fungal "would pathogens" such as *Mycospherella* spp. could lead to greater infection efficiency and a more aggressive pathogen.

Another potential strategy for increasing pathogen aggressiveness is by mutation or deletion of specific genes or nucleic acids that limit pathogenesis. Viruslike particles (VLPs) and dsRNA are common in fungal pathogens. Although their involvement in pathogenicity or aggressiveness in most instances is unknown, in a small number of cases (e.g., *Cryphonectria parasitica*) these elements have been demonstrated to result in reduced aggressiveness, a phenomenon called hypovirulence (1). The presence of the *C. parasitica* hypovirulence dsRNA appears to suppress specific, highly expressed poly(A)$^+$RNAs of unknown function that may account for their ability to cause hypovirulence (43). Growth of hypovirulent fungi on media containing cycloheximide has resulted in elimination of the dsRNA and conversion to a more aggressive form (20). Similar strategies to cure VLPs and dsRNAs from potential biocontrol fungi may be worthwhile, considering the prevalence of these elements.

Fungi are also known to contain dominant traits (called avirulence genes) that result in resistant interactions when challenged plants have genes specific for recognition of the corresponding avirulence genes of the pathogen. These gene-for-gene interactions were first described by Harold Flor in the early 1940s for flax rust caused by *Melampsora lini* (see ref. 32) and later have been described for numerous other obligately parasitic fungi. Since resistance results from the action of a dominant gene in the pathogen, mutants may be obtained that overcome resistance. For example, Gabriel et al. (21) have obtained chemically induced mutants of the wheat powdery mildew pathogen (*Erysiphe graminis* f.sp. *tritici* with increased virulence toward wheat plants of a particular genotype. However, the use of mutation to increase the ability of fungi to cause disease in weeds is likely to be unsuccessful. This is due to the fact that natural populations of weed plants are apt to be more genetically diverse than agricultural monocultures. The chances are slim for finding mutations in the pathogen population that match every potential gene for resistance in a genetically diverse plant population.

Improvement in the Ability to Disseminate

Little is known about the genetic control of fungal dispersal. Most plant pathogenic fungi disseminate passively in the form of spores. Unlike plant viruses that in some instances rely on an intimate relationship with a vector for dispersal, spread of fungal pathogens is generally left to chance distribution by the physical (wind, rain) or biotic environment. How the phenotypic and genotypic makeup of a fungus affect the ability to be dispersed in this manner is a subject in need of greater research.

However, some fungi do not fit the preceding generalizations, and these exceptions may allow for genetic analysis of the role of specific traits in fungal dispersal and perhaps lead to principles for improvement in the ability of pathogens to disseminate. The Dutch elm disease pathogen, *Ophiostoma ulmi,* for example, produces conidia that are sticky and adhere to bark beetles of the family Scolytidae. These beetles include elm-feeding species that, because of their selectivity in feeding and reproductive activity in elm bark and sapwood, serve as highly efficient vectors of the pathogen (53). The genetic and biochemical control for the mechanism of adhesion is unknown, but such a mechanism, if transferred to a heterologous fungus, might also aid in its insect-mediated dispersal. Targeting of dispersal thus could be in control of the feeding and reproductive behavior of the insect.

Similar factors for fungal adhesion to plants may affect the ability of a fungus to persist at the site of plant infection (16,23). Hamer et al. (23) have recently described a mechanism by which spores of *Magnaporthe grisea* adhere to the surface of leaves. The genetic control of this mechanism is likely to be complex, however, since it apparently not only involves production of an adhesive mucilage in a specialized conidial structure but also may involve the morphology of the conidium itself.

Other spores are actively as well as passively distributed. Many lower fungi have motile flagellated propagules known as zoospores. These zoospores are important for dispersal especially in aquatic environments or under flooding conditions and may also be important for establishing secondary infections in roots of the same or closely located plants. Selection for more persistent zoospores or zoospores with modified trophic responses may increase the likelihood of long-range dispersal of these organisms.

Improvement in Competitive Ability

Prior to infection, fungal plant pathogens are subjected to the same demands of the environment as other plant-associated microorganisms. Resources available for these microorganisms are limited; simultaneous demand for these resources among microbe populations is competition (11). It is still unclear how a fungal pathogen may be improved in competitive ability. Presumably, it will involve an enhanced ability to utilize existing nutrients or to overcome antagonistic and/or antibiotic barriers.

Enhanced ability to use available nutrients doubtlessly occurs by complex mechanisms. Yet individual components of nutrient acquisition may be isolated and tested for their ability to affect relative competitive ability. For example, a detailed genetic analysis of the high-affinity iron acquisition (siderophore) system of *Ustilago maydis* has recently been undertaken (34). The ability to test the effect of this system on the population dynamics of the pathogen not only will give information concerning the potential selective advantage of such a system but also will test whether, in general, modifications in nutrient acquisition systems will be useful in programs for pathogen improvement.

Antibiotic production by rhizosphere bacteria has recently been determined to be a component of their ability to suppress disease in plants caused by fungi (54). Competitive advantage may thus be accrued by selecting fungal pathogens insensitive to the antibiotic or antagonistic activity of the microbial flora normally associated with the target plant or plants. Screening wild-type or mutagenized strains for their ability to overcome antagonist inhibition in vitro could be a first step in increasing competitive ability. Along this line, a second, related strategy may be employed. It has been noted in several instances where fungal diseases on crop plants actually increase significantly after fungicide treatment (4). This phenomenon has been attributed to the increase in populations of previously insignificant pathogens after the fungicide-induced decrease in the normally antagonistic flora of the plant. Pathogens of weeds growing in fields of crops sprayed with fungicides may be greatly improved in their competitive ability and ability to control weeds if they are first selected or otherwise modified for fungicide resistance.

Improvement In Safety

An overriding concern to workers developing biological control agents should be the safety of the organism they produce and deploy. Biocontrol is meant to supplement or replace traditional methods of weed control with safe and effective alternatives. Public health and environmental concerns should be addressed in research aimed at developing these novel forms of weed control.

Three major concerns for the safety of biocontrol agents are (1) the potential for epidemics in nontarget plant species, (2) the potential of fungal pathogens with novel genotypes to persist and displace native microbial species with adverse consequences, and (3) the potential of novel pathogens to produce undesirable by-products such as mycotoxins. These concerns may also be addressed using genetic technology.

The potential for modification of host range of a genetically modified fungus to nontarget plants must be assessed. Where genetic modifications are small and well characterized, as by the precise addition or deletion of a single gene of known function, the likelihood of unpredicted pathogenic effects is reduced. More uncontrolled modification through chemical mutation, protoplast fusion, and so on, may have greater potential for creating unanticipated changes in pathogenic phenotype. However, given the difficulty of even intentionally modifying host range, unintended modification of this type seems unlikely. Nevertheless, host range studies of genetically modified fungi ought to be undertaken to determine the probability of this occurring. As stated previously, genetic manipulation may be used to reduce host range for nontarget plants just as easily as to increase host range. Deletion of genes for biosynthesis of potentially hazardous fungal metabolites could also be readily accomplished by mutation or gene eviction.

Another concern with genetically modified plant pathogens is their potential to persist and cause unpredictable ecological problems as a result. Presumably, modified pathogens released into the environment in large numbers have the capability of displacing native microflora. Although persistence in the environment and competitive advantage may be beneficial as a basis for aggressiveness, in terms of safety assurance and regulatory policy it may be more important to debilitate a pathogen so it will not survive in the environment for any prolonged period of time. With this in mind it is interesting to note that Garber et al. (22) have described conditional epidemics on maize caused by *Cochliobolus heterostrophus*. Auxotrophic strains of this pathogen are less fit than wild-type ones and do not survive well under field conditions in the presence or absence of wild-type *C. heterostrophus*. However, the frequency of the auxotrophic strain in the pathogen population can be increased by supplementing field plots with dilute histidine, the nutritional requirement of the auxotroph. When the auxotroph alone was introduced to fields, a much greater amount of disease was caused when it was supplemented with histidine than when it was not supplemented. Wild-type strains did not cause more disease when the histidine treatment was added. Therefore, the epidemic was dependent upon input of the amino acid supplement, and the frequency of the mutant form in the pathogen population was controllable.

Other genes also can assure that pathogens will not persist or become a dominant form under field conditions. For example, albino mutants of *C. heterostrophus*, virulent in laboratory tests, do not survive in the field (70), presumably because of sensitivity to UV irradiation. Likewise, a strain of *C. heterostrophus* having a gene *(TOX1)* for T-toxin production is less fit than a near isogeneic strain lacking the gene for toxin-producing ability (29) when allowed to infect toxin-insensitive maize. Therefore, the introduction of selected genes into a fungal pathogen, even ones increasing aggressiveness such as *TOX*1, may ultimately reduce the fitness of a pathogen in a predictable manner. Epidemics caused by debilitated strains would depend upon the initial primary inoculation of the pathogen and not upon rounds of secondary infection arising from the primary infection. In this regard, pathogen strains highly aggressive but deficient in the ability to sporulate may also be beneficial. The initial inoculant (perhaps in the form of fragments of mycelium) could be applied to create sufficient disease in the weed population, but the resulting pathogens would fail to be dispersed because of the inability to sporulate.

Conclusions

What are the prospects for genetic improvement of biocontrol fungi? Writing as recently as 1985, Lindemann (37) suggested that genetic systems of "fungi are so poorly characterized that genetic improvements in fungi which act as biological control agents are not likely to be developed in the near future." This assessment probably holds true today as well, but not for the reason that Lindemann suggests. Great advances have been made in the past 3 years in the ability to manipulate fungal pathogens genetically. Transformation, site-directed integration, and gene replacement manipulations all have been accomplished for fungal pathogens. These methods, along with the ability to transform with autonomously replicating plasmids, allow for the most sophisticated types of genetic manipulations, all of which now are becoming available for use with fungal plant pathogens. The major impediments for development of improved biological control fungi today are the lack of fundamental knowledge concerning fungal pathogenesis and the lack of clear regulatory policy concerning this development.

Fungal pathogenesis is polyphyletic. Since the vast majority of fungi are not plant pathogens and since nearly all major types of fungi have representative species that are plant pathogens, it seems clear that the ability to infect and cause disease on plants evolved independently, probably several different times. Because of these independent origins, mechanisms for pathogenesis are apt to be diverse. Only a few disease interactions have been studied to even a slight degree, and these have been for diseases of crop plants. To exploit our newly developed abilities to manipulate fungal pathogens genetically, we require more knowledge concerning biochemical mechanisms of pathogenesis and the molecular basis for pathogen survival, dispersal, and host defense. Only further research will allow us in a rational way to design pathogens improved in their ability to control weeds.

Implicit within a review of this nature is the idea that if genetic improvement can be done, it ought to be done. In fact, the genetic improvement and environmental release of weed pathogens is not something that can be taken for granted by today's regulatory standards. Many pathogens subjected to the genetic manipulations suggested in this review would not now, or possibly ever, be sanctioned for environmental release, and perhaps rightly so. The decision for environmental release must be made by regulatory bodies and an informed public. The responsibility of the scientific community in these decisions seems clear: to keep the public informed and knowledgeable with our most thoughtful opinions concerning environmental risk and to provide a data base from which risk can be assessed.

Literature Cited

1. Anagnostakis, S. L. 1988. *Cryphonectria parasitica,* cause of chestnut blight. Pages 123–136, *in:* Genetics of Plant Pathogenic Fungi. Advances in Plant Pathology, Vo. 6. Academic Press, San Diego.

2. Beckman, C. 1987. The Nature of Wilt Diseases of Plants. APS Press, St. Paul.

3. Bennett, J. W., and L. Lasure, eds. 1985. Gene Manipulations in Fungi. Academic Press, Orlando, FL

4. Blakeman, J. P., and N. J. Fokkema. 1982. Potential for biological control of plant diseases on the phylloplane. Ann. Rev. Phytopathol. 20:167–192.

5. Blanch, P. A., M. J. C. Asher, and J. H. Burnett. 1981. Inheritance of pathogenicity and cultural characters in *Gaeumannomyces graminis* var. *tritici*. Trans. Br. Mycol. Soc. 77:391–399.

6. Brasier, C. M., and J. N. Gibbs. 1976. Inheritance of pathogenicity and cultural characters in *Ceratocytis ulmi:* hybridization of aggressive and non-aggressive strains. Ann. Appl. Biol 83:31–37.

7. Caten, C. E., C. Person, J. V. Groth, and S. J. Dhahi. 1984. The genetics of pathogenic aggressiveness in three dikaryons of *Ustilago hordei*. Can. J. Bot. 62:1209–1219.

8. Chumley, F., K. Parsons, and B. Valent. 1988. Genes that control host cultivar specificity in the rice blast fungus. J. Cell. Biochem. Suppl. 12c:241.

9. Ciuffetti, L. M., K. M. Weltring, B. G. Turgeon, O. C. Yoder, and H. D. VanEtten. 1988. Transformation of *Nectria haematococca* with a gene for pisatin demethylating activity and the role of pisatin detoxification in virulence. J. Cell. Biochem. Suppl. 12C:278.

10. Cooley, R. N., R. K. Shaw, F. C. H. Franklin, and C. E. Caten. 1988. Transformation of the phytopathogenic fungus *Septoria nodorum* to hygromycin B resistance. Curr. Genet. 13:383–389.

11. Cullen, D., and J. H. Andrews. 1984. Epiphytic microbes as biological control agents. Pages 381–399, *in:* Plant–Microbe Interactions: Molecular Genetic Perspectives, Vol. 1; T. Kosuge and E. W. Nester, eds. Macmillan, New York.

12. Défago, G., H. Kern, and L. Sedlar. 1983. Genetic analysis of tomatine insensitivity, sterol content and pathogenicity for green tomato fruits in mutants of *Fusarium solani*. Physiol. Plant Pathol. 22:39–43.

13. Défago, G., and H. Kern. 1983. Induction of *Fusarium solani* mutants insensitive to tomatine, their pathogenicity and aggressiveness to tomato fruits and pea plants. Physiol. Plant Pathol. 22:29–37.

14. Dickman, M. B., S. S. Patil, and P. E. Kolattukudy. 1982. Purification, characterization and rôle in infection of an extracellular cutinolytic enzyme from *Colletotrichum gloeosporioides* Penz. on *Carica papaya* L. Physiol. Plant Pathol. 20:333–347.

15. Droby, S., D. Prusky, B. Jacoby, and A. Goldman. 1987. Induction of antifungal resorcinols in flesh of unripe mango fruits and its relation to latent infection by *Alternaria alternata*. Physiol. Molec. Plant Pathol. 30:285–292.

16. Epstein, L., L. B. Laccetti, R. C. Staples, and H. C. Hoch. 1987. Cell–substratum adhesive protein involved in surface contact responses of the bean rust fungus. Physiol. Molec. Plant Pathol. 30:373–388.

17. Ettinger, W. F., S. K. Thukral, and P. E. Kolattukudy. 1987. Structure of cutinase gene, cDNA, and the derived amino acid sequence from phytopathogenic fungi. Biochemistry 26:7883–7892.

18. Farman, M. L., and R. P. Oliver. 1988. The transformation of protoplasts of *Leptosphaeria maculans* to hygromycin B resistance. Curr. Genet. 13:327–330.

19. Fuchs, A., F. W. deVries, and M. Platero Sanz. 1980. The mechanism of pisatin degradation by *F. oxysporum* f.sp. *pisi*. Physiol. Plant Pathol. 16:119–133..

20. Fulbright, D. W. 1984. Effect of eliminating dsRNA in hypovirulent *Endothia parasitica*. Phytopathology 74:722–724.

21. Gabriel, D. W., N. Lisker, and A. H. Ellingboe. 1982. The induction and analysis of two classes of mutations affecting pathogenicity in an obligate parasite. Phytopathology 72:1026–1028.

22. Garber, R. C., W. E. Fry, and O. C. Yoder. 1983. Conditional field epidemics on plants: a resource for research in population biology. Ecology 64:1653–1655.

23. Hamer, J. E., R. J. Howard, F. G. Chumley, and B. Valent. 1988. A mechanism for surface attachment in spores of a plant pathogenic fungus. Science 239:288–290.

24. Henson, J. M., N. K. Blake, and A. L. Pilgeram. 1988. Transformation of *Gaeumannomyces graminis*. J. Cell. Biochem Suppl. 12C:287.

25. Keller, N. P., G. C. Bergstrom, and O. C. Yoder. 1988. Stability of foreign DNA in the *Cochliobolus heterostrophus* genome. J. Cell. Biochem. Suppl. 12C:282.

26. Kelly, J. M., and M. J. Hynes. 1985. Transformation of *Aspergillus niger* by the *amd*S gene of *Aspergillus nidulans*. EMBO J. 4:475–479.

27. Kistler, H. C., and U. K. Benny. 1988. Genetic transformation of the fungal wilt pathogen *Fusarium oxysporum*. Curr. Genet. 13:145–149.

28. Kistler, H. C., and H. D. VanEtten (1984) Regulation of pisatin demethylation in *Nectria haematococca* and its influence on pisatin tolerance and virulence. J. Gen. Microbiol. 130:2605–2613.

29. Klittich, C. J. R., and C. R. Bronson. 1986. Reduced fitness associated with TOX1 of *Cochliobolus heterostrophus*. Phytopathology 76:1294–1298.

30. Kolattukudy, P. E. 1985. Enzymatic penetration of the plant cuticle by fungal pathogens. Ann. Rev. Phytopathol. 23:223–250.

31. Kronstad, J., J. Wang, S. Covert, and S. Leong. 1988. Isolation of metabolic and pathogenicity genes from *Ustilago maydis*. J. Cell. Biochem. Suppl. 12C:283.

32. Lawrence G. J. 1988. *Melampsora lini,* rust of flax and linseed. Pages 313–331, *in:* Genetics of Plant Pathogenic Fungi. Advances in Plant Pathology, Vol. 6. Academic Press, San Diego.

33. Leong, S. A. 1988. Recombinant DNA research in phytopathogenic fungi. Pages 1–26, *in:* Genetics of Plant Pathogenic Fungi. Advances in Plant Pathology. Vol. 6. Academic Press, San Diego.

34. Leong, S. A., J. Wang, A. Budde, D. Holden, T. Kinscherf, and T. Smith. 1987. Molecular strategies for the analysis of the interaction of *Ustilago maydis* and maize. Pages 95–106, *in:* Molecular Strategies for Crop Protection. Alan R. Liss, New York.

35. Leung, H., F. S. Borromeo, and M. A. Bernardo. 1988. Genetic control of cultivar specificity on rice in the blast fungus. J. Cell. Biochem. Suppl. 12C:269.

36. Leung, H., and M. Taga. 1988. *Magnaporthe grisea (Pyricularia* species): the blast fungus. Pages 175–188, *in* Genetics of Plant Pathogenic Fungi. Advances in Plant Pathology, Vol. 6. Academic Press, San Diego.

37. Lindemann, J. 1985. Genetic manipulation of microorganisms for biological control. Pages 116–130, *in:* Biological Control on the Phylloplane. C. E. Windels and S. E. Lindow, eds. APS Press, St. Paul.

38. Michelmore, R. W., and S. H. Hulbert. 1987. Molecular markers for genetic analysis of phytopathogenic fungi. Ann. Rev. Phytopathol 25:383–404.

39. Mullin, P. G., B. G. Turgeon, R. C. Garber, and O. C. Yoder. 1988. Integration of transforming DNA by homologous recombination allows gene replacement in *Cochliobolus heterostrophus*. J. Cell. Biochem. Suppl. 12C:285.

40. Oliver, R. P., I. N. Roberts, H. Harling, L. Kenyon, P. J. Punt, M. A. Dingemans, and C. A. M. J. J van den Hondel. 1987. Transformation of *Fulvia fulva*, a fungal pathogen of tomato, to hygromycin B resistance. Curr. Genet. 12:231–233.

41. Panaccione, D. G., M. McKiernan, and R. M. Hanau. 1988. *Colletotrichum graminicola* transformed with homologous and heterologous benomyl-resistance genes retains expected pathogenicity to corn. Mol. Plant–Microbe Interactions 1:113–120.

42. Parsons, K. A., F. G. Chumley, and B. Valent. 1987. Genetic transformation of the fungal pathogen responsible for rice blast disease. Proc. Natl. Acad. Sci. USA 84:4161–4165.

43. Powell, W. A., and N. K. Van Alfen. 1987. Differential accumulation of poly (A)$^+$ RNA between virulent and double-stranded RNA-induced hypovirulent strains of *Cryphonectria (Endothia)* parasitica. Mol. Col. Biol. 7:3688–3693.

44. Prusky, D., N. T. Keen, and I. Eaks. 1983. Further evidence for the involvement of a preformed antifungal compound in the latency of *Colletotrichum Gloeosporioides* on unripe avocado fruits. Physiol. Plant Pathol. 22:189–198.

45. Rambosek, J., and J. Leach. 1987. Recombinant DNA in filamentous fungi: progress and prospects. CRC Crit. Rev. Biotechnol. 6:357–393.

46. Revuelta, J. L., and M. Jayaram. 1986. Transformation of *Phycomyces blakesleeanus* to G–148 resistance by an autonomously replicating plasmid. Proc. Natl. Acad. Sci. USA 83:7344–7347.

47. Rodriguez, R. J., and O. C. Yoder. 1987. Selectable genes for transformation of the fungal plant pathogen *Glomerella cingulata* f.sp. *phaseoli (Collectotrichum lindemuthianum)* Gene 54:73–81.

48. Rowe, R. C. 1980. Comparative pathogenicity and host ranges of *Fusarium oxysporum* isolates causing crown and root rot of greenhouse and field-grown tomatoes in North America and Japan. Phytopathology 70:1143–1148.

49. Salch, Y. P., and M. N. Beremand. 1988. Development of a transformation system for *Fusarium sambucinum*. J. Cell. Biochem. Suppl. 12C:290.

50. Schäfer, W., H. D. VanEtten, and O. C. Yoder. 1988. Transformation of the maize pathogen *Cochliobolus heterestrophus* to a pea pathogen by insertion of a PDA gene of *Nectria haematococca*. J. Cell. Biochem. 12C:291.

51. Scheffer, R. P., and R. S. Livingston. 1984. Host selective toxins and their role in plant diseases. Science 223:17–22.

52. Sidhu, G. S., ed. 1988. Genetics of Plant Pathogenic Fungi. Advances in Plant Pathology, Vol. 6. Academic Press, San Diego.

53. Stipes, R. J., and R. J. Campana, (eds). 1981. Compendium of elm diseases. APS Press, St. Paul.

54. Thomashow, L. S., and D. M. Weller. 1988. Role of a phenazine antibiotic from *Pseudomonas fluorescens* in biological control of *Gaeumannomyces graminis* var. *tritici*. J. Bacteriol. 170:3499–3508.

55. Timberlake, W. E., M. T. Boylan, M. B. Cooley, P. M. Mirabito, E. B. O'Hara, and C. E. Willett. 1985. Rapid identification of mutation–complementing restriction fragments from *Aspergillus nidulans* cosmids. Exp. Mycol. 9:351–355.

56. Tsukuda, T., S. Carleton, S. Fotheringham, and W. K. Holloman. 1988. Isolation and characterization of an autonomously replicating sequence from *Ustilago maydis*. Mol. Cell. Biol. 8:3703–3709.

57. Turgeon, B. G., R. C. Garber, and O. C. Yoder. 1985. Transformation of the fungal maize pathogen *Cochliobolus heterostrophus* using the *Aspergillus nidulans amd*S gene. Mol. Gen. Genet. 201:450–453.

58. Turgeon, B. G., R. C. Garber, and O. C. Yoder. 1987. Development of a fungal transformation system based on selection of sequences with promoter activity. Mol. Cell. Biol. 7:3297–3305.

59. Typas, M. A. 1983. Heterokaryon incompatibility and interspecific hybridization between *Verticillium albo-atrum* and *Verticillium dahliae* following protoplast fusion and microinjection. J. Gen. Microbiol. 129:3043–3056.

60. Valent, B., M. S. Crawford, C. G. Weaver and F. G. Chumley. 1986. Genetic studies of fertility and pathogenicity in *Magnaporthe grisea (Pyricularia oryzae)*. Iowa State J. Res. 60:569–594.

61. VanEtten, H. D., and H. C. Kistler. 1988. *Nectria haematococca* mating populations I and VI. Pages 189–206, *in:* Genetics of Plant Pathogenic Fungi. Advances in Plant Pathology, Vol. 6. Academic Press, San Diego.

62. VanEtten, H. D., D. E. Matthews, and S. F. Mackintosh. 1987. Adaptation of pathogenic fungi to toxic chemical barriers in plants: the pisatin demethylase of *Nectria haematococca* as an example. Pages 59–70, *in:* Molecular Strategies for Crop Protection. Alan R. Liss, New York.

63. Walton, J. D. 1987. Two enzymes involved in biosynthesis of the host-selective phytotoxin HC-toxin. Proc. Natl. Acad. Sci. USA 84:8444–8447.

64. Walton, J. D., and F. R. Holden. 1988. Properties of two enzymes involved in the biosynthesis of the fungal pathogenicity factor HC-toxin. Mol. Plant–Microbe Interactions 1:128–134.

65. Wang, J., D. W. Holden, and S. A. Leong. 1988. Gene transfer system for the phytopathogenic fungus *Ustilago maydis*. Proc. Natl. Acad. Sci. 85:865–869.

66. Ward, M., B. Wilkinson, and G. Turner. 1986. Transformation of *Aspergillus nidulans* with a cloned, oligomycin-resistant ATP synthase subunit 9 gene. Mol. Gen. Genet. 202:265–270.

67. Weltring, K. M., B. G. Turgeon, O. C. Yoder, and H. D. VanEtten. 1988. Cloning a phytoalexin detoxification gene from the plant pathogenic fungus *Nectria haematococca* by expression in *Aspergillus nidulans*. Gene (in press). 68:335–344.

68. Wernars, K., T. Goosen, B. M. J. Wennekes, K. Swart, C. A. M. J. J. vanden Hondel, and H. W. J. vanden Broek. 1987. Cotransformation of *Aspergillus nidulans:* a tool for replacing fungal genes. Mol. Gen. Genet. 209:71–77.

69. Wessel, W. L., K. A. Clare, and W. A. Gibbons. 1988. Purification and characterization of Hc toxin synthetase from *Helminthosporium carbonum*. Biochem. Soc. Trans. 16:401–402.

70. Yoder, O. C. 1988. *Cochliobolus heterostrophus* cause of southern corn leaf blight. Pages 93–112, *in:* Genetics of Plant Pathogenic Fungi. Advances in Plant Pathology, Vol. 6. Academic Press, San Diego.

10

Protoplast Fusion for the Production of Superior Biocontrol Fungi

G. E. Harman and T. E. Stasz

Biological control (biocontrol) of plant pathogens is becoming an important component of weed and plant disease management practices. Biocontrol potentially offers answers to many persistent problems in agriculture, including problems of resource limitations, nonsustainable agricultural systems, and overreliance on pesticides (7). Much emphasis is now being placed on the use of specific biocontrol agents, as opposed to modification of the environment, to improve the level of naturally occurring biocontrol (7). Many fungi and other microorganisms have been shown to control various plant pathogens and weeds; products sold commercially include the mycoherbicides COLLEGO and DeVINE.

Although some successes are evident, methods to prepare superior strains of these beneficial fungi are required. Strains require a number of attributes, including wide adaptation to environmental conditions and superior competitive ability relative to other microflora.

Several general approaches to preparation of superior strains can be envisioned. These include mutation (as induced by irradiation, chemical mutagenesis, or plasmid insertion), sexual or asexual recombination of two parental strains, or transformation. Chemical or radiation mutation is the simplest procedure but is restricted to changes in single genes. Additional unwanted mutations may occur elsewhere in the genome. Insertional mutation is a precise method for deletion or modification of activity of a single gene.

Sexual or asexual recombination allows the recombination of genomes of two strains. This may be accomplished through anastomosis, protoplast fusion, or sexual processes. These techniques can be used to induce variability among progeny or to recombine useful complementary properties from two different strains into single superior biocontrol strains.

Transformation systems are useful for addition of specific genes. When specific genes critical in biocontrol are identified, they can be cloned and transferred to otherwise superior strains even across wide taxonomic barriers. Such techniques allow deletion or addition of specific genes and can be used to determine and quantify the role of specific genes in biocontrol processes.

Therefore, various approaches to production of genetically superior biocontrol strains have their own advantages and disadvantages. Sexual or asexual recombination of entire genomes is the method of choice if one desires to combine useful properties of two strains into single superior strains, particularly if polygenic traits are to be combined or if the genetic bases of the desirable properties are not understood. Many fungi used in biocontrol do not have sexual stages, so asexual hybridization must be utilized. The natural process of anastosmosis is relatively inefficient, but protoplast fusion can be utilized to give large numbers of fused cells (32).

This chapter describes methodology for utilizing protoplast fusion, together with steps for avoiding pitfalls. It also describes postfusion genetic events in selected fungal systems. Finally, it describes successful utilization of protoplast fusion to produce superior strains of *Trichoderma* for biocontrol.

General Methods for Protoplast Fusion

Protoplast Preparation

The first requirement for protoplast fusion is the ability to produce large numbers (approximately 10^8) of protoplasts rapidly and reliably. Specific methods must be empirically refined for each fungal system tested. However, general procedures can be described. The review by Davis (9) provides considerable information on the intricacies of specific systems.

The first consideration is the manner in which biomass for protoplasting is produced. The age and the nature of this material will strongly affect the quantity and quality of protoplasts obtained. Toyama et al. (42) produced protoplasts from immature conidia, but most workers have used actively growing cells (9). In general, young hyphae or cells in the log phase of growth are more suitable than are older ones (9). In our work with *Trichoderma*, we prepare approximately 100 small (1–4 mm^2) squares of agar from a young nonsporulating culture of the strain to be tested. These are incubated in 200 ml of potato dextrose broth plus 1.5% yeast extract. The mixture is incubated 16–18 hours (depending on the strain) until small spheres of hyphae 4–6 mm in diameter are produced (39). Care is taken to exclude spores from the preparation, since conidia are the same size as protoplasts and interfere in subsequent procedures. If older hyphae are used, not only are fewer protoplasts obtained, but they are more highly vacuolated and fragile.

Inclusion of certain compounds that affect cell wall composition may be helpful or required for the routine production of large numbers of protoplasts. Such amendments usually are added to the medium used for production of hyphae to be protoplasted. Reducing agents (e.g., thiol compounds), detergents (e.g., triton X–100), chelating compounds (e.g., EDTA) or materials interfering with cell wall synthesis (e.g., polyoxin antibiotics or 2-deoxyglucose) have been used with various fungi to aid release of protoplasts (9). In our work with *Trichoderma* (Stasz et al., 38), some strains gave very poor protoplast yields when grown in potato dextrose plus yeast extract broth. However, the addition of 100–250 mg/ L 2-deoxyglucose (2-DG) to this medium resulted in high protoplast yields, regardless of the strain used. The concentration of 2-DG required for each strain was optimal over a narrow range, and this optimal concentration had to be determined empirically for each strain.

Once a useful biomass is obtained, protoplasts can be produced. Such production requires a suitable cell-wall-degrading enzyme contained in an osmoticant of appropriate strength to avoid bursting of protoplasts.

Several enzymes have been used with different fungi. Early studies used preparations from various sources that were prepared by the individual investigators; usual sources of enzymes were from snails or from filamentous fungi (9). Recently, several effective commercial enzyme preparations have become available, including β-glucuronidase, chitinase, and driselase (all available from Sigma Chemicals, St. Louis). However, the most widely used enzyme for this purpose now is Novozym 234 (Novo Labs, Wilton, CT). For preparation of *Trichoderma* protoplasts, young thalli prepared as previously described are incubated with shaking in a mixture containing Novozym 234 in 0.7 M NaCl (39). Incubation time in enzyme solutions should be as short as possible; prolonged incubation results in reductions in protoplast viability.

Following protoplast release, residual hyphae must be removed by filtration (we use Miracloth, Calbiochem, La Jolla, CA), and the protoplasts must then be recovered in the enzyme mixture. Protoplasts are removed from the enzyme osmoticant mixture by gentle centrifugation ($\sim 200 \times g$) (the enzyme solution can be reused); the protoplasts are then resuspended in fresh osmotic medium. Choice of osmoticants depends on the fungi being used; Davis (9) lists several salts, sugars, or sugar alcohols that are commonly used. Ones giving optimal results must be determined by experiment with the organism in question. In addition, a buffer (usually near pH 7) is used, and, at least for *Trichoderma*, Ca^2 is required for protoplast stability (39). We routinely use a mixture (STC) containing 0.6 M sorbitol, 10 mM tris–HCl, and 10 mM CaCl$_2$, pH 7.5.

Protoplast Fusion

The procedures for protoplast fusion are well described, with only minor variations present between labs. Polyethylene glycol (PEG) at a final concentration of approximately 30% in the presence of Ca^2 is the usual fusogen (31,39). Protoplasts must remain in contact with PEG for several minutes to accomplish fusion, but must be removed from this solution within 30 minutes to avoid reductions in protoplast viability. Other procedures (e.g., electroporation) are available but are less widely used.

Visualization of Heterofusants

Frequently it is useful to be able to monitor fusion visually. The most useful method of accomplishing this is to label protoplasts of the parental strains with complementary vital fluorescent stains. Thus, prior to fusion, protoplasts of one strain can be labeled with a stain fluorescing one color, whereas those of the other can be labeled with a stain fluorescing a distinctly different color. The two stains should be capable of being excited and fluoresce when irradiated with a common wavelength of light, so that the two colors can be visualized at the same time in a single preparation.

Such staining systems have been used in other organisms for several years. Rhodamine isothiocyanate (red fluorescence) and fluorescein isothiocyanate (green fluorescence) are complementary vital stains useful for nonfungal systems (21). However, with *Trichoderma* protoplasts, we found that these materials stain only a small percentage of protoplasts and therefore are not useful. We found, however, that rhodamine 6G (green fluorescence) and hydroethidine (red fluorescence) give good staining of protoplasts, are nontoxic when used correctly, and can be easily used to monitor heterofusant formation (15).

Selection of Heterofusants

Once protoplast fusion is accomplished, methods must be utilized that permit selection of heterofusants from the mixture of homofusants and nonfused cells.

With other systems, cell sorting (flow cytometry) can be used to select only cells that contain two colors from mixtures arising from complementarily stained cells (21). We were able to do the same with *Trichoderma* protoplasts stained with rhodamine 6G and hydroethidine. However, this technique was not useful in obtaining heterokaryons; physically selected cell mixtures gave rise to the parental types in the absence of continued selection pressure (Stasz, Harman, and Harris, unpublished).

Heterofusants usually are selected by forced nutritional complementation. Parental strains are selected or mutated to be deficient in their ability to synthesize a particular nutrient required for growth (e.g., an amino acid or vitamin). When two strains with dissimilar nutritional requirements are fused and then plated on a medium deficient in the specific nutrients (minimal medium) required for growth of either parent, only heterofusants should grow. It is essential, however, to have appropriate controls (i.e., each parent fused with itself) to determine that growth resulting on the minimal medium indeed is a consequence of protoplast fusion.

Other methods of selection of heterofusants are possible (29). Instead of nutritional complementation, complementary resistance to toxicants can be used. Thus, if one parent is resistant to compound x and the second is resistant to compound y, the progeny might be expected to be resistant to both x and y. However, this approach has difficulties. For example, resistance to the toxicants needs to be a dominant character for expression to be exhibited in the heterokaryon. Second, resistance is rarely complete; most resistant strains grow more slowly in the presence of the toxicant than in its absence. Fused protoplasts themselves initially grow slowly, and the combination of a low level of residual toxicity plus damage to protoplasts during the fusion process may give rise to few or no somatic hybrid colonies even after a successful fusion. This difficulty may be overcome by initially plating the fused protoplasts on a medium deficient in the toxicants and incubating such plates until thalli appear. The toxicants then can be added as any overlay and somatic hybrids can be selected by their ability to grow through this toxic layer.

Complementary mutants affecting colony pigments also may be useful in selection of heterofusants. Thus, if a fungus normally produces a blue or green pigment, mutants may be found that are white or yellow. When such mutants are fused, normal color may be restored (5), presumably because each parent is deficient in a different enzyme in the pigment synthetic pathway. This selection differs from either nutritional or toxin selection procedures in that the thallus is not subjected to constant selection pressure for a specific character but is instead subject only to mechanical selection if and when complementation is discerned.

Two other approaches are possible. First, protoplasts of one parental strain can be killed (e.g., by heat or irradiation) in such a way that the nuclei remain largely intact. These killed cells then can be fused with living protoplasts containing a recessive allele of a selectable marker (e.g., auxotrophy or toxicant susceptibility) and the heterofusants selected by growth on an appropriate medium (29).

Finally, nuclei can be isolated from a wild strain, fused with protoplasts with a selectable marker, and selected as described earlier (11,36). This procedure has the advantage of adding only nuclei. Mitochondria and, if present, plasmids and/ or mycoviruses also contain genetic material that may confuse the genetic analysis of progeny.

Genetic Markers

Genetic markers are required to determine the genetic nature of progeny from protoplast fusion. Markers used to date are nearly invariably those used for selection of heterofusants (e.g., nutritional deficiencies, toxicant resistance, and/ or colony pigmentation). The markers are ones constantly or intermittently being selected for, and they are few in number.

Not only are a larger number of markers necessary to cover a larger portion of the genome, but, as we will show, behavior of genes not under constant selection pressure in progeny may behave differently from those being selected for. Thus, we need to make a distinction at this point between *selected* genes and *nonselected* genes. Molecular biology gives us two general methods for choosing and identifying large numbers of nonselected genes.

Isozyme analysis is the first of these. Genes produce specific products, many of which have enzymic functions that can be detected. When homogenates of any organism are subjected to starch gel electrophoresis, these enzymes are separated according to their electrical charge at the particular pH of the electrophoresis buffer. Activity of a large number of specific enzymes can be detected. By appropriate testing, isozymes critical to primary metabolism can be detected that can be unequivocally interpreted. For *Trichoderma* we tested 63 enzyme assays in four buffer systems. Of these, 16 were found to give very useful, reproducible, and easily interpreted profiles (41). Moreover, there is a great deal of polymorphism in isozyme mobility among strains. We detected 118 distinct alleles in 16 loci in 87 *Trichoderma* and *Gliocladium* strains. Thus, for each strain it was possible to construct a profile of isozyme alleles (40). This permitted strains to be easily identified. Moreover, most pairs of strains differed in electrophoretic mobility of 4–16 isozymes, and the differences provided nonselected gene markers.

Isozyme analysis has one further advantage. Some enzymes are dimeric (i.e., two subunits associate to form an active enzyme). If the subunits produced by the parental strains differ in electrophoretic mobility, then heterokaryotic progeny strains will produce three bands. The first band will be the homodimer from one parental gene, the second will be the homodimer from the second parent, and the third will be a novel band (the heterodimer) of intermediate mobility that is composed of one monomer from each parental gene. Thus, if genes coding for both parental enzymes are produced in close proximity (i.e., the same cell), a characteristic heterodimeric enzyme will be produced (30). This gives a powerful tool to demonstrate heterokaryosis. Such heterodimeric bands have been found in fused *Agaricus* thalli and have proved the existence of a true somatic hybrid (28).

The second powerful method of generating a nearly infinite number of specific genetic markers for genetic analysis is restriction length polymorphism (RFLP) analysis. In this procedure, DNA is isolated from the organism in question and is cut into reproducible-sized pieces using specific restriction enzymes (27). The resulting mixtures of DNA of varying lengths are then electrophoresed to give a banding pattern based on size. After separation they are probed with DNA coding for specific genes, or representing a specific segment of the genome, using Southern analysis (27). Since each restriction enzyme will cleave different genomes at different sites, each fractionation of different genomes will be unique. The pattern of DNA banding with homology to a specific probe may be similar or dissimilar after cutting with a specific restriction enzyme. However, by using different restriction enzymes (i.e., varying the sites at which the DNA will be cut), distinctly different patterns can be obtained for most gene probes (4). Each gene probe will reveal, therefore, the origin of any specific DNA segment in a progeny from protoplast fusion. RFLP analysis can also be utilized to determine whether nonparental progeny differ in their genomic genetic components or whether the differences exist because of extrachromosomal characters.

Distinguishing Between Heterokaryons and Recombinants

Progeny frequently are obtained with characters different from those of the parents. These differences, if controlled by nuclear genes, could result from heterokaryosis (different nuclei) or from recombination within single nuclei. Many fungi have a mechanism to separate such differences. When conidia form, they receive only a single nucleus from the parental thallus. Different conidia receive different nuclei if the thallus is heterokaryotic. Therefore, numbers and types of nuclei within a particular thallus can be sorted into their component mixtures by isolation of single conidia (e.g., 39). If one is unsure whether this property occurs in the fungus of interest, this can be tested fairly simply. Two complementary auxotrophs are prepared from the same parental strain and fused. Use of two auxotrophs of the same strain eliminate the problems of vegetative incompatibility discussed later. Such fusions should give rise to prototrophic heterokaryotic strains. When these sporulate, conidia can be plated on appropriate media and the number of prototrophic and auxotrophic progeny can be determined. Since karyogamy is rare (see later), most of the conidia should give rise to thalli auxotrophic for one or the other required nutrients if each conidium initially receives only a single nucleus. Thus, single conidial analysis quantitatively distinguishes between heterokaryons and various types of recombinant nuclei in appropriate fungi.

Postfusion Genetic Events

The assumption underlying nearly all interpretations of the genetic nature of nonparental progeny resulting from protoplast fusion is that they result from the operation of the parasexual cycle, as defined by Pontecorvo (34). In this process, plasmogamy (cell fusion) occurs to give rise to a heterokaryon. Subsequently (and rarely) two nuclei can fuse (karyogamy) to give rise to a diploid nucleus. Diploid nuclei frequently are unstable, and loss of chromosomes gives rise to various aneuploid and recombinant haploid nuclei. Such events occur in some fungi following protoplast fusion. However, even in these fungi, analysis of the genetic nature of progeny may be complex; as noted earlier, protoplast fusion gives rise to mixtures of all organelles within the cell. Since mitochondria, plasmids, and mycoviruses, if present, all contain transmissible genetic elements, some of the variation that occurs may be due to these extrachromosomal elements. For example, progeny may inherit nuclei from one parent and mitochondria from the other. In yeast, novel mitochondria have been detected that resulted from recombination between two mitochondrial genomes (10). In addition, as we will see, the inheritance of selected markers may be quite different from inheritance of genetic markers not under selection pressure. The data in *Trichoderma* clearly show that the classic parasexual cycle does not operate, even though protoplast fusion results in progeny with a very wide range of variation that has been successfully used to obtain superior biocontrol strains.

Vegetative Incompatibility

The primary barrier to operation of the parasexual cycle is the existence of vegetative incompatibility. The nature and consequences of such incompatibility recently have been summarized by Croft for *Aspergillus* (8). Two types of incompatibility are known; the first is a barrier to anastomosis, and this type of incompatibility may be overcome by protoplast fusion. However, even if protoplast fusion is utilized, postfusion vegetative incompatibility may occur. Incompatibility or a low level of compatibility does not affect the frequency of fusion, but results in low levels of recovery of progeny and in slow growth of progeny (8,38,39).

Generally, the greater the degree of unrelatedness between two strains, the greater the level of incompatibility. For example, when two auxotrophs of the same strain are fused, the resulting colonies are fully compatible and colonies develop rapidly and at high frequencies. Frequently, prototrophic progeny may total as much as 10% of the total viable propagules (8,39). This number, the percentage complementation, is frequently used as a measure of compatibility. When fusions between more distantly related strains are attempted, compatibility is decreased markedly. With *Aspergillus*, strains belonging to the same species or species group were still compatible (8). Within at least some formae speciales of *Fusarium oxysporum*, only those strains belonging to the same formae speciales were compatible (22,23). In *Trichoderma*, no fusions within or between species were found to be of very limited compatibility (39), so patterns of compatibility grouping differ among various genera of imperfect fungi.

Within fungi, where such information has been determined, incompatibility has been found to be heterogenic, and strains with dissimilar alleles at any of several loci are incompatible (1,8,25,33).

Even though specific genes confer incompatibility, this property is not an absolute. Degrees of incompatibility occur. Fortunately, even in relatively incompatible fusions, some progeny will be obtained (37), and these can be used as sources of improved strains for biocontrol.

Systems Where Parasexuality Occurs

Some fungal strains, when fused, unquestionably give rise to progeny in which the classical parasexual cycle operates. The first step in parasexuality is the formation of a heterokaryon. Such heterokaryons usually are selected on the basis of prototrophy when auxotrophic parental strains are fused. The heterokaryotic nature of strains can be proved by recovery of both parental strains from a thallus of the progeny. As noted earlier, in fungi where each conidium receives only a single nucleus from its conidiophore, single-spore analysis may be used to deter-mine both the existence and the frequency of occurrence of each nuclear type. Closely related strains of many fungi, including *Aspergillus, Penicillium, Sac-charomyces*, and *Fusarium*, readily produce heterokaryons (32) following proto-plast fusion. Heterokaryons may be balanced (i.e., each nuclear type is present in equal numbers) or imbalanced (i.e., one nuclear type becomes much more numerous than its opposite type). In most cases, each hyphal cell is multinucleate, so each cell is in a heterokaryotic state.

However, other conditions exist. In *Verticillium dahliae* each hyphal cell contains only one nucleus. Following anastomosis of mutants of the same strain (i.e., dissimilar strains were not studied), heterokaryotic cells are formed, but heterokaryosis is restricted to the anastomosed cells. Anastomosis gives a more or less stable mosaic of homokaryotic and isolated heterokaryotic cells (18,35).

In *Agaricus*, the balance of heterokaryosis was influenced by the ability of particular nuclei to migrate through the thallus (19). Nuclei from one strain were able to proliferate and migrate through the cytoplasm of a second strain, and thus become the prevalent nuclear type. This process gave rise to cybrid progeny strains with the nucleus of one strain and the mitochondria of a second strain.

When heterokaryons are formed, karyogamy (nuclear fusion) is possible. Kary-ogamy is a rare event, occurring at a rate of 3×10^{-4} in a heterokaryon of *Aspergillus nidulans* (6), and possibly less frequently in other fungi (18,43). Thus, formation of stable balanced heterokaryons enhances the likelihood that such an event may occur. To the extent that vegetative incompatibility limits perpetuation of balanced heterokaryons, it also limits the probability of formation of diploid heterologous nuclei. *Verticillium* appears to be a special case in this regard; as noted earlier, heterokaryotic cells only result directly from fusion, and all other cells remain uninucleate. Nonetheless, convincing evidence for the presence for recombinant nuclei has been presented for this genus (35,43); pre-sumably, karyogamy occurs relatively frequently in the few heterokaryotic cells in these fungi.

Evidence for the occurrence of karyogamy leading to the recovery of diploid or aneuploid progeny has taken several forms. First is the recovery of recombinant progeny. Thus, if one parent, for example, has the genotype $aux1^-$, $aux2^-$, $aux3^-$, and res^+ and a second strain contains $aux4^-$ and $aux5^-$ genes (where $aux1$–5 are deficiencies for synthesis of different nutritional factors and res^+ is a resistance to a specific toxicant), several kinds of recombinant progeny are possible (e.g., $aux1^-$, $aux2^-$, and $aux5^-$; prototrophic strains with and without res^+; $aux2^-$, $aux4^-$; etc.). Unequivocal evidence for recombination in progeny requires many gene markers. Recombination following fusion has been demon-strated in related strains of *Aspergillus, Penicillium, Verticillium, Saccharo-myces, Cephalosporium*, and others (32,43). Other evidence for karyogamy, as indicated by the presence of diploids, comes from addition of chemicals inducing haploidization to nonparental progeny. Certain chemicals (e.g., benomyl, pfluorphenylalanine or other treatments [3,17,26]) induce the formation of haploids or aneuploids from diploids. Thus, if diploid nonparental progeny are obtained, treatment of these with haploidization agents may induce a number of novel genotypes. Finally, diploids or aneuploids should be detectable by measurement of the quantity of DNA/nucleus. This approach has been used to suggest the level of ploidy of progeny resulting from parasexuality in the fungi noted earlier. Thus, parasexuality does occur in many fungi and can be an important tool in obtaining novel genotypes for biocontrol.

Fungi in Which Parasexuality Does Not Occur

In our work with *Trichoderma*, we have found a distinctly different pattern of postgenetic events. Fusions of distantly related strains of *Aspergillus* or *Penicillium* (8,24) give at least superficially similar results to the ones we will describe.

We did intra- and interstrain as well as intra- and interspecific fusions (15). When two auxotrophs of the same strain were fused, about 1–10% of the resulting colonies were prototrophic. These prototrophs grew rapidly, were morphologically similar to the wild-type parental strain, and were balanced heterokaryons containing approximately equal numbers of the two types of auxotrophic nuclei, as determined by single-conidium isolation. Karyogamy was not detected in these balanced heterokaryons even after prolonged growth (38,39).

When interstrain fusions were made, results were significantly different. Typically, numbers of prototrophic progeny recovered were much fewer, with only about one prototrophic progeny isolated from about 10^5 viable protoplasts. However, monitoring of protoplast fusions by complementary staining with hydroethidine and rhodamine 6G revealed that at least 10% of the protoplasts were involved in an interstrain fusion event (38). Taxonomic position did not indicate the degree of compatibility; all intra- or interspecific fusions produced similar low numbers of progeny (41).

Initial prototrophic colonies grew very slowly, even on permissive media. When these were analyzed using isozymes, the profile was of one parent or the other. Even though the initial fused cells were by definition heterokaryons, no heterodimeric bands were detected for dimeric enzymes. Evidently, functioning balanced heterokaryons were not produced. As slow-growing colonies were cultured further, they were found to be highly unstable. More rapidly growing sectors appeared. These sectors were of either isozyme pattern, even though only one parental isozyme pattern was detected in the initial colony. Morphologies of strains derived from sectors were extremely variable (Figure 10.1). Results similar to those obtained with sectors could also be obtained when progeny were derived from single conidia from colonies with nonparental properties.

These data showed that nuclei are present in thalli of progeny strains that could not be detected by isozyme analysis. There are two possibilities for this. Either dissimilar nuclei were present, but only one type was expressed (as occurred in

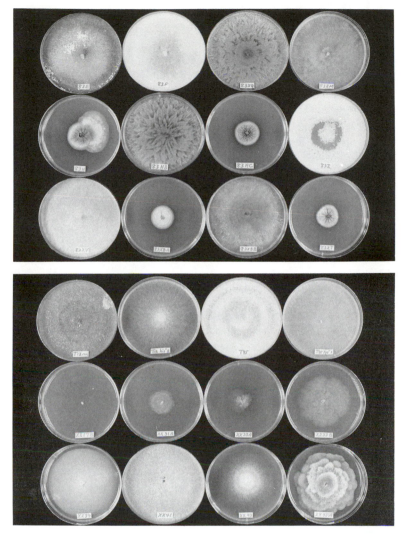

Figure 10.1. Cultures of protoplast fusion progeny grown on potato dextrose agar showing the extreme range of variability induced by this procedure. All strains shown were derived from a single thallus after fusing *Trichoderma harzianum* strains T12 and T95. The original thallus grew slowly and was morphologically similar to strain 83 XT (lower right). This original progeny strain was isozymically identical to the T12 parent and sectored to give rise to the variation shown here. The isozyme pattern of the progeny, proceeding from left to right, of these progeny is T12 (83E), T95 (83F), T12 (83XQ), T12 (83XM), T12 (83H), T12 (83HB), T12 (83HC), T95 (83I), T95 (83XV), T12 (83XBA), T12 (83XBB), and T12 (83XT). Some progeny still are unstable even after cultivation for approximately 2 years; note the sector in strain 83H. The variants shown here are not due to heterokaryosis, as shown by the fact that single spore isolations from those tested give rise to similar isozyme and morphological phenotypes. Similar data have been presented elsewhere (41).

Bacillus subtilus [20]), or else some nuclear types are present much less frequently than others. We examined this question from two approaches. First, we isolated conidia from presumptive heterokaryons to determine relative numbers of specific nuclear types. Second, we isolated DNA and conducted RFLP analysis. Results from both approaches were consistent. Single-spore isolation indicated that nuclei isozymically similar to one parental type may be present in 10,000-fold greater numbers than were nuclei similar to the opposite parental type. Within each parental isozyme class are a wide range of morphotypes (37,39).

From these data, it was possible to determine some of the genetic events following protoplast fusion in *Trichoderma*. When protoplasts were fused, cell contents mixed, forming heterokaryotic cells that contained cytoplasm and organelles of both strains. In fusions between two auxotrophs of the same strain, prototrophic, relatively rapid-growing progeny developed. These were balanced heterokaryons, as determined by analysis of subprogeny arising from single conidia.

In fusions between dissimilar strains, heterokaryotic cells again were formed. However, developing progeny rapidly became heavily imbalanced, and nuclei of one parental strain prevailed, whereas nuclei of the other remained at much lower numbers—differences were as great as 10,000 to 1. However, both nuclear types were present, and even after extensive periods of cultivation, sectors appeared that expressed the phenotype of the opposite parent (37,39). The mechanism whereby one nuclear type prevailed over the other is unknown, as are the events that permitted the spontaneous appearance of sectors that expressed the opposite isozyme phenotype.

Heterokaryosis did not, however, explain much of the variation in progeny following protoplast fusion. For example, all the progeny shown in Figure 10.1 are derived from a single slow-growing prototroph that initially expressed the phenotype of parental strain Tl2. The derivative progeny exhibited various isozyme phenotypes, and all morphotypes were stable through conidiation. Since they all were stable through conidiation, they were homokaryotic. However, they differed markedly in morphology, growth rate, and degree of prototrophy. Sectors frequently expressed a higher degree of prototrophy than did the original progeny. These differences, therefore, cannot be due to heterokaryosis. Isozyme analysis indicates that only one parental nuclear type may be present, the observed variation must be due to other factors (37,39).

These observations and conclusions were inconsistent with the classical model of parasexuality. If heterokaryons are produced in interstrain fusions in which both nuclei are functional, prototrophy should be fully restored and growth, whether rapid or slow, should be equal on permissive and nonpermissive media. This surely would be the case when diploids are produced by karyogamy. Instead we find no evidence for formation of heterokaryons or diploids. Significantly, recombination of nonselected (i.e., isozyme) characters was not detected despite extensive testing. We do, however, obtain a series of progeny in which prototrophy (a character under constant selection pressure) is incrementally but completely restored. In addition, families of progeny are obtained that are homokaryotic (stable through conidiation) and genotypically identical to each other and one or the other parent but that are tremendously different in morphotype (39).

A possible interpretation of the results obtained with *Trichoderma* is that integration of genetic material from one nucleus into another (transformation) occurs following protoplast fusion. We have shown that dicell and multicell fusions occur between dissimilar protoplasts. These fusions result in the formation of multinucleate heterofusants containing 2–25 or more nuclei of each parental type (39). One hypothesis of subsequent events is as follows. As the fusant regenerates a cell wall and hyphal growth commences, or very soon afterward, nuclei of one parental type persist, whereas nuclei of the other parental type degenerate. The prevailing nuclei direct cell functions and undergo division; because we fuse protoplasts of auxotrophic mutants, these nuclei are deficient in a gene for the production of an amino acid. Fragments of DNA from the degenerating nuclei are produced and persist. Fragments coding for genetic sequence conferring a selective advantage are retained. Among auxotrophic strains on minimal media, these are genes for synthesis of the deficient nutrient. Some fragments may become self-replicating. Fragments containing the gene missing in the prevailing nuclei would be strongly selected for. Initially, this gene is expressed, but poorly. This poor expression results in very slow hyphal growth. Fragments with the needed gene may lack effective promotor sequences, or they may be located in regions of the cell where molecular machinery for transcription is lacking. Subsequently, DNA fragments transform nuclei of the prevailing type. Stable integration of a DNA fragment containing the needed gene could allow its full expression, and the transformed nucleus would be prototrophic. The level of prototrophy acquired would depend on the number and sites of integration events with respect to expression of the needed gene.

Preparation of Superior Biocontrol Strains

To the best of our knowledge, only one group of closely related superior biocontrol strains have been prepared using protoplast fusion. However, many other candidate strains with apparently superior properties are being tested by us. These superior strains resulted from fusion of two strains of *Trichoderma harzianum* and are biocontrol agents of soil-borne plant pathogenic fungi. The two parental strains themselves had desirable features; parental strain T12 was able to control *Pythium* and *Alternaria* spp. and possessed good ability to compete with soil bacteria, particularly pseudomonads (13). The second strain, T95, also was able to control *Pythium* spp. and *Rhizoctonia solani* and was also able to colonize plant roots (e.g., was rhizosphere competent) (2,14).

When these two strains were fused, initial slow-growing unstable thalli were obtained. Subsequently, these sectored to give several hundred progeny. Most of these were inferior biocontrol strains, compared with the wild (prototrophic) parents. However, one class of progeny, representing less than 2% of the total, grew more rapidly than the wild parents at either 11° or 25°C. These were found to be superior bioprotectants (16).

One of these, strain 1295-22 (ATCC 20847), has been chosen for further development and testing. It is more effective than either parent at protecting seeds of various vegetable and agronomic seeds and seedlings against a range of pathogenic fungi. At least as important, it is more strongly rhizosphere competent than either parent (16).

This strain arose from protoplast fusion, but the parasexual cycle probably was not involved in its production. In essence, protoplast fusion was used in this case to generate a large number of widely divergent progeny strains. When these were obtained, they were divided into classes based on the nutritional, morphological, isozyme, and growth rate properties. These classes were then screened using bioassay systems, and ones that appeared more promising than the parental strains were selected for further testing. Stability of these strains was obtained by culturing until variation ceased and then determining that single-spore progeny all were similar. Although this procedure has been successful, an understanding of the basic genetic events following protoplast fusion in taxons of interest to biological control specialists will make possible greater progress in the production of superior biocontrol strains.

Literature Cited

1. Adams, G., N. Johnson, J. F. Leslie, and L. P. Hart. 1987. Heterokaryons of *Gibberella zeae* formed following hyphal anastomosis or protoplast fusion. Exp. Mycol. 11:339–353.

2. Ahmad, J. S., and R. Baker. 1987. Competitive saprophytic ability and cellulolytic activity of rhizosphere-competent mutants of *Trichoderma harzianum*. Phytopathology 77:358–362.

3. Anderson, J. B. 1983. Induced somatic segregation in *Armillaria mellea* diploids. Exp. Mycol. 7:141–147.

4. Beckmann, J. S., and M. Soller. 1986. Restriction fragment length polymorphisms in plant genetic improvement. Pages 196–250, *in*: Miflin, B. J., ed., Oxford Surveys of Plant Molecular and Cell Biology, Vol. 3. Oxford University Press, Oxford, U.K.

5. Bojnanska, A., M. Sipiczki, and L. Ferenczy. 1980. Characterization of conidiation mutants in *Trichoderma viride* by hyphal anastomosis and protoplast fusion. Acad. Sci. Hung. 27:305–307.

6. Caten, C. E. 1981. Parasexual processes in fungi. Pages 191–214, *in*: Gull, K., and G. S. Oliver, ed. The Fungal Nucleus. Cambridge University Press, Cambridge, U.K.

7. Cook, R. J., and K. F. Baker. 1983. The Nature and Practice of Biological Control of Plant Pathogens. The American Phytopathological Society, St. Paul, MN.

8. Croft, J. H. 1985. Protoplast fusion and incompatibility in *Aspergillus*. Pages 225–240, *in*: Perberdy, J. F., and L. Ferenczy, eds., Fungal Protoplasts: Applications in Biochemistry and Genetics. Dekker, New York.

9. Davis, B. 1985. Factors influencing protoplast isolution. Pages 45–71, *in*: Peberdy, J. F., and L. Ferenczy, eds., Fungal Protoplasts: Applications in Biochemistry and Genetics. Dekker, New York.

10. Ferenczy, L. 1985. Transfer of cytoplasmic genetic elements by protoplast fusion. Pages 307–321, *in*: Peberdy, J. F., and L. Ferenczy, eds., Fungal Protoplasts: Applications in Biochemistry and Genetics. Dekker, New York.

11. Ferenczy, L., and M. Pesti. 1982. Transfer of isolated nuclei into protoplasts of *Saccharomyces cerevisae*. Curr. Microbiol. 7:157–160.

12. Garber, R. C., and O. C. Yoder. 1983. Isolation of DNA from filamentous fungi and separation into nuclear, mitochondrial, ribosomal, and plasmid components. Anal. Biochem. 135:416–422.

13. Hadar, Y., G. E. Harman, and A. G. Taylor. 1984. Evaluation of *Trichoderma koningii* and *T. harzianum* from New York soils for biological control of seed rot caused by *Pythium* spp. Phytopathology 74:106–110.

14. Harman, G. E., I. Chet, and R. Baker. 1980. *Trichoderma hamatum* effects on seed and seedling disease induced in radish and pea by *Pythium* spp. or Rhizoctonia solani. Phytopathology 70:1167–1172.

15. Harman, G. E., and T. E. Stasz. 1988. Fluorescent vital stains for complementary labelling of protoplasts from *Trichoderma* spp. Stain Technol. 63:241–247.

16. Harman, G. E., A. G. Taylor, and T. E. Stasz. 1989. Combining effective strains of *Trichoderma harzianum* and solid matrix priming to provide improved biological seed treatment systems. Plant Dis. (in press).

17. Hastie, A. C. 1970. Benlate-induced instability of *Aspergillus* diploids. Nature 226:771.

18. Hastie, A. C. 1981. The genetics of conidial fungi. Pages 511–547, *in*: Cole, G. T., and B. Kendrick, eds., Biology of Conidial Fungi, Vol. 2. Academic Press, New York.

19. Hintz, W. E. A., J. B. Anderson, and P. A. Horgen. 1988. Nuclear migration and mitochondrial inheritance in the mushroom *Agaricus bitorgius*. Genetics 119:35–41.

20. Hotchkiss, R. D., and M. H. Gabor. 1980. Biparental products of bacterial protoplast fusion showing unequal parental chromosome expression. Proc. Natl. Acad. Sci. USA 77:3553–3557.

21. Junker, S., and S. Pedersen. 1981. A universally applicable method of isolating somatic cell hybrids by two-color flow sorting. Biochem. Biophys. Res. Commun. 102:977–984.

22. Katan, T., E. Hadar, and J. Katan. 1988. Vegetative compatibility of *Fusarium oxysporum* f.sp. *dianthi* from carnation in Israel. Phytopathology (in press).

23. Katan, T., and J. Katan. 1988. Vegetative compatibility grouping of *Fusarium oxysporum* f.sp. *vasinfectum* from tissue and rhizosphere of cotton plants. Phytopathology 78:852–855.

24. Kevei, F., and J. F. Peberdy. 1985. Interspecies hybridization after protoplast fusion in *Aspergillus*. Pages 241–257, *in*: Peberdy, J. F., and L. Ferenczy, eds., Fungal Protoplasts: Applications in Biochemistry and Genetics. Dekker, New York.

25. Leach, J., and O. C. Yoder. 1983. Heterokaryon incompatibility in the plant-pathogenic fungus, *Cochliobolus heterostrophus*. J. Hered. 74:149–152.

26. Lhoas, P. 1961. Mitotic haploidization by treatment of *Aspergillus niger* diploids with para-fluorophenylalanine. Nature 190:744.

27. Maniatis, T., E. F. Fritsch, and J. Sambrook. 1982. Molecular Cloning: A Laboratory Manual. Cold Spring Harbor Laboratory, Cold Spring Harbor, NY.

28. May, B., and D. J. Royse. 1982. Confirmation of crosses between lines of *Agaricus brunnescens* by isozyme analysis. Exper. Mycol. 6:283–292.

29. Mellon, F. M. 1985. Protoplast fusion and hybridization in *Penicillium*. Pages 69–82, *in*: Timberlake, W. E., ed., Molecular Genetics of Filamentous Fungi. Alan R. Liss, New York.

30. Micales, J. A., M. R. Bonde, and G. L. Peterson. 1986. The use of isozyme analysis in fungal taxonomy and genetics. Mycotaxon 27:405–449.

31. Peberdy, J. F. 1980. Protoplast fusion–a new approach to interspecies genetic manipulation and breeding in fungi. Pages 63–72, *in*: Ferenczy, L., and G. L. Farkas, eds., Advances in Protoplast Research. Pergamon Press, Oxford, U.K.

32. Peberdy, J. F., and L. Ferenczy, eds. 1985. Fungal Protoplasts: Applications in Biochemistry and Genetics. Dekker, New York.

33. Perkins, D. D., and B. C. Turner. 1988. *Neurospora* from natural populations: toward the population biology of a haploid eukaryote. Exp. Mycol. 12:91–131.

34. Pontecorvo, G. 1956. The parasexual cycle in fungi. Ann. Rev. Microbiol. 10:393–400.

35. Puhalla, J. E., and J. E. Mayfield. 1974. The mechanism of heterokaryotic growth in *Verticillium dahliae*. Genetics 76:411–422.

36. Sivan, A., G. E. Harman, and T. E. Stasz. 1988. Fusion of isolated nuclei with protoplasts of *Trichoderma harzianum*. Phytopathology 78:1549.

37. Stasz, T. E., and G. E. Harman. 1989. Nonparental progeny resulting from protoplast fusion of *Trichoderma* in the absence of parasexuality. Exp. Mycol. 14 (in press).

38. Stasz, T. E., G. E. Harman, and M. L. Gullino. 1989. Limited vegetative compatibility following intra- and interspecific protoplast fusion in *Trichoderma*. Exp. Mycol. 13:364–371.

39. Stasz, T. E., G. E. Harman, and N. F. Weeden. 1988. Protoplast preparation and fusion in two biocontrol strains of *Trichoderma harzianum*. Mycologia 88:141–150.

40. Stasz, T. E., K. Nixon, G. E. Harman, N. F. Weeden, and G. A. Kuter. 1989. Evaluation of phenetic species and phylogenetic relationships in the genus *Trichoderma* by eladistic analysis isozyme polymorphism. Mycologia 81:391–403.

41. Stasz, T. E., N. F. Weeden, and G. E. Harman. 1988. Methods of isozyme electrophoresis for *Trichoderma* and *Gliocladium* species. Mycologia 80:870–874.

42. Toyama, H. Y., A. Yamaguchi, A. Shinmyo, and H. Okada. 1984. Protoplast fusion of *Trichoderma reesei* using immature conidia. Appl. Environ. Microbiol. 47:363–368.

43. Typas, M. A. 1983. Heterokaryon incompatibility and interspecific hybridization between *Verticillium albo-atrum* and *Verticillium dahliae* following protoplast fusion and microinjection. J. Gen. Microbiol. 129:3043–3056.

Application Technology

11

Integration of Biological Control Agents with Chemical Pesticides

Roy J. Smith, Jr.

Introduction

Integrated pest management for crops is a concept that combines pest control principles, practices, materials, and strategies to maintain plant health by minimizing damage from pests (38,57,58). Components of integrated pest management systems vary according to the presence of different modifying factors. Strategies include minimum use of chemical pesticides to maintain pests below economic thresholds, use of biological control agents for specific pests, use of resistant crop cultivars, modification of cultural practices to prevent or reduce pest infestations, and use of any inputs to prevent the successful interaction among pests and the crop.

Integrated weed management is a viable component of integrated pest management (57,58,59). The weed management system combines use of multiple-pest-resistant, high-yielding, well-adapted crop cultivars that also resist weed competition and precise placement and timing of fertilizers to give the crop a competitive advantage. Seedbed tillage and seeding methods that enhance crop growth while minimizing weed growth, optimum crop plant population, and use of crop cultivars that form a canopy for shading early season weed growth are viable parts of the system. Such systems also include the use of judicious irrigation practices; timely and appropriate cultivations; carefully planned crop rotations; field sanitation; harvesting methods that do not spread weed seeds; use of biological control agents, such as pathogens, insects, and nematodes; and employment of effective chemical weed control methods. Also, preventive weed control practices to reduce the numbers of weed seeds and other propagules in the soil is a component of the system.

Many of the elements needed for effective integrated weed management systems are currently limited or unavailable (44,45,57,58). Some important needs include plant pathogens and insects to control weeds selectively, highly competi-

Cooperative contribution of the Agricultural Research Service, U.S. Department of Agriculture and University of Arkansas Agricultural Experiment Station, Stuttgart, Arkansas.

tive crop cultivars, cultivars that are highly tolerant of herbicides and their residues, and information on chemical and biological interactions that occur among pesticides and living organisms used in the crop production system.

Although weed control technology integrates preventive, cultural, mechanical, chemical, and biological practices, the use of herbicides is probably the most important component of the weed management system for most major crops produced in the United States (35,44,45,58,59). Because more than 140 herbicides are available presently for control of weeds in cropping systems (3), numerous herbicide programs that combine preplant, preemergence, and postemergence treatments are available for control of important weeds in most crops. In contrast, only two microbial herbicides are registered for use in crop sites: an endemic fungal pathogen, *Colletotrichum gloeosporioides* (Penz.) Sacc. f.sp. *aeschynomene* (hereafter referred to as C.g.a.[1],[2]) for control of northern joint vetch [*Aeschynomene virginica* (L.) B.S.P.] in rice (*Oryza sativa* L.) and soybeans (*Glycine max* L.) in the South (14,15,61) and a soil-borne fungus, *Phytophthora palmivora* (Butler) Butler, for control of strangler vine [*Morrenia ordorata* (H. & A.) Lindl.] in citrus groves in Florida (39,53). For example, in Arkansas in 1988, 10 chemical herbicides and only one microbial herbicide were recommended for weed control in rice (5). Likewise, 19 specific herbicides, alone or combined, were registered for use in soybeans in Arkansas in 1988 whereas only one biological herbicide was registered (5,61,78). Consequently, chemical herbicides are combined into numerous tank mixtures and sequential treatments to control complexes of weeds in rice and soybeans. COLLEGO, however, is limited to only a few combinations because tank mixtures of COLLEGO and some herbicides as well as other pesticides used in rice or soybeans inhibit germination and growth of the pathogen (42,43,61).

The use of plant pathogens as well as other biological agents to control weeds offers a comparatively untapped source of technology for selective weed control in crops (51,58). Although little research has been conducted on the discovery and development of microbial agents for weed control compared with chemical herbicides, additional scientific and financial resources should identify new pathogens with mycoherbicide potential and permit successful use of additional biological weed control agents in integrated pest management systems (44).

The purpose of this chapter is to discuss the research and development of microbial herbicides for weed control in crops and their integration into weed control and pest management programs in crop production systems. Emphasis is on interactions of C.g.a. and other fungal pathogens with microbial herbicide potential and chemical herbicides, fungicides, and insecticides.

Discussion

Research, Development, and Integration of C.g.a.

A disease of northern joint vetch was discovered in 1969 at the University of Arkansas Rice Research and Extension Center at Stuttgart in field plot experiments conducted to determine the interference of northern joint vetch on rice (61). The pathogen, C.g.a., infected and killed weed seedlings in preliminary growth chamber, greenhouse, and field experiments. Host range research indicated that the fungus infested and killed only northern joint vetch in rice and soybeans field environments of the South. Extensive information on the discovery, development, and deployment of C.g.a. as a biological control agent has been published (12,14,15,21,31,59,60,62–64,66–77,80–84).

Small-plot replicated field experiments and aerial applications of C.g.a. to flooded rice fields in nonreplicated field trials developed precise and accurate information on times, rates, volumes, methods of application, water management, and crop production practices (61). Although in early experiments and field trials fresh spore formulations controlled northern joint vetch, C.g.a. dry-formulated products developed by industry also consistently controlled northern joint vetch in rice and soybeans. Pilot tests in commercial rice and soybeans field environments were conducted under experimental use permits granted by the U.S. Environmental Protection Agency (EPA). During this period, research and development continued in close cooperation with scientists and technical people in industry while they developed a suitable dry-formulated product, defined registration requirements with EPA, assessed market potential, and conducted required safety tests. Furthermore, after industry assembled all the efficacy, pathogen biology, and toxicology data on C.g.a. in a coherent document, they developed a marketable two-component product of dry-formulated spores and a fungal spore-rehydrating agent, which is a sugar solution. EPA granted a conditional registration for COLLEGO in June 1982, and following submission of additional data EPA granted the full registration in October 1982 (15). The first commercially registered mycoherbicide for use on an annual weed in annual agronomic crops was applied to farmers' rice and soybeans fields in 1982. During the 7 years (1982–1988) that COLLEGO has been available farmers have treated approximately 8000 hectares annually of rice and soybeans.

Intensive research was conducted on the integration of C.g.a. with chemical herbicides, fungicides, and insecticides, each alone and combined. Many chemical pesticides applied in tank mixtures or sequentially with C.g.a. have inhibited C.g.a. infection and control of northern joint vetch while others applied in specific management programs have had a synergistic effect or no effect on C.g.a. activity. Consequently, interactions between chemical pesticides and plant pathogens with microbial herbicide potential range from enhancement to suppression of disease incidence.

In greenhouse experiments, tank mixture treatments of C.g.a. with the herbicides propanil (Figure 11.1) at 2.2 kg/ha and 2,4,5-T at 0.28 kg/ha, and the fungicides benomyl or fentin hydroxide each at 0.56 kg/ha, inhibited fungal infection of northern joint vetch (42).

Propanil, 2,4,5-T, fentin hydroxide, and benomyl, however, applied within days after C.g.a. treatments, did not inhibit disease infection and development

Figure 11.1. Common and Chemical names and Classification of Pesticides

Common name	Chemical name	Classification*
Acifluorfen	5-[2-chloro-4-(trifluoromethyl)phenoxy]-2-nitrobenzoic acid	H
Atrazine	6-chloro-N-ethyl-N'-(1-methylethyl)-1,3,5-triazine-2,4-diamine	H
Benomyl	methyl 1-(butylcarbamoyl)-2-benzimidazolecarbamate	F
Bentazon	3-(1-methylethyl)-(1H)-2,1,3-benzothiadiazin-4-(3H)-one 2,2-dioxide	H
Bromacil	5-bromo-6-methyl-3-(1-methylpropyl)-2,4(1H,3H)pyrimidinedione	H
Carbofuran	2,3-dihydro-2,2-dimethyl-7-benzofuranyl methylcarbamate	I
Chlorimuron	2-[[[[(4-chloro-6-methoxy-2-pyrimidinyl)amino]carbonyl]amino]sulfonyl]benzoic acid	H
Chlorothalonil	2,4,5,6-tetrachloro-1,3-benzenedicarbonitrile	F
Dicamba	3,6-dichloro-2-methoxybenzoic acid	H
Diclofop	(±)-2-[4-(2,4-dichlorophenoxy)phenoxy]propanoic acid	H
Diquat	6,7-dihydrodipyrido[1,2-α:2',1'-c]pyrazinediium ion	H
Diuron	N'-(3,4-dichlorophenyl)-N,N-dimethylurea	H
Fenoxaprop	(±)-2-[4-[(6-chloro-2-benzoxazolyl)oxy]phenoxy]propanoic acid	H
Fentin hydroxide	triphenyltin hydroxide	F
Fluazifop	(±)-2-[4-[[5-(trifluoromethyl)-2-pyridinyl]oxy]phenoxy]propanoic acid	H
Fomesafen	5-[2-chloro-4-(trifluoromethyl)phenoxy]-N-(methylsulfonyl)-2-nitrobenzamide	H
Glyphosate	N-(phosphonomethyl)glycine	H
Imazaquin	2-[4,5-dihydro-4-methyl-4-(1-methylethyl)-5-oxo-1H-imidazol-2-yl]-3-quinolinecarboxylic acid	H
Imazethapyr	(±)-2-[4,5-dihydro-4-methyl-4-(1-methylethyl)-5-oxo-1H-imidazol-2-yl]-5-ethyl-3-pyridinecarboxylic acid	H
Iprodione	3-(3,5-dichlorophenyl)-N-(1-methylethyl)-2,4-dioxo-1-imidazolinecarboxamide	F
Lactofen	(±)-2-ethoxy-1-methyl-2-oxoethyl 5-[2-chloro-4-(trifluoromethyl)phenoxy]2-nitrobenzoate	H
Malathion	(O,O,-dimethyl phosphorodithioate of diethyl mercaptosuccinate)	I
Mefluidide	N-[2,4-dimethyl-5-[[(trifluoromethyl)sulfonyl]amino]phenyl]acetamide	H
Metalaxyl	methyl N-(2,6-dimethylphenyl)-N-(methoxyacetyl)-DL-alaninate	F
Metribuzin	4-amino-6-(1,1-dimethylethyl)-3-(methylthio)-1,2,4-triazin-5(4H)-one	H
Molinate	S-ethyl hexahydro-1H-azepine-1-carbothioate	H
Oryzalin	4-(dipropylamino)-3,5-dinitrobenzenesulfonamide	H

(continued)

192

Figure 11.1. (continued) Common and Chemical names and Classification of Pesticides

Common name	Chemical name	Classification*
Paraquat	1,1′-dimethyl-4,4′-bipyridinium ion	H
Pencycuron	N-[(4-chlorophenyl)methyl]-N-cyclopentyl-N-phenyl urea	F
Pendimethalin	N-(1-ethylpropyl)-3,4-dimethyl-2,6-dinitrobenzenamine	H
Propanil	N-(3,4-dichlorophenyl)propanamide	H
Propiconazol	2-[[2(2,4-dichlorophenyl)-4-propyl-1,3-dioxalan-2-yl]methyl]-1-H-1,2,4-triazole	F
Sethoxydim	2-[1-(ethoxyimino)butyl]-5-[2-(ethylthio)propyl]-3-hydroxy-2-cyclohexen-1-one	H
Simazine	6-chloro-N,N′-diethyl-1,3,5-triazine-2,4-diamine	H
SN-84364	3′-isopropoxy-2-(trifluoromethyl)benzanilide	F
Thiabendazole	2-(thiazol-4-yl)benzimidazole	F
Thiobencarb	S-[(4-chlorophenyl)methyl]diethylcarbamothioate	H
Trifluralin	2,6-dinitro-N,N-dipropyl-4-(trifluoromethyl)benzenamine	H
2,4-D	(2,4-dichlorophenoxy)acetic acid	H
2,4-DB	4-(2,4-dichlorophenoxy)butanoic acid	H
2,4,5-T	(2,4,5-trichlorophenoxy)acetic acid	H

*F = fungicide; H = herbicide; I = insecticide.

193

in greenhouse experiments (42). The insecticides malathion and carbofuran, each at 0.56 kg/ha, or the herbicides acifluorfen at 0.07–0.56 kg/ha or bentazon at 0.56–1.1 kg/ha applied in tank mixture or sequential treatments with C.g.a. did not inhibit infection or disease development on the weed.

Many replicated small-plot field experiments and nonreplicated large-plot aerial trials have been conducted on the integration of C.g.a. with chemical pesticides. In a small-plot experiment conducted for 3 years from 1981 to 1983, tank mixtures of C.g.a. and acifluorfen controlled northern joint vetch and hemp sesbania [*Sesbania exaltata* (Raf.) Rydb. *ex* A. W. Hill] in rice (61). In replicated small-plot experiments conducted in 1982 to 1984, tank mixtures of C.g.a. and acifluorfen controlled northern joint vetch and hemp sesbania in soybeans (41). Aerial applications in 1981 of a tank mixture of C.g.a. and acifluorfen controlled both weeds in a commercial rice field (42), and a similar treatment also controlled both weeds in a commercial soybean field. Neither crop was injured in the tests.

In replicated small-plot experiments, treatments with propanil or acifluorfen applied 1 week following C.g.a. application did not reduce disease development on northern joint vetch in rice (Figure 11.2). Likewise, two sequential applications of fentin hydroxide beginning 1–2 weeks after an application of C.g.a. did not reduce disease development. Fentin hydroxide, however, applied 1 week before C.g.a., reduced activity of the pathogen on northern joint vetch (61). Unlike the

Figure 11.2.   Interaction of C.g.a. and Various Pesticide Treatments on Northern Joint Vetch in Rice, Colt and Stuttgart, AR, 1981 and 1982*

Pesticide Treatment†	Rate (billion spores or kg/ha)	Control‡ (%)
Untreated check	—	0 d
C.g.a.	190	72 b
Benomyl	0.56	2 d
Fentin hydroxide	0.56	0 d
Propanil	2.2	91 ab
Acifluorfen	0.14	8 d
C.g.a. fb benomyl§	190, 0.56	49 c
C.g.a. fb fentin hydroxide§	190, 0.56	81 b
C.g.a. fb propanil‖	190, 2.2	100 a
C.g.a. fb acifluorfen‖	190, 0.14	84 b

*Adapted from ref. 42.
†C.g.a. treatments applied when northern joint vetch plants emerged through the rice canopy. Northern joint vetch and rice plants ranged from 46 to 63 cm and 46 to 65 cm tall, respectively. fb = followed by.
‡Average control for four experiments based on dry weights of weeds harvested at rice maturity. Values followed by the same letter are not significantly different at the 5% level by Duncan's multiple range test.
§Benomyl and fentin hydroxide were applied twice at indicated rate at each time with the first application 1–2 weeks after C.g.a. treatment and the second application 2 weeks after the first.
‖Propanil and acifluorfen were applied 1 week after C.g.a.

greenhouse results, two sequential applications of benomyl beginning 1–2 weeks after C.g.a. reduced mycoherbicide activity on the weed (see Figure 11.2). Carbofuran at 0.56 kg/ha, when applied as a standard early-season treatment, had no effect on the activity of C.g.a. applied conventionally (42).

In replicated small-plot experiments conducted in 1985 and 1986, propiconazol reduced efficacy of C.g.a. treatments similar to benomyl in rice (Figure 11.3). Two applications of propiconazol applied 1 and 2 weeks after C.g.a. inhibited activity of C.g.a. compared with the mycoherbicide alone. Propiconazol applied 1 week before and 1 week after C.g.a. treatment did not inhibit activity of C.g.a. on northern joint vetch. Likewise, the fungicides pencycuron applied before or after C.g.a. or SN–84364 applied twice after C.g.a. did not lower the activity of C.g.a. on northern joint vetch in rice. Also, when C.g.a. was tank-mixed with acifluorfen and followed with sequential applications of propiconazol or pency-curon, 73–80% of the northern joint vetch plants were controlled in rice. Propico-nazol, however, applied after a C.g.a. plus acifluorfen treatment reduced weed control. The fungicide iprodione applied before or after C.g.a. treatments or after C.g.a. plus acifluorfen did not significantly reduce activity of C.g.a. on northern joint vetch in rice (Figure 11.4).

In a replicated small-plot field experiment conducted in 1982, applications of

Figure 11.3.   Interaction of C.g.a. and Various Pesticide Treatments on Northern Joint Vetch in Rice, Stuttgart, AR, 1985 and 1986*

Pesticide treatment[†]	Rate[§] (billion spores or kg/ha)	Control[‡] (%)
Untreated check	—	0 c
C.g.a.	190	91 a
C.g.a. fb benomyl	190, 0.56	67 b
C.g.a. fb propiconazol	190, 0.5	67 b
Propiconazol fb C.g.a. fb propiconazol‖	0.5, 190, 0.5	81 ab
C.g.a. fb pencycuron	190, 0.56	87 ab
Pencycuron fb C.g.a. fb pencycuron‖	0.56, 190, 0.56	86 ab
C.g.a. fb SN-84364	190, 0.4	89 a
C.g.a. + acifluorfen fb propiconazol	190 + 0.22, 0.5	73 ab
C.g.a. + acifluorfen fb pencycuron	190 + 0.22, 0.56	80 ab

*Adapted from ref. 40.

†C.g.a. treatments applied when northern joint vetch emerged through the rice canopy. Northern joint vetch and rice plants ranged from 55 to 70 cm and 50 to 65 cm tall, respectively. fb = followed by.

‡Average control for two experiments based on visual ratings. Values followed by the same letter are not significantly different at the 5% level by Duncan's multiple-range test.

§Benomyl, propiconazol, pencycuron, and SN-84364 were applied twice as indicated in column with the first application 1 week after C.g.a. or C.g.a. + acifluorfen treatment and the second application 1 week after the first.

‖Propiconazol and pencycuron were applied twice at indicated rate at each time with the first application 1 week before and the second application 1 week after C.g.a. treatment.

Figure 11.4.   Interaction of C.g.a. and Various Pesticide Treatments on Northern Joint Vetch in Rice, Stuttgart, AR, 1986 and 1987*

Pesticide Treatment	Rate§	Control‡ 1986	1987
	(billion spores of kg/ha)	(%)	
Untreated check	—	0 c	0 c
C.g.a.	190	93 a	85 a
C.g.a. fb benomyl	190, 0.56	57 b	42 b
C.g.a. fb iprodione	190, 0.56	92 a	71 a
Iprodione fb C.g.a. fb iprodione‖	0.56, 190, 0.56	—	74 a
C.g.a. + acifluorfen fb iprodione§	190 + 0.11, 0.56	—	89 a

*R. J. Smith, Jr., and L. Q. Yu, unpublished.

†C.g.a. treatments were applied when nothern joint vetch emerged through the rice canopy. Northern joint vetch and rice plants ranged from 50 to 60 cm and 70 to 75 cm tall, respectively. fb = followed by.

‡Average control of three replications based on visual ratings. Values followed by the same letter are not significantly different at the 5% level by Duncan's multiple-range test.

§Benomyl and iprodione were applied twice at indicated rate at each time with the first application 1 week after C.g.a. or C.g.a. + acifluorfen treatments and the second application 1 week after the first.

‖Iprodione was applied twice at the indicated rate at each time with the first application 1 week before and the second 1 week after C.g.a. treatment.

C.g.a. at 190 billion spore/ha alone or as a tank mixture with bentazon at 1.1 kg/ ha controlled 93% of the northern joint vetch plants compared to only 60% for bentazon alone (42). In the same experiment a three-way tank mixture of C.g.a. plus bentazon plus acifluorfen (190 billion spores/ha + 0.56 + 0.28 kg/ha) controlled 87% of the weed plants. In 1986 a tank mixture of C.g.a. plus bentazon (190 billion spores/ha + 1.1 kg/ha) followed by the fungicide propiconazol at 0.5 kg/ha applied sequentially 1 and 2 weeks after the weed treatment controlled 80% of the northern joint vetch (R. J. Smith, Jr., unpublished).

In pilot-test rice plots and commercially treated rice fields, C.g.a. has been integrated successfully with chemical herbicides. Early-season treatments of registered rice herbicides such as propanil, molinate, thiobencarb, pendimethalin, fenoxaprop, or combinations have not inhibited the efficacy of C.g.a. applied at midseason for control of northern joint vetch in rice (5).

The previously cited research indicated that C.g.a. can be integrated with other pesticides in effective pest management programs. Herbicides used for controlling weeds in rice, including acifluorfen, bentazon, and propanil, as well as others, were integrated into the weed management programs effectively with C.g.a. Indeed, herbicides, including acifluorfen and bentazon, were applied in tank mixtures with C.g.a. to broaden the spectrum of weed species controlled in rice and soybeans. Insecticides, including carbofuran for early season control of rice water weevil (*Lissorhoptrus oryzophilus* Kuschel) or malathion for control of fall armyworm (*Spodoptera frugiperda*, J. E. Smith) and rice stink bug (*Oebalus pugnax* Fabricius) (6) can be integrated with C.g.a. without affecting the activity of the mycoherbicide.

Several registered fungicides are used to control rice and soybean foliar diseases. Benomyl and thiabendazole are registered to control rice blast (*Pyricularia oryzae* Cav.) and several foliar diseases of soybean (7,8). The registered fungicides benomyl, thiabendazole, and propiconazol and the experimental fungicides iprodione, pencycuron, and SN–84364 control rice sheath blight incited by *Rhizoctonia oryzae* Ryker and Gooch. Benomyl and thiabendazole may be applied 3 weeks after C.g.a. treatments for successful control of northern joint vetch and the rice diseases blast and sheath blight. Propiconazol may be applied 2 weeks after C.g.a. treatments with successful control of northern joint vetch and sheath blight. (This information comes from personal communication with B. A. Huey of University of Arkansas Cooperative Extension Service on April 15, 1988.) The experimental fungicides iprodione, pencycuron, and SN–84364 applied as early as 1 week after C.g.a. treatment have successfully controlled northern joint vetch and rice sheath blight in replicated experiments and nonreplicated field trials.

Benomyl or thiabendazole applications in soybeans disease control programs would probably have to follow a C.g.a. treatment by at least 3 weeks to prevent interference with mycoherbicide activity on northern joint vetch.

Adverse interactions of C.g.a. and fungicides on rice have been prevented by using a computerized system based on accumulated heat units (degree-day at a base temperature of 10°C) to time applications of C.g.a. and fungicides (37). After the farmer supplies information on the rice cultivar and the data of rice emergence, the program returns information to the farmer on timing sequences for applying C.g.a. with benomyl, thiabendazole, and propiconazol or other pesticides. A similar program is not available for soybeans.

New strains of C.g.a., tolerant to benomyl, are currently under investigation to evaluate their virulence, host specificity, and stability. TeBeest (65) induced benomyl tolerance in C.g.a. by treating spores with ethylmethanesulfonate and selecting for in vitro benomyl tolerance. Four mutants, selected for laboratory testing, appear to be genetically stable, host specific and potentially useful as biological control agents for northern joint vetch. The four strains were repeatedly cultured and passed through the host without loss of benomyl tolerance. Two benomyl-tolerant strains, field tested in replicated small-plot experiments from 1983 to 1987, controlled northern joint vetch in rice when benomyl was applied before or after treating with the C.g.a. strains (79). Charudattan (27) reviewed the potential of genetically altered strains of pathogen for biological control of weeds.

Research, Development and Integration of *Phytophthora palmivora*

A disease of stranglervine was discovered in 1972 in Orange County, Florida (53). A fungus, *P. palmivora*, was isolated and identified as the causal agent in

1973 (39). Laboratory, greenhouse, and field research indicated that the pathogen consistently controlled stranglervine with adequate safety to citrus and nontarget plants in the environment where the fungus would be applied as a mycoherbicide (53). Some plants, susceptible to *P. palmivora*, are not grown in citrus groves. Extensive research on susceptibility of citrus species and cultivars showed that *P. palmivora* was not a threat to citrus trees in natural environments in Florida. *P. palmivora* was tested in 1978–1980 as a successful mycoherbicide for control of stranglervine. DeVine, a commercial formulation of *P. palmivora*, was registered in 1981 as a mycoherbicide for stranglervine (67). It proved to be an effective mycoherbicide because it controlled initial weed infestations and provided residual control for several years after only one application (39).

P. *palmivora* was inhibited by several chemical pesticides. Germination of *P. palmivora* chlamydospores was inhibited when the pathogen was applied in tank mixtures with bromacil, diuron, glyphosate, paraquat, and simazine (27). However, herbicides applied sequentially with *P. palmivora* did not inhibit its activity. Glyphosate and *P. palmivora* were both active on susceptible weeds when the chemical herbicide was applied 3 weeks before the mycoherbicide. The fungicide metalaxyl inhibited the activity of *P. palmivora* when applied after the mycoherbicide (53).

Research, Development, and Integration of Potential Mycoherbicides

Numerous other fungi have been identified recently as inciting diseases on many important weed species (Figure 11.5). Templeton (71) and Charudattan (26) list microbial weed control agents with corresponding target weeds. After research is completed to assess efficacy and commercial potential, additional research will be required to integrate them into weed and pest management programs with chemical pesticides. Many potential mycoherbicides have been studied to determine their interactions with chemical pesticides for integration into pest management programs with chemical pesticides.

Alternaria cassiae on Sicklepod

The fungus *Alternaria cassiae* Jurair and Khan, identified as a potential mycoherbicide in 1981 (86), has been researched and developed for integration into weed and pest management programs for control of sicklepod (*Cassia obtusifolia* L.) (4,87,88). *A. cassiae* is currently under development as a mycoherbicide for control of sicklepod and coffee senna (*Cassia occidentales* L.) in soybeans and peanuts (*Arachis hypogaea* L.) (9). *A. cassiae* applied as single and sequential treatments consistently controlled sicklepod in soybeans in replicated small-plot experiments (87).

Industry has developed an *A. cassiae* product for field testing under an experimental use permit (4). These trials revealed that *A. cassiae* controlled sicklepod

Figure 11.5. Weeds Susceptible to Selected Fungal Pathogens Recently Identified as Potential Mycoherbicides

Weed	Pathogen	Location	Reference
Anoda, spurred *Anoda cristata* (L.) Schlect.	*Puccinia heterospora* B. & C.	South Carolina	(54)
Beggarweed, Florida *Desmodium tortuosum* (Sw.) DC.	*Colletotrichum truncatum* (Schw.) Andrus & Moore	Georgia	(25)
Bindweed, field *Convolvulus arvensis* L.	*Phomopsis convolvulus* Ormeno	Canada	(47)
Cocklebur, common *Xanthium strumarium* L.	*Alternaria helianthi* (Hansq.) Tubaki & Nishihara	Mississippi	(11)
Crotalaria, showy *Crotalaria spectabilis* Roth	*Colletotrichum dematium* f.sp. *crotalariae; Fusarium udum* f.sp. *crotalariae*	Florida	(29)
Dodder, swamp *Cuscuta gronovii* Willd. ex R. & S.	*Alternaria* sp. *Fusaruim* sp.	Wisconsin	(13)
Dock, broadleaf *Rumex obtusifolius* L.	*Uromyces rumices* (Schum.) Wint.	Switzerland	(56)
Goosegrass *Eleusine indica* (L.) Gaertn.	*Bipolaris setariae* (Saw.) Shoem. *Pircularia grisea* (Cke.) Sacc.	South Carolina	(33)
Jimsonweed *Datura stramonium* L.	*Alternaria crassa* (Sacc.) Rands	Mississippi	(16, 22, 85)
Knapweed, diffuse, spotted, Russian *Centaurea difussa* Lam. *C. maculosa* Lam. *C. repens* L.	*Puccinia* spp. (several species)	Canada	(89)
Nightshade, eastern black *Solanum ptycanthum* Dun.	*Colletotrichum coccodes* (Walr.) Hughes	Minnesota	(2)
Nut sedge, purple *Cyperus rotundus* L.	*Balansia cyperi* Edg.	Louisiana	(30)
Ragwort, tansy *Senecia jacobaea* L.	*Puccinia expansa* Link	Switzerland	(1)
Sesbania, hemp *Sesbania exaltata* (Raf.)	*Alternaria crassa* (Sacc.) Rands	Mississippi	(17)
Rydb. ex A. W. Hill	*Colletotrichum truncatum*	Mississippi	(18)
Sicklepod *Cassia obtusifolia* L.	*Pseudocercospora nigricans* (Cooke) Deighton	Florida	(36)
Signal grass, broadleaf *Brachiaria platyphylla* (Griseb.) Nash	*Bipolaris setariae*	North Carolina	(52)
Spurge, cypress *Euphorbia cyparissias* L.	*Uromyces scutellatus* (Pers.) Lev.	Switzerland	(32)

in soybeans, peanuts, and cotton (*Gossypium hirsutum* L.) when applied at precise rates and plant growth stages. In favorable environments the formulated product consistently controlled sicklepod without infecting crops and most nontarget plants. Only sicklepod, coffee senna, and showy crotalaria (*Crotalaria spectabilis* Roth) were susceptible to the mycoherbicide.

Tank mixture treatments of formulated *A. cassiae* with the herbicides acifluorfen, chlorimuron, and imazaquin controlled sicklepod better than either *A. cassiae* alone or chemical herbicides alone in field trials (4). Herbicides with synergistic activity in greenhouse experiments when tank-mixed with *A. cassiae* included bentazon, diclofop, fluazifop, mefluidide, metribuzin, oryzalin, and sethoxydim. The formulated product of *A. cassiae* tank-mixed with 2,4-DB or lactofen controlled sicklepod and coffee senna in peanuts in favorable field environments (9). In dry environments a three-way tank mixture of *A. cassiae*, 2,4-DB and gibberellic acid controlled sicklepod season long in peanuts.

In previous research, activity of the formulated product of *A. cassiae* was reduced on sicklepod by adding several surfactants but was unaffected by nonphytotoxic crop oils (10). Fungicides may inhibit activity of *A. cassiae*, and the experimental use permit states that fungicides not be applied for 3 weeks following *A. cassiae* treatments (4).

Cercospora rodmanii on Water Hyacinth

The fungal pathogen *Cercospora rodmanii* Conway, discovered in 1973 in Florida, induced leaf spots, leaf necrosis, and root rot on water hyacinth [*Eichhornia crassipes* (Mart.) Solms] (28). After greenhouse and field testing, industry tested a formulated product under an experimental use permit to evaluate *C. rodmanii* as a microbial herbicide. Results indicated that this fungus may be a good candidate for registration as a mycoherbicide. The toxicity of 2,4-D and diquat to *C. rodmanii* was rate dependent; however, reduced rates of both herbicides when tank-mixed with the fungus did not reduce growth of the fungus (28). 2,4-D at a reduced rate of only 6.4% of a standard rate applied 3 weeks after *C. rodmanii* controlled water hyacinth better than other treatments, but even this treatment did not control the weed sufficiently for commercialization of the fungal product. Charudattan (28) concluded that additional research is required to integrate *C. rodmanii* and chemical herbicides for effective control of water hyacinth in natural environments.

Fusarium solani f.sp. cucurbitae on Texas Gourd

An indigenous soil-borne fungus, *Fusarium solani* (Mart.) App. & Wr. f.sp. *cucurbitae* (FSC) was isolated from infected seeds and seedlings of Texas gourd [*Cucurbita texana* (Scheele) Gray] and evaluated as a mycoherbicide (20). In controlled-environment and field experiments, microconidial suspensions of FSC consistently controlled Texas gourd. Small-plot experiments indicated that FSC controlled Texas gourd in soybeans without injury to the crop or nontarget plants in southwest Arkansas. Tank mixture treatments of FSC conidia and trifluralin applied preplant incorporated controlled Texas gourd in soybeans (91). This combination reduced seedling emergence more than either FSC or trifluralin alone. Yu and Templeton (94) reported that trifluralin at field rates was not toxic to FSC and that tank mixture or sequential applications of FSC and trifluralin caused earlier disease incidence and higher seedling mortality than that achieved with FSC alone. The compatibility of FSC with trifluralin suggests that FSC can be integrated into existing weed management strategies to broaden the spectrum of weeds controlled in soybean fields infested with Texas gourd and other weeds (91).

Puccinia canaliculata on Yellow Nut Sedge

The rust fungus *Puccinia canaliculata* (Schw.) Lagerh. has potential for controlling yellow nut sedge (*Cyperus esculentus* L.) (23,48). Release of the pathogen early in the spring on seedling yellow nut sedge reduced plant populations by 46% and tuber formation by 66% and completely inhibited flowering (48). Disease development of a *P. canaliculata* strain from Salisbury, Maryland, on yellow nut sedge was reduced by treatments of bentazon at 0.3 or 0.6 kg/ha applied postemergence sequentially after spores (23). Percent disease necrosis on yellow nut sedge foliage was less than half in plots treated with bentazon and *P. canaliculata* compared with plots treated with the rust fungus only. Chlorothalonil reduced disease development of *P. canaliculata* on yellow nut sedge when the fungicide was applied after the fungus was established (49).

Cochliobolus lunatus on Barnyardgrass

The fungus *Cochliobolus lunatus* Nelson and Haasis incited leaf necrosis on barnyardgrass (*Echinochloa crus-galli* [L.] Beauv.) in the Netherlands and caused death of seedling plants in the one- to two-leaf stages of growth (55). Larger barnyardgrass plants, 22 and 30 days old, were killed by combined treatments of *C. lunatus* and sublethal rates of atrazine, compared with only 60% leaf necrosis on plants treated with either *C. lunatus* or atrazine. Plants that were 47 days old exhibited 75% necrosis from a combined treatment of *C. lunatus* and atrazine compared to 15% and 3%, respectively, for treatments of either atrazine or *C. lunatus*. In a second experiment, combined treatments of *C. lunatus* and atrazine killed all barnyardgrass seedlings with only slight (6%) necrosis to corn (*Zea mays* L.). Either *C. lunatus* or atrazine alone caused 20–23% necrosis on barnyardgrass and 3–5% on corn. In host range experiments, tomato (*Lycopersicon esculentum* Mill.) and bean (*Phaseolus vulgaris* L.) were immune to *C. lunatus*, whereas rye (*Secale cereale* L.), wheat (*Triticum aestivum* L.), oat (*Avena sativa* L.), and barley (*Hordeum vulgare* L.) exhibited slight necrosis of leaf tips. Scheepens (55) concluded that combined treatments of *C. lunatus* and atrazine have potential for selective control of barnyardgrass in corn.

Other Fungal Pathogen Interactions with Chemical Herbicides

Research suggests that the activity of several potential mycoherbicides is influenced by chemical pesticides, especially herbicides.

Acifluorfen enhanced the activity of *Fusarium lateritium* (Nees) emend. Snyder and Hansen on prickly sida (*Sida spinosa* L.) in Mississippi (50). In growth chamber experiments, a tank mixture of *F. lateritium* plus acifluorfen controlled 100% of prickly sida plants compared with 8% and 68% for only *F. lateritium* or acifluorfen, respectively. Quimby (50) concluded that a tank mixture of the mycoherbicide and acifluorfen had synergistic activity on prickly sida.

In growth chamber, greenhouse, and field experiments in Canada, acifluorfen, bentazon, and chlorimuron enhanced the activity of *Colletotrichum coccodes* (Wallr.) Hughes, a potential mycoherbicide on velvetleaf (*Abutilon theophrasti* Medic.) in soybeans (34,90,92). Tank mixtures of these herbicides and *C. coccodes* controlled velvetleaf better than either chemical or biological herbicides only. In the laboratory, bentazon did not affect germination of spores or fungal growth of *C. coccodes* (34). In greenhouse experiments dicamba, 2,4-D, and imazethapyr interacted synergistically with *C. coccodes* on velvetleaf when the chemical and biological herbicides were applied in tank mixtures (93). Conversely, fomesafen inhibited activity of *C. coccodes* on velvetleaf. Synergistic activity of herbicides such as bentazon, dicamba, 2,4-D, imazathapyr, and others, and *C. coccodes* as a mycoherbicide is essential for use in control programs for velvetleaf in soybeans and corn.

In greenhouse and field experiments in Mississippi, 2,4-DB and *Fusarium lateritium* applied in tank mixtures or 2,4-DB applied after the fungus inhibited infection and reduced control of velvetleaf (19). However, 2,4-DB applied before *F. lateritium* infected and controlled velvetleaf. It was concluded that timely applications of 2,4-DB and *F. lateritium* controlled velvetleaf better than either the chemical or biological herbicide alone.

Acifluorfen and imazaquin applied at one third to one half of standard rates enhanced activity of *Colletotrichum truncatum* (Schw.) Andrus and Moore on Florida beggarweed (*Desmodium tortuosum* [Sw.] DC) in experiments in Georgia (24).

Colletotrichum gloeosporioides Penz. Sacc. f.sp *malvae* (C.g.m.) is being developed for control of round-leaved mallow (*Malva pusilla* Sm.) in wheat and other crops in Canada (46). A formulated spore product, produced by the Philom Bios Co., was field-tested from 1986 to 1988. Restricted and full registrations are expected for this mycoherbicide in Canada in 1989 and 1990, respectively. Although research has indicated significant interactions of C.g.m. and chemical pesticides, the data have not been published. Integration of C.g.m. with chemical pesticides, however, will be essential for control of round-leaved mallow in cropping systems. (This information comes from personal communication with

K. Mortensen of the Canadian Agricultural Research Station on October 18, 1988.)

Conclusions

Integration of microbial herbicides and chemical pesticides is essential to judicious use of plant pathogens in weed and pest management programs in crop production systems. Because microbial herbicides control a comparatively narrow spectrum of weed species, chemical herbicides are generally required to control the complex of weed species that infest crops. Also, few microbial herbicides are available compared with the many chemical herbicides available for weed control in major crops. Therefore, mycoherbicides must be integrated with chemical herbicides for effective weed management. In addition, diseases and insects that infest crops must be controlled with timely applications of fungicides and insecticides. Consequently, microbial herbicides must be integrated with the numerous chemical fungicides and insecticides used for pest management.

Research and development of registered and experimental mycoherbicides indicate that mycoherbicides can be integrated successfully with chemical pesticides into effective pest management programs. For example, C.g.a. has been integrated successfully with chemical pesticides for control of northern joint vetch as well as other weed species, diseases, and insects with timely, efficacious treatments of chemical herbicides, fungicides, and insecticides. Experiments with other microbial herbicides also indicate that they are compatible with chemical pesticides for managing pests in profitable crop production systems while ensuring a quality environment. Indeed, chemical herbicides combined with microbial herbicides in tank mixture or timely sequential applications frequently increase the activity of each type of herbicide on target weeds.

As new, improved chemical pesticides are developed for control of weeds, diseases, and insects, continued research will be required to determine the effect they have on microbial herbicides and integration into pest management programs.

Development of new pathogen strains resistant to improved pesticides offers opportunity to reduce the adverse impact pesticides have on microbial herbicides. Also, research is required to develop genetically altered pathogen strains that have increased pathogenicity on target weeds and are compatible with chemical pesticides used in pest management programs.

Integration of microbial herbicides as viable components of weed management programs will be a challenge to researchers and organizations concerned with pest management sciences. Costs, benefits, and risks of all components of integrated weed and pest management programs must be examined carefully. Weed pathogens offer opportunities for development of improved weed control practices that will be compatible with all components of integrated pest management systems.

Literature Cited

1. Alber, G., G. Defago, H. Kern, and L. Sedlar. 1986. Host range of *Puccinia expansa* Link (= *P. glomerata* Grev.), a possible fungal biocontrol agent against *Senecio* weeds. Weed Res. 26:69–74.

2. Anderson, R. N., and H. L. Walker. 1985. *Colletotrichum coccodes*: a pathogen of eastern black nightshade (*Solanum ptycanthum*). Weed Sci. 33:902–905.

3. Anonymous. 1983. Herbicide Handbook, 5th ed. Weed Science Society of America, Champaign, IL.

4. Anonymous. 1986. *Alternaria cassia* mycoherbicide for control of sicklepod. Mycogen Corporation, San Diego, CA.

5. Anonymous. 1988a. Recommended chemicals for weed and brush control. Misc. Publ. No. 44. University Arkansas Cooperative Ext. Serv., Little Rock, AR.

6. Anonymous. 1988b. Insecticide recommendations for Arkansas. Misc. Publ. No. 144. University Arkansas Cooperative Ext. Serv., Little Rock, AR.

7. Anonymous. 1988c. Abstract of pesticides recommended for plant disease control—Arkansas. Misc. Publ. No. 154. University Arkansas Coop. Ext. Serv., Little Rock, AR.

8. Anonymous. 1988d. Rice diseases and their control. Ext. Leaflet No. 198. University Arkansas Coop. Ext. Serv., Little Rock, AR.

9. Bannon, J. S., R. A. Hudson, L. Stowell, and J. Glatzhofer. 1988. Combinations of herbicides/plant growth regulators with CASST™. Proc. South. Weed Sci. Soc. 41:268.

10. Bannon, J. S., and H. L. Walker. 1987. Influence of non-ionic surfactants and non-phytotoxic crop oils on control of sicklepod by *Alternaria cassiae*. Proc. South. Weed Sci. Soc. 40:288.

11. Bassi, A., and P. C. Quimby, Jr. 1985. Infection of cocklebur by *Alternarai helianthi*. Proc. South. Weed Sci. Soc. 38:373.

12. Beasley, J. N., L. T. Patterson, G. E. Templeton, and R. J. Smith, Jr. 1975. Responses of animals to a fungus used as a biological herbicide. Arkansas Farm Res. 24(6):16.

13. Bewick, T. A., L. K. Binning, and W. R. Stevenson. 1986. Abstr. Weed Sci. Soc. Am. 26:55.

14. Bowers, R. C. 1982. Commercialization of microbial biological control agents. Pages 157–173, *in*: R. Charudattan and H. L. Walker, eds., Biological Control of Weeds with Plant Pathogens. Wiley, New York.

15. Bowers, R. C. 1986. Commercialization of Collego™—an industrialist's view. Weed Sci. 34 (Suppl. 1):24–25.

16. Boyette, C. D. 1986. Host range evaluations of *Alternia crassa* for biocontrol of jimsonweed (*Datura stramonium* L.). Abstr. Weed Sci. Soc. Am. 26:52.

17. Boyette, C. D. 1987. Biocontrol of hemp sesbania [*Sesbania exaltata* (Raf.) Cory.] by an induced host range alteration of *Alternaria crassa*. Abstr. Weed Sci. Soc. Am. 27:48.

18. Boyette, C. D. 1988. Efficacy and host range of a recently discovered fungal pathogen for biocontrol of hemp sesbania. Proc. South. Weed Sci. Soc. 41:267.

19. Boyette, C. D., and P. C. Quimby. 1988. Interaction of *Fusarium lateritium* with 2,4-DB for control of velvetleaf (*Abutilon theophrasti* Medik.). Abstr. Weed Sci. Soc. Am. 28:82.

20. Boyette, C. D., G. E. Templeton, and L. R. Oliver. 1984. Texas gourd (*Cucurbita texana*) control with *Fusarium solani* f.sp. *cucurbitae*. Weed Sci. 32:649–655.

21. Boyette, C. D., G. E. Templeton, and R. J. Smith, Jr. 1979. Control of winged waterprimrose (*Jussiaea decurrens*) and northern jointvetch (*Aeschynomene virginica*) with fungal pathogens. Weed Sci. 27:497–501.

22. Boyette, C. D., G. J. Weidemann, D. O. TeBeest, and L. B. Turfitt. 1986. Biocontrol of jimsonweed in the field with *Alternaria crassa*. Proc. South. Weed Sci. Soc. 39:388.

23. Bruckart, W. L., D. R. Johnson, and J. R. Frank. 1988. Bentazon reduces rust-induced disease in yellow nutsedge, *Cyperus esculentus*. Weed Tech. 2:299–303.

24. Cardina, J., and R. H. Littrell. 1986. Enhancement of anthracnose severity on Florida beggarweed [*Desmodium tortuosum* (Sw.) DC]. Abstr. Weed Sci. Soc. Am. 26:51.

25. Cardina, J., R. H. Littrell, and R. T. Hanlin. 1988. Anthracnose of Florida beggarweed (*Desmodium tortuosum*) caused by *Colletotrichum truncatum*. Weed Sci. 36:329–334.

26. Charudattan, R. 1984. Microbial control pathogens and weeds. J. Georgia Entomol. Soc. 19(3)2nd Suppl.:40–62.

27. Charudattan, R. 1985. The use of natural and genetically altered strains of pathogens for weed control. Pages 347–372, *in*: M. A. Hoy and D. C. Herzog, eds., Biological Control in Agricultural IPM Systems. Academic Press, New York.

28. Charudattan, R. 1986. Integrated control of waterhyacinth (*Eichhornia crassipes*) with a pathogen, insects, and herbicides. Weed Sci. 34(Suppl. 1):26–30.

29. Charudattan, R. 1986. Biological control of showy crotalaria (*Crotalaria spectabilis* Roth) with two fungal pathogens. Abstr. Weed Sci. Soc. Am. 26:51.

30. Clay, K. 1986. New disease (*Balansia cyperi*) of purple nutsedge (*Cyperus rotundus*). Plant Dis. 70:597–599.

31. Daniel, J. T., G. E. Templeton, R. J. Smith, Jr., and W. T. Fox. 1973. Biological control of northern jointvetch in rice with an endemic fungal disease. Weed Sci. 21:303–307.

32. Defago, G., H. Kern, and L. Sedlar. 1985. Potential control of weedy spurges by the rust *Uromyces scutellatus*. Weed Sci. 33:857–860.

33. Figliola, S., N. D. Camper, and W. H. Ridings. 1988. Potential biological control agents for goosegrass. Proc. South. Weed Sci. Soc. 41:269.

34. Gotlieb, A. R., A. K. Watson, and L. A. Wymore. 1986. Synergism between bentazon and the mycoherbicide, *Colletotrichum coccodes*, for control of velvetleaf (*Abutilon theophrasti* Medik.). Abstr. Weed Sci. Soc. Am. 26:53.

35. Hill, G. D. 1982. Herbicide technology for integrated weed management systems. Weed Sci. 30(Suppl.):35–39.

36. Hofmeister, F. M., and R. Charudattan. 1987. *Pseudocercospora nigricans*, a pathogen of sicklepod (*Cassia obtusifolia*) with biocontrol potential. Plant Dis. 71:44–46.

37. Huey, B. A. 1987. How to use the DD50 computer printout. Misc. Publ. No. 211. Univ. Arkansas Coop. Ext. Serv., Little Rock, AR.

38. Kendrick, J. B., Jr. 1988. A viewpoint on integrated pest management. Plant Dis. 72:647.

39. Kenney, D. S. 1986. DeVine®—the way it was developed—an industrialist's view. Weed Sci. 34(Suppl. 1):15–16.

40. Khodayari, K., and R. J. Smith, Jr. 1988. A mycoherbicide integrated with fungicides in rice, *Oryza sativa*. Weed Tech. 2:282–285.

41. Khodayari, K., R. J. Smith, Jr., J. T. Walker, and D. O. TeBeest. 1987. Applicators for a weed pathogen plus acifluorfen in soybean. Weed Tech. 1:37–40.

42. Klerk, R. A. 1983. Use of a microbial herbicide for weed control in rice (*Oryza sativa*). M.S. Thesis, University of Arkansas, Fayetteville, AR, May, 1983.

43. Klerk, R. A., R. J. Smith, Jr., and D. O. TeBeest. 1985. Integration of a microbial herbicide into weed and pest control programs on rice (*Oryza sativa*). Weed Sci. 33:95–99.

44. McWhorter, C. G. 1984. Future needs in weed science. Weed Sci. 32:850–855.

45. McWhorter, C. G., and W. C. Shaw. 1982. Research needs for integrated weed management systems. Weed Sci. 30(Suppl.):40–45.

46. Mortensen, K. 1988. The potential of an endemic fungus, *Colletotrichum gloeosporioides*, for biological control of round-leaved mallow (*Malva pusilla*) and velvetleaf (*Abutilon theophrasti*). Weed Sci. 36:473–478.

47. Ormeno-Nunez, J., R. D. Reeleder, and A. K. Watson. 1988. A foliar disease of field bindweed (*Convolvulus arvensis*) caused by *Phomopsis convolvulus*. Plant Dis. 72:338–342.

48. Phatak, S. C., M. B. Callaway, and C. S. Vavrina. 1987. Biological control and its integration in weed management systems for purple and yellow nutsedge (*Cyprus rotundus* and *C. esculentus*). Weed Tech. 1:84–91.

49. Phatak, S. C., D. R. Sumner, H. D. Wells, D. K. Bell, and N. C. Glaze. 1983. Biological control of yellow nutsedge with the indigenous rust fungus *Puccinia canaliculata*. Science 219:1446–1447.

50. Quimby, P. C. 1985. Pathogenic control of prickly sida and velvetleaf: an alternate technique for producing and testing *Fusarium lateritium*. Proc. South. Weed Sci. Soc. 38:365–371.

51. Quimby, P. C., Jr., and H. L. Walker. 1982. Pathogens as mechanisms for integrated weed management. Weed Sci. 30(Suppl.):30–34.

52. Ravenell, D. I., and C. G. Van Dyke. 1986. Efficacy and host range of *Bipolaris setariae* as a potential biocontrol pathogen of broadleaf signalgrass [*Brachiaria platyphylla* (Griseb.) Nash]. Abstr. Weed Sci. Soc. Am. 26:54.

53. Ridings, W. H. 1986. Biological control of stranglervine in citrus—a researcher's view. Weed Sci. 34(Suppl. 1):31–32.

54. Ridings, W. H., P. H. Hoyer, and L. E. Schimmel. 1986. Biological control of spurred anoda [*Anoda cristata* (L.) Schlecht.] using teleospores of *Puccinaia heterspora*. Abstr. Weed Sci. Soc. Am. 26:54–55.

55. Scheepens, P. C. 1987. Joint action of *Cochliobolus lunatus* and atrazine on *Echinochloa crusgalli* (L.) Beauv. Weed Res. 27:43–47.

56. Schubiger, F. X., G. Defago, H. Kern, and L. Sedlar. 1986. Damage to *Rumex crispus* L. and *Rumex obtusifolius* L. caused by the rust fungus *Uromyces rumicis* (Schum.) Wint. Weed Res. 26:347–350.

57. Shaw, W. C. 1982. Integrated weed management systems technology for pest management. Weed Sci. 30(Suppl.):2–12.

58. Shaw, W. C. 1984. The new weed science—a view of the 21st century. Proc. West. Weed Sci. Soc. 37:8–29.

59. Smith, R. J., Jr. 1982. Integration of microbial herbicides with existing pest management programs. Pages 189–203, *in*: R. Charudattan and H. L. Walker, eds., Biological Control of Weeds with Plant Pathogens. Wiley, New York.

60. Smith, R. J., Jr. 1983. Integrated weed management in rice in the USA. Korean J. Weed Sci. 3(1):1–13.

61. Smith, R. J., Jr. 1986. Biological control of northern jointvetch (*Aeschynomene virginica*) in rice (*Oryza sativa*) and soybeans (*Glycine max*)—a researcher's view. Weed Sci. 34(Suppl. 1):17–23.

62. Smith, R. J., Jr., J. T. Daniel, W. T. Fox, and G. E. Templeton. 1973. Distribution in Arkansas of a fungus disease used for biocontrol of northern jointvetch in rice. Plant Dis. Rep. 57:695–697.

63. Smith, R. J., Jr., W. T. Fox, J. T. Daniel, and G. E. Templeton. 1973. Can plant diseases be used to control weeds? Arkansas Farm Res. 22(4):12.

64. TeBeest, D. O. 1982. Survival of *Colletotrichum gloeosporioides* f.sp. *aeschynomene* in rice irrigation water and soil. Plant Dis. 66:469–472.

65. TeBeest, D. O. 1984. Induction of tolerance to benomyl in *Colletotrichum gloeosporioides* f.sp. *aeschynomene* by ethyl methanesulfonate. Phytopathology 74:864.

66. TeBeest, D. O., and J. M. Brumley. 1978. *Colletotrichum gloeosporioides* borne with the seed of *Aeschynomene virginica*. Plant Dis. Rep. 62:675–678.

67. TeBeest, D. O., and G. E. Templeton. 1985. Mycoherbicides: progress in the biological control of weeds. Plant Dis. 69:6–10.

68. TeBeest, D. O., G. E. Templeton, and R. J. Smith, Jr. 1978. Temperature and moisture requirements for development of anthracnose on northern jointvetch. Phytopathology 68:389–393.

69. TeBeest, D. O., G. E. Templeton, and R. J. Smith, Jr. 1978. Histopathology of *Colletotrichum gloeosporioides* f.sp. *aeschynomene* on northern jointvetch. Phytopathology 68:1271–1275.

70. TeBeest, D. O., G. E. Templeton, and R. J. Smith, Jr. 1978. Decline of a biocontrol fungus during winter. Arkansas Farm Res. 27(1):12.

71. Templeton, G. E. 1982. Status of weed control with plant pathogens. Pages 29–44, *in*: R. Charudattan and H. L. Walker, eds., Biological Control of Weeds with Plant Pathogens. Wiley, New York.

72. Templeton, G. E. 1982. Biological herbicides: discovery, development, deployment. Weed Sci. 30:430–433.

73. Templeton, G. E. 1983. Integrating biological control of weeds in rice into a weed control program. Pages 219–225, *in*: Proc. Conf. on Weed Control in Rice (Aug. 31–Sept. 4, 1981). Int. Rice Res. Inst., Los Banos, Laguna, Philippines.

74. Templeton, G. E. 1986. Mycoherbicide research at the University of Arkansas—past, present, and future. Weed Sci. 34(Suppl. 1):35–37.

75. Templeton, G. E., and R. J. Smith, Jr. 1977. Managing weeds with pathogens. Pages 167–176, *in*: J. G. Horsfall and E. B. Cowling, eds., Plant Disease: An Advanced Treatise, Vol. 1. Academic Press, New York.

76. Templeton, G. E., R. J. Smith, Jr., and W. Klomparens. 1980. Commercialization of fungi and bacteria for biological control. Biocontrol News and Information 1(4):291–294.

77. Templeton, G. E., R. J. Smith, Jr., and D. O. TeBeest. 1986. Progress and potential of weed control with mycoherbicides. Rev. Weed Sci. 2:2–14.

78. Templeton, G. E., R. J. Smith, Jr., and D. O. TeBeest. 1988. Perspectives on mycoherbicides two decades after discovery of the Collego™ pathogen. Proc. VII Int. Symp. Biol. Contr. Weeds, E. S. Delfosse, ed. (March 6–11, 1988) Rome, Italy. In press.

79. Templeton, G. E., R. J. Smith, Jr., D. O. TeBeest, and J. N. Beasley. 1988. Rice research overview—mycoherbicides. Arkansas Farm Res. 37(2):7.

80. Templeton, G. E., R. J. Smith, Jr., D. O. TeBeest, J. N. Beasley, and R. A. Klerk. 1981. Field evaluation of dried fungus spores for biocontrol of curly indigo in rice and soybeans. Arkansas Farm Res. 30(6):8.

81. Templeton, G. E., D. O. TeBeest, and R. J. Smith, Jr. 1976. Development of an endemic fungal pathogen as a mycoherbicide for biocontrol of northern jointvetch in rice. Pages 214–216, *in*: T. E. Freeman, ed. Proc. IV Int. Symp. Biol. Control Weeds. University of Florida, Gainesville.

82. Templeton, G. E., D. O. TeBeest, and R. J. Smith, Jr. 1979. Biological weed control with mycoherbicides. Annu. Rev. Phytopathol. 17:301–310.

83. Templeton, G. E., D. O. TeBeest, and R. J. Smith, Jr. 1984. Biological weed control in rice with a strain of *Colletotrichum gloeosporioides* (Penz.) Sacc. used as a mycoherbicide. Crop Protection 3:409–422.

84. Templeton, G. E., and E. E. Trujillo. 1981. The use of plant pathogens in the biological control of weeds. Pages 345–350, *in*: D. Pimentel, ed., CRC Handbook of Pest Management in Agriculture, Vol. 2. CRC Press, Boca Raton, FL.

85. Turfitt, L. B., and C. D. Boyette. 1986. Factors influencing biocontrol of jimsonweed with *Alternaria crassa*. Proc. Soc. Weed Sci. Soc. 39:389.

86. Walker, H. L. 1982. Seedling blight of sicklepod caused by *Alternaria cassiae*. Plant Dis. 66:426–428.

87. Walker, H. L., and D. Boyette. 1985. Biocontrol of sicklepod (*Cassia obtusifolia*) in soybeans (*Glycine max*) with *Alternaria cassiae*. Weed Sci. 33:212–215.

88. Walker, H. L., and J. A. Riley. 1982. Evaluation of *Alternaria cassiae* for the biocontrol of sicklepod (*Cassia obtusifolia*). Weed Sci. 30:651–654.

89. Watson, A. K., and M. Clement. 1986. Evaluation of rust fungi as biological control agents of weedy *Centaurea* in North America. Weed Sci. 34(Suppl. 1):7–10.

90. Watson, A. K., A. R. Gotlieb, and L. A. Wymore. 1986. Interactions between a mycoherbicide *Colletotrichum coccodes* and herbicides for control of velvetleaf (*Abutilon theophrasti* Medik.). Abstr. Weed Sci. Soc. Am. 26:52–53.

91. Weidemann, G. J., and G. E. Templeton. 1988. Control of Texas gourd, *Cucurbita texana*, with *Fusarium solani* f.sp. *cucurbitae*. Weed Tech. 2:271–274.

92. Wymore, L. A., C. Poirier, and A. K. Watson. 1988. *Colletotrichum coccodes*, a potential bioherbicide for control of velvetleaf (*Abutilon theophrasti*). Plant Dis. 72:534–538.

93. Wymore, L. A., and A. K. Watson. 1988. Interaction between the mycoherbicide [*Colletotrichum coccodes* (Wallr.) Hughes] and selected herbicides for velvetleaf (*Abutilon theophrasti* Medik.). Abstr. Weed Sci. Soc. Am. 28:50.

94. Yu, S. M., and G. E. Templeton. 1983. The relationship of trifluralin to collar rot of texas gourd caused by *Fusarium solani* f.sp. *cucurbitae*. Phytopathology 73:823.

12

Progress in the Production, Formulation, and Application of Mycoherbicides

C. Douglas Boyette, P. Charles Quimby, Jr., William J. Connick, Jr., Donald J. Daigle, and Floyd E. Fulgham

Introduction

Chemical herbicides have been the mainstay for weed control practices in the United States since the end of World War II and are responsible for much of the unparalleled increased crop productivity that has occurred during this period (38). The high costs involved in developing and registering chemical herbicides and recent trends in environmental awareness have prompted researchers to investigate alternative systems of weed control. Ideally, such a system would control target weeds at or near the same levels as that achieved with chemical herbicides, while at the same time not posing a threat to either the environment or nontarget organisms.

To date, the most biologically effective alternatives to chemical weed control agents that have been extensively evaluated are the plant pathogens, more specifically, plant pathogenic fungi. There are two broad approaches to biological weed control using plant pathogens: (1) the classical approach and (2) the mycoherbicide (or bioherbicide) approach (57). In the classical approach the pathogen (generally an exotic) is introduced into a susceptible weed population and allowed to spread unchecked by the lack of any natural resistance in the population. In the mycoherbicidal approach the pathogen (generally an endemic) is applied to target weeds using techniques and methodology similar to those used with chemical herbicides. This approach has proved to be the more effective of the two in obtaining the rapid, high levels of weed control in row crops that are desired in current socio-agronomic practices (56–58).

This chapter discusses recent developments and advancements in mycoherbicide production, formulation, and application technology. Commercially produced mycoherbicides are also discussed.

Inoculum Production

Infective Units

From a standpoint of practicality and economics, the infective units of the candidate mycoherbicide must be produced in a timely and cost-effective manner.

With few exceptions, the most suitable infective units are fungal spores. There are several types of spores, but asexually produced spores, or conidia, are generally the easiest to produce under experimental conditions, and since spores are the most common mechanisms for natural disease dispersal, they should, logically, serve as the best candidates as infective units of mycoherbicides. Some fungi, such as *Rhizoctonia* spp., normally do not produce spores. Because many of these types of fungi are rather nonspecific in host preference, they have not been extensively evaluated as mycoherbicides (58).

In some instances, mycelial fragments may be used in place of spores for some fungi that do not produce spores or produce them sparingly (25,51,60). However, mycelial fragments are harder to quantitate than spores, less readily separated from the culture medium, and often less infective than spores (60). Additionally, the durability, longevity, and viability of mycelium is generally much less than that of spores (23).

Culture Medium Selection

For small-scale inoculum production where economics are not a primary concern, relatively expensive materials, such as V–8 vegetable juice and agar culture, have been used successfully to induce sporulation and obtain inoculum of several mycoherbicides (9,10,28).

However, mass production of mycoherbicides on a larger scale, such as in pilot test studies and by industry, requires that the candidate mycoherbicide be produced as cost efficiently as possible while the quality and quantity of the final product is retained. Materials for scaled-up testing include crude agricultural products that are readily available at low costs and in unlimited quantities. Protein sources, such as soybean flour, corn steep liquor, distiller's solubles, brewer's yeast, autolyzed yeast, milk solids, cottonseed flour, linseed meal, corn protein, and a variety of fish meal are some of the materials that may be used. The carbon sources that are commonly tested include cornstarch, corn flour, glucose, hydrolyzed-corn-derived materials, glycerol, and sucrose (23).

A growth medium with a balanced ratio of carbon and nitrogen generally produces the best vegetative (mycelial) growth. However, to induce sporulation it may be necessary to alter this ratio or to amend the medium with other nutrients such as calcium, chelating agents, and various amino acids (60).

Often, carbon sources that do not yield maximum vegetative growth will enhance sporulation. To optimize growth and sporulation, the carbon, nitrogen, and mineral levels may require precise balancing (23). In addition to the effect on growth and sporulation, the carbon–nitrogen ratio may also affect the viability, longevity, and virulence of the fungus being cultured. For example, Toussoun et al. (60) demonstrated that the vegetative growth of *Fusarium solani* f.sp. *phaseoli* in vitro was increased by a high carbon–low nitrogen ratio, whereas virulence of the fungus on *Phaseolus vulgaris* was decreased. Conversely, a low carbon–high nitrogen growth medium resulted in decreased vegetative growth and increased virulence.

Culturing Mycoherbicides

Methodology has been developed to produce mycoherbicides using solid substrate fermentations, submerged culture fermentations, or combinations of both at different stages of the growth cycle of the fungus. Some specific examples follow:

Solid Substrate Fermentation

The technology of culturing fungi using a solid substrate is not as well developed in the West (with the exception of mushroom spawn production) as it is in Japan (31). There are several inherent problems associated with solid substrate fermentations. Among these are high labor costs, difficulties in maintaining sterility, lack of controls on fermentation conditions, and recovery of the spores from the substrate (23). These problems are in large part responsible for the development of a highly developed submerged culture fermentation industry in the West (31). Because not all fungi will produce spores in submerged culture, solid substrate fermentations, or variations thereof, may sometimes offer the only method of spore production. Various cereal grains and straws can be used to produce inoculum of a number of plant pathogenic fungi (60). It is relatively simple and inexpensive to produce fungal inoculum using these substrates. Additionally, the cereal grains or straw pieces make quantification and dispersal relatively easy and accurate. Hildebrand and McCain (32) used wheat straw that was infested with *Fusarium oxysporum* f.sp. *cannabis* to control marijuana (*Cannabis sativa*), and Boyette (5) used oat seed infested with *F. solani* f.sp. *cucurbitae* to control Texas gourd (*Cucurbita texana*). It is, however, difficult to sterilize, inoculate, and store these types of bulky materials until they are ready to be used in the field.

Combined Solid Substrate and Submerged Fermentation

Several mycoherbicides have been produced using a combination of solid substrate and submerged fermentations techniques. *Alternaria macrospora* for controlling spurred anoda (*Anoda cristata*) was first mass produced by culturing mycelium of the fungus for 48 hours in V–8 juice medium (62). The mycelium was collected, blended, and mixed with 1000 g of vermiculite, spread into foil-lined pans, and exposed to fluorescent light for a period of 7 hours, or to direct sunlight for 20–30 minutes to induce sporulation. After 24 hours, the mycelium began to produce conidiophores and conidia of *A. macrospora*, and after air-drying at 35°C for another 48 hours, the mixture was sieved, packaged, and stored at 4°C. Each gram of culture medium produced a dry weight of approximately 4 g of spores. Spore yields were approximately 1×10^5 spores/g of dried mycelium (63). This procedure has also been used to produce inoculum of other mycoherbicides, such as *Colletotrichum malvarum*, a mycoherbicide for prickly sida (*Sida spinosa*), and *Fusarium lateritium*, a mycoherbicide for spurred anoda, prickly sida, and velvetleaf (*Abutilon theophrasti*) (64).

Modifications of this methodology were made to produce spores of *A. cassiae* for use as a mycoherbicide against sicklepod (67). The growth medium consisted of 15 g/L soyflour, 15 g/L cornmeal, 30 g/L sucrose, and 3 g/L calcium carbonate. The fungus was grown in submerged culture at 25°C for 24 hours. The mycelia were then collected, homogenized, and poured into foil-lined trays and subjected to 10 minutes of ultraviolet light treatment every 12 hours for a period of 3–5 days to induce sporulation. The surface of the mycelium sporulated profusely, and after 72 hours the spores were collected by a vacuum, dried over $CaSO_4$, and stored at 4°C. Approximately 8 g of spores were produced per liter of growth medium, with each gram of spores containing approximately 1×10^8 spores.

Sufficient quantities of *A. cassiae* spores were produced using this technique to conduct field efficacy tests for a five-state regional test and for a 2-year pilot test study (22,65). This fungus is currently being developed as a commercial mycoherbicide by Mycogen Corporation, San Diego, CA, and the commercial product will be discussed in a later section.

This technique has also been used to produce spores of *A. crassa* for jimsonweed (*Datura stramonium*) control (12,13), *A. helianthi* for cocklebur (*Xanthium strumarium*) and wild sunflower (*Helianthus annuus*) control (2,44), and *Bipolaris sorghicola* for Johnson grass (*Sorghum halepense*) control (61).

Submerged Culture Fermentations

From a practical and economic standpoint, fungi that sporulate in liquid culture are favored over those that require additional steps to induce sporulation, and this factor may prove to be an essential requirement for the commercial development of a fungus as a mycoherbicide (3,4).

For early evaluations and in small-scale experiments, such as virulence and host range experiments, ample inoculum can usually be produced in shake flasks. However, with shake flasks it is difficult to control many of the fungal growth parameters that influence mycelial growth or sporulation, such as pH, temperature, agitation, and aeration. For larger quantities of inoculum and more precise growth parameter control, laboratory model fermenters are essential. Some models monitor and control several environmental factors such as temperature, agitation, dissolved oxygen, and pH, all of which may affect growth and sporulation of the organism being cultured.

The commercially produced mycoherbicides COLLEGO (Ecogen, Inc., Langhorne, PA) and DeVine (Abbott Laboratories, North Chicago, IL) are both produced using submerged culture techniques. The formulated products of each are discussed in a later section.

Commercially Produced Mycoherbicides

Phytophthora palmivora was the first fungus to be marketed as a mycoherbicide. The fungus infects and kills stranglervine (*Morrenia odorata*), a problem weed in Florida citrus groves. For early field evaluations, chlamydospores of the fungus were produced in V–8 juice medium contained in shake flasks (52). Abbott Laboratories, North Chicago, IL, in a cooperative research effort with Florida researchers, developed *P. palmivora* as a mycoherbicide and marketed it under the trade name DeVine. The infective units are chlamydospores and are not highly stable (51). The material has a shelf life of only about 6 weeks and must be handled like fresh milk (51,53). However, the marketing area is small enough to make the refrigerated distribution and custom-order sales possible (51).

Abbott Laboratories also developed an experimental formulation of *Cercospora rodmanii* for controlling water hyacinth (*Eichhornia crassipes* in Florida waterways. The experimental formulation, called ABG–5003, consisted of mycelial fragments and spores and was applied as a wettable powder formulation (21,29). Although biological control of the weed was achieved, the efficacy was less than is required for commercialization, largely because of the restrictive environmental requirements of the fungus (21,25). More recently, it has been found that weed control by the fungus can be enhanced by using it in combination sublethal doses of various chemical herbicides or with certain insects (20).

Colletotrichum gloeosporioides f.sp. *aeschynomene* (CGA), for northern joint vetch control, was the first fungus to be evaluated as a mycoherbicide (28). In collaboration with researchers at the University of Arkansas and the U.S. Department of Agriculture, the Upjohn Company was able to mass-produce CGA and marketed it under the trade name COLLEGO (Ecogen, Inc., Langhorne, PA) for use as a mycoherbicide to control northern joint vetch (*Aeschynomene virginica*) in Arkansas and Louisiana rice fields. The formulated material consists of an active component (dried CGA spores) and an inert rehydrating agent used to wet the spores and plant surfaces and to improve germination. These components are packaged separately and are added to the desired volume of water just prior to application (3,4).

The Mycogen Corporation (San Diego, CA) developed an experimental formulation called CASST to control sicklepod in soybeans and peanuts. It is a two-component product consisting of an emulsifiable paraffinic oil "adjuvant" (designated MYD 751M) and spores of *Alternaria cassiae* (MYX 104, 100% ai). To prepare the spray mixture, the oil component is first emulsified in water at a 1% (v/v) concentration and then the spores are added and dispersed by stirring. This formulation is applied at a rate of about 1 lb spores/acre to sicklepod in the cotyledon to second-leaf stage of development (26).

Another mycoherbicide that is nearing commercialization is BioMal (Philom Bios, Saskatoon, Sask., Canada), which contains the spores of fungus *Colletotrichum gloeosporioides* f.sp. *malvae*. The fungus is pathogenic to round-leaved mallow (*Malva pusilla*). An experimental formulation of BioMal using a silica gel carrier has routinely provided over 90% control of this weed in the field. The wettable powder formulation of this hydrophilic fungus disperses easily in water and is applied as a spray (30).

Formulation and Application

Formulation is the blending of active ingredients, such as fungal spores, with inert carriers, such as diluents and surfactants, in order to alter the physical characteristics to a more desirable form. This may include diluting to a common potency, enhancing stability and/or biological activity, improving mixing and sprayability, and possibly integrating the mycoherbicide into a pest management system (3,27).

In many cases the method of production of mycoherbicides determines the method of application (46). As stated previously, mycoherbicides are applied in much the same manner as are chemical herbicides, often with the same equipment. However, mycoherbicides are living organisms and must be treated as such. For example, chemical residue in spray tanks may be detrimental to the fungus, and the applicator should ensure that the tanks and lines on the spraying system are clear of these residues. A slurry of activated charcoal and liquid detergent is recommended for this cleansing (46). Similarly, when making field applications, it would not be prudent to apply other pesticides, especially fungicides, to mycoherbicide-treated areas unless they are compatible. For example, the fungicides benomyl and propiconazol applied sequentially 7 and 14 days after COLLEGO was applied suppressed disease development on northern joint vetch (35,37). Similarly, the efficacy of DeVine was reduced if the fungicides Aliette and Ridomil were used within 45 days following application of the mycoherbicide (37).

Liquid-Based Formulations

Just as water is considered to be the "universal solvent" for chemicals, it may also be the "universal carrier" for mycoherbicides. This is because water is inexpensive, readily available, easy to handle, and necessary to maintain life for all plant pathogens.

During initial evaluations, spray volumes and inoculum concentrations are very high, usually about 900–1000 L/ha with 10^5–10^7 spores/ml. These volumes and rates can generally be reduced to more practical levels by conducting experiments to determine threshold levels.

The simplest mycoherbicide delivery system is the fungus contained in and sprayed in water. Many weeds are covered with a waxy cuticle that prevents water from spreading evenly, and this prevents an equal distribution when aqueous materials, such as mycoherbicides, are applied. Surfactants aid in wetting the plants and in dispersing the fungal spores throughout the spray mix. Because spores of mycoherbicides are finite units, it is of paramount importance that the surface area be covered with the material as evenly and equally as possible. A number of surfactants have been used in mycoherbicide research (Figure 12.1). Since some surfactants may be detrimental to the growth and/or germination of fungi, preliminary experiments should be conducted to determine the effect of the surfactant on the candidate mycoherbicide. For example, *Alternaria cassiae* spores do not germinate consistently well in either Tween–20 or Tween–80 surfactants, but in 0.02–0.04% nonionic nonoxynol surfactant spore germination is not reduced, and spray coverage is increased resulting in enhanced control of sicklepod (53). Figure 12.1 lists some of the various mycoherbicides formulated in a liquid-based formulation.

Solid-Based Formulations

With some exceptions, liquid formulations of mycoherbicides are best suited for use as postemergence sprays and are used primarily to incite leaf and stem diseases. Conversely, pathogens that infect at or below the soil line are probably delivered best in a solid, or granular, formulation. Granular formulations are better suited for use as preplant or preemergence mycoherbicides than are spray formulations because (1) the granules provide a buffer from environmental extremes; (2) the granules can serve as a food base for the fungus, resulting in a longer period of persistence; (3) the granules are less likely to be washed away from the treated areas than are spores. Figure 12.2 lists some of the pathogens that have been formulated in solid-based formulations. Some specific examples follow:

A cornmeal–sand formulation of *Fusarium solani* f.sp. *cucurbitae* (FSC) produces mycelium and a mixture of microconidia, macroconidia, and chlamydospores (10). The ratio of these spore types can be altered by addition of various nutrients to the basal medium (71). Almost complete control (96% avg) of Texas gourd was achieved with preplant and preemergence applications of this granular formulation of FSC (10,69,70).

Another solid substrate that has been effectively used is vermiculite. Walker (62,64) grew mycelium of *Alternaria macrospora* in liquid shake culture and mixed the mycelium with vermiculite. The fungus sporulated profusely, and after air-drying the mixture was applied both preemergence and postemergence to spurred anoda (*Anoda cristata*), yielding 75–95% weed control. The control achieved preemergence with the fungus-infested vermiculite formulation was as good as the control achieved postemergence with foliar sprays of *A. macrospora*.

Granular formulations of several biocontrol fungi have also been made using

Figure 12.1. Liquid-Based Mycoherbicide Formulations

Weed host	Pathogen	Formulation	Reference
Velvetleaf (*Abutilon theophrasti* Medik.)	*Fusarium lateritium* (Nees.) ex. Fr.	Water + Tween-20 surfactant (0.02%)	11, 63
		Experimental formulation— water;	28
Northern joint vetch [*Aeschynomene virginica* (L.) B.S.P.]	*Colletotrichum gloeosporioides* (Penz.) Sacc. f.sp. *aeschynomene*	commercial formulation— component A: dried spores, component B: rehydrating agent + surfactant	23
Spurred anoda (*Anoda cristata* Schlecht.)	*Alternaria macrospora* Zimm.	Water + nonoxynol surfactant (0.02%); sucrose (5% w/v)	62, 64
Giant ragweed (*Ambrosia trifida* L.)	*Protomyces gravidus* Davis	Water	19
Field bindweed (*Convolvulus arvensis* L.)	*Phomopsis convolvulus* Ormeno	Water + gelatin (0.1%)	73
Jimsonweed (*Datura stramonium* L.)	*Alternaria crassa* (Sacc.) Rands	Water + nonoxynol surfactant (0.04%)	6, 13
Florida beggarweed (*Desmodium tortuosum* (Sw.) DC.	*Colletotrichum truncatum* (Schw.) Andrus & Moore		17, 18
Sicklepod (*Cassia occidentalis* L.)	*Alternaria cassiae* Jurair & Khan	Water + nonoxynol surfactant (0.04%); paraffin wax mineral oil, soybean oil, corn syrup, lecithin	22, 65, 68; 27, 47, 48
Velvetleaf (*Abutilon theophrasti* Medik.)	*Colletotrichum coccodes* (Walk.) Hughes	Water + sorbitol (0.75%)	73
Common purslane (*Portulaca oleracea* L.)	*Dichotomophthora portulaceaceae* Mehrlich and Fitz. ex. M. D. Ellis	Water + Tween-20 surfactant (0.02%)	40
Hemp sesbania [*Sesbania exaltata* (Raf.) Cory]	*Colletotrichum truncatum* (Schw.) Andrus & Moore	Water + nonoxynol surfactant (0.02%); paraffin wax; mineral oil, soybean oil lecithin	7
Eastern black nightshade (*Solanum ptycanthum* Dunn.)	*Colletotrichum coccodes* (Walk.) Hughes	Water + Tween-80 surfactant (0.02%)	1
Strangler vine (*Morrenia ordorata* L.)	*Phytopthora palmivora* (Butl.) Butl	Commercial formulation: chlamydospores in water	50
Horse purslane (*Trianthema portulacastrum* L.)	*Gibbago trianthemae*	Water + Tween-20 surfactant (0.02%)	41

Figure 12.2.   Solid-Based Mycoherbicide Formulations

Weed host	Pathogen	Formulation	Reference
Velvetleaf (*Abutilon theophrasti* Medik.)	*Fusarium lateritium* Nees. ex. Fr.	Sodium alginate–kaolin granules	11, 43
Spurred anoda (*Anoda cristata* Schlecht.)	*Alternaria macrospora* Zimm.	Vermiculite	62, 64
Texas gourd [*Cucurbita texana* (A.) Gray]	*Fusarium solani* App. + Wr. f.sp. *cucurbitae* Snyd. + Hans.	Fungus-infested oats; cornmeal/sand; sodium alginate–raolin granules	5
Marijuana (*Cannabis sativa* L.)	*Fusarium oxysporum* var. *cannabis* Snyd.	Fungus-infested wheat straw	31

sodium alginate (66,68). This process was adapted from work with time-released herbicide formulations (24). In this process, fungal mycelium is mixed with sodium alginate and various fillers, such as kaolin clay, and dripped into a 0.25 *M* solution of calcium chloride. The Ca^{2+} ions react immediately with the sodium alginate to form gel beads. The beads are allowed to harden in the calcium chloride solution for a few minutes and can then be collected, rinsed, and air-dried. The granules, of fairly uniform size and shape, can then be used in a manner similar to preplant or preemergence herbicides, or rehydrated and exposed to UV light. The rehydration and UV light treatment induces the fungus to produce spores, which can then be collected and used as postemergence sprays (11,12,43,65). Figure 12.2 lists some solid-substrate mycoherbicides that have been formulated.

Formulations That Improve Mycoherbicide Efficacy

Various adjuvants and amendments have been used to either improve or modify spore germination, pathogen virulence, or environment requirements, all of which greatly determine the bioherbicidal potential of a candidate microorganism. Some specific examples follow:

The addition of sucrose to spray mixes of *A. macrospora* resulted in increased disease severity on spurred anoda (63,65). Also, increased spore germination and disease severity of Florida beggarweed (*Desmodium tortuosum*) anthracnose was reported when small quantities of sucrose and gum xanthan were added to aqueous spore suspensions of *Colletotrichum truncatum* (17).

Disease severity of Johnson grass (*Sorghum halepense*) by (*Bipolaris sorjicola*) was significantly increased by adding 1% Soy-Dex to the fungus spray mix (71). Similarly, when sorbitol was added to spray mixes of *C. coccodes*, there was a 20-fold increase of viable spores reisolated from inoculated velvetleaf. When this amendment was used, three 9-hour dew periods on consecutive nights were as effective as a single 18-hour dew treatment (73).

Recent work has indicated that invert (water-in-oil) emulsions may be a potential method to retard evaporation and trap water in the spray mixture, thereby decreasing the amount of dew required for spore germination and infection to occur (47,48). In these studies lecithin was used as the emulsifying agent, and paraffin oil and wax were used to retard evaporation further and help retain droplet size (27). Specialized spraying equipment was developed to deliver this viscous material. Greenhouse and field results indicated that biocontrol of sicklepod could be achieved with little or no dew (47,48). Recently, this system was used to enhance control of hemp sesbania (*Sesbania exaltata*) in the field with *Colletotrichum truncatum* (C. D. Boyette, 1989, unpublished data).

Another constraint of most mycoherbicides is their limited host range. Probably the simplest method to overcome this limitation is to apply a mixture of pathogens to mixed weed populations. For example, the rice weeds northern joint vetch and winged water primrose, *Jussiaea decurrens*, can be simultaneously controlled with a single application of CGA and *Colletotrichum gloeosporiodes* f.sp. *jussiaea* (CGJ) (9), and a mixture of these two pathogens along with *C. malvarum* effectively controlled northern joint vetch, winged water primrose, and prickly sida (56).

Disease severity of several plant pathogens are known to increase in response to various chemical herbicides (33). The addition of sublethal rates of the herbicides linuron, imaziquin, and lactofen to *Alternaria cassiae* spores in an invert formulation resulted in significantly increased control of sicklepod (45). Control of velvetleaf was significantly improved by sequential applications of the herbicide 2,4-DB and spores of *Fusarium lateritium*. However, fungal germination and disease severity were greatly reduced when the fungus and herbicide were tank-mixed (11). Biocontrol of velvetleaf was also improved significantly by adding field rates of the cotton defoliant thiadiazuron to an aqueous spray mixture of *C. coccodes* (72).

The rust fungus *Puccinia canaliculata*, parasitic to yellow nut sedge (*Cyperus esculentus*), is erratic in infecting and controlling this weed when applied alone even under optimal environmental conditions (15,16). However, sequential applications of the herbicide paraquat, followed by *P. canaliculata* spores, resulted in a synergistic disease interaction, with almost complete control of yellow nut sedge, compared to 10% and 60% control, respectively, for paraquat alone and the fungus alone (14).

Conclusions

From the examples discussed in this chapter it is clear that the key to the successful use of mycoherbicides depends upon a number of factors. Of paramount importance is the creation and maintenance of a favorable microenvironment in which the pathogens can thrive and manifest disease. Through innovative approaches to production, formulation, and application of mycoherbicides, much progress in this area has been made. Closer cooperative research efforts among plant pathologists, formulation chemists, fermentation scientists, and agricultural engineers are vital for the continued success of microbiological weed control.

Literature Cited

1. Andersen, R. N., and Walker, H. L. 1985. *Colletotrichum coccodes*: a pathogen of eastern black nightshade (*Solanum ptycanthum*). Weed Sci. 33:902–905.

2. Bassi, A., Jr., and Quimby, P. C., Jr. 1985. Infection of cocklebur by *Alternaria helianthi*. Proc. South. Weed Sci. 38:373.

3. Bowers, R. C. 1982. Commercialization of microbial biological control agents. Pages 157–173, *in*: R. Charudattan and H. L. Walker, eds., Biological control of weeds with plant pathogens. Wiley, New York.

4. Bowers, R. C. 1986. Commercialization of Collego™: an industrialist's view. Weed Sci. 34s:24–25.

5. Boyette, C.D. 1982. Evaluation of *Fusarium solani* f.sp. *cucurbitae* as a potential bioherbicide for controlling Texas gourd. PhD. Dissertation, University of Arkansas, Fayetteville.

6. Boyette, C. D. 1986. Evaluation of *Alternaria crassa* for biological control of *Datura stramonium*: host range and virulence. Plant Sci. Lett. 45:223–228.

7. Boyette, C. D. 1988. Efficacy and host range of a recently discovered fungal pathogen for biocontrol of hemp sesbania. Proc. South. Weed Sci. Soc. 41:267.

8. Boyette, C. D., and Quimby, P. C., Jr. 1988. Interaction of *Fusarium lateritium* and 2,4 DB for control of velvetleaf (*Abutilon theopharasti*). Proc. Weed Sci. Soc. 28:232.

9. Boyette, C. D., Templeton, G. E., and Smith, R. J., Jr. 1979. Control of winged waterprimrose (*Jussiaea decurrens*) and northern jointvetch (*Aeschynomene virginica*) with fungal pathogens. Weed Sci. 27:497–501.

10. Boyette, C. D., Templeton, G. E., and Oliver, L. R. 1985. Texas gourd (*Cucurbita texana*) control with *Fusarium solani* f.sp. *cucurbitae*. Weed Sci. 32:649–654.

11. Boyette, C. D., and Walker, H. L. 1985. Evaluation of *Fusarium lateritium* as a biological herbicide for controlling velvetleaf (*Abutilon theophrasti*) and prickly sida (*Sida spinosa*). Weed Sci. 34:106–109.

12. Boyette, C. D., and Walker, H. L. 1985. Production and storage of *cercospora kikuchii* for field studies. Phytopathology 75:183–185.

13. Boyette, C. D., Weidemann, G. J., TeBeest, D. O., and Turfitt, L. B. 1986. Biocontrol of jimsonweed in the field with *Alternaria crassa*. Proc. South. Weed Sci. 39:149.

14. Bruckart, W. L., Johnson, D. R., and Frank, J. R. 1988. Bentazon reduces rust-induced disease in yellow nutsedge, *Cyperus esculentus*. Weed Technol. 2:299–303.

15. Callaway, M. B., Phatak, S. C., and Wells, H. D. 1985. Effect of rust and rust–herbicide combinations on yellow nutsedge. Proc. South. Weed Sci. Soc. 38:31.

16. Callaway, M. B., Phatak, S. C., and Wells, H. D. 1987. Interactions of *Puccinia canaliculata* (Schw.) Lagerh. with herbicides on tuber production and growth of *Cyperus esculentus* (L.) Trop. Pest Manag. 33:22–26.

17. Cardina, J. 1986. Enhancement of anthracnose severity on Florida beggarweed. Proc. Weed Sci. Soc. 26:138 (Abstr.).

18. Cardina, J., Littrell, R. H., and Hanlin, R. T. 1988. Anthracnose of Florida beggarweed (*Desmodium tortuosum*) *caused by Colletotrichum truncatum*. Weed Sci. 36:329–334.

19. Cartwright, R. D., and Templeton, G. E. 1988. Biological limitations of *Protomyces gravidus* as a mycoherbicide for giant ragweed, *Ambrosia trifida*. Plant Dis. 72:580–582.

20. Charudattan, R. 1986. Integrated control of waterhyacinth (*Eichhornia crassipes*) with a pathogen, insects, and herbicides. Weed Sci. 34 (Suppl. 1):26–30.

21. Charudattan, R., Lenda, S. B., Kluepfel, M., and Osman, Y.A. 1985. Biocontrol efficacy of *Cercospora rodmanii* on waterhyacinth. Phytopathology 75:1263–1269.

22. Charudattan, R., Walker, H. L., Boyette, C. D., Ridings, W. H., TeBeest, D. O., Van Dyke, C. G., and Worsham, A. D. 1986. Evaluation of *Alternaria cassiae* as a mycoherbicide for sicklepod (*Cassia obtusifolia*) in regional field tests, South. Coop. Series Bull. 317, Alabama Agric. Exp. Stn., Auburn University.

23. Churchill, B. W. 1982. Mass production of microorganisms for biological control. Pages 139–156, *in*: R. Charudattan and H. L. Walker, eds., Biological control of weeds with plant pathogens. Wiley, New York.

24. Connick, W. J., Jr. 1982. Controlled release of the herbicides 2,4-D and dichlobenil from alginate gels. J. Appl. Polym. Sci. 27:3341–3348.

25. Conway, K. E. 1976. Evaluation of *Cercospora rodmanii* as a biological control of waterhyacinth. Photopathology 66:914–917.

26. Crowley, H. D. 1989. Personal communication. Mycogen Corporation, Ruston, LA.

27. Daigle, D. J., Connick, W. J., Jr., Quimby, P. C., Jr., Evans, J. P., Trask-Merrell, B., and Fulgham, F. E. 1990. Invert emulsions: delivery system and water source for the mycoherbicide, *Alternaria cassiae*. Weed Technol. (2:In press).

28. Daniel, J. T., Templeton, G. E., Smith, R. J., Jr., and Fox, W. T. 1973. Biological control of northern jointvetch in rice with an endemic fungal disease. Weed Sci. 21:303–307.

29. Freeman, T. E., and Charudattan, R. 1984. *Cercospora rodmanii* Conway—a biocontrol agent for waterhyacinth. Bulletin 842 (technical), Agricultural Experiment Station, Institute of Food and Agricultural Sciences, University of Florida, Gainesville, FL.

30. Gantotti, B. V. 1989. Personal communication. Philom Bios, Saskatoon, Sask., Canada.

31. Gray, P. P. 1986. Mass production of microorganisms, fermentation techniques, and equipment. Workshop Proceedings: Potential for Mycoherbicides in Australia, pp. 4–7. Agric. Research and Veterinary Center, Orange, New South Wales, Australia.

32. Hildebrand, D. C., and McCain, A. H. 1978. The use of various substrates for large-scale production of *Fusarium oxysporum* f.sp. *cannabis* inoculum. Phytopathology 68:1099–1101.

33. Katan, J., and Eshel, Y. 1973. Interactions between herbicides and plant pathogens. Residue Rev. 45:145–177.

34. Kenney, D. S. 1986. DeVine®—the way it was developed—an industrialist's view. Weed Sci. 34s:15–16.

35. Khodayari, K., and Smith, R. J., Jr. 1988. A mycoherbicide integrated with fungicides in rice, *Oryza sativa*. Weed Technol. 2:282–285.

36. Khodayari, K., Smith, R. J., Jr., Walker, J. T., and TeBeest, D. O. 1987. Applicators for a weed pathogen plus acifluorfen in soybeans. Weed Technol. 1:37–40.

37. Klerk, R. A., Smith, R. J., Jr., and TeBeest, D. O. 1985. Integration of a microbial herbicide into weed and pest control programs in rice (*Oryza sativa*). Weed Sci. 33:95–99.

38. McWhorter, C. G., and Chandler, J. M. 1982. Conventional weed control technology. Pages 5–27, *in*: R. Charudattan and H. L. Walker, eds., Biological control of weeds with plant pathogens. Wiley, New York.

39. McWhorter, C. G., Fulgham, F. E., and Barrantine, W. L. 1988. An air-assist spray nozzle for applying herbicides in ultra low volume. Weed Sci. 36:118–121.

40. Mitchell, J. K. 1986. *Dichotomophthora portulacae* causing black stem rot on common purslane in Texas. Plant Dis. 70:603.

41. Mitchell, J. K. 1988. *Gibbago trianthemae*, a recently described hyphomycete with bioherbicide potential for control of horse purslane (*Trianthema portulacastrum*). Plant Dis. 72:354–355.

42. Ormeno-Nunes, J., Reeleder, R. D., and Watson, A. K. 1988. A foliar disease of field bindweed (*Convolvulus arvensis*) caused by *Phomopsis convolvulus*. Plant Dis. 72:338–342.

43. Quimby, P. C., Jr. 1985. Pathogenic control of prickly sida and velvetleaf: an alternate technique for producing and testing *Fusarium lateritium*. Proc. South. Weed Sci. Soc. 38:365–371.

44. Quimby, P. C., Jr. 1989. Response of common cocklebur (*Xanthium strumarium*) to *Alternaria helianthi*. Weed Technol. 3:177–181.

45. Quimby, P. C., Jr., and Boyette, C. D. 1986. *Alternaria cassiae* can be integrated with selected herbicides. Proc. South. Weed Sci. Soc. 38:389.

46. Quimby, P. C., Jr., and Boyette, C. D. 1987. Production and application of biocontrol agents. Pages 265–280, *in*: C. G. McWhorter and M. R. Gebhardt, eds., *Methods of Applying Herbicides*. Weed Science Society, Champaign, IL.

47. Quimby, P. C., Jr., Fulgham, F. E., Boyette, C. D., and Connick, W. J., Jr. 1988. An invert emulsion replaces dew in biocontrol of sicklepod—a preliminary study. Pages 264–270, *in*: Pesticide Formulations and application System, Vol. 8, D. A. Hovde and G. B. Beestman, eds., ASTM-STP 980. American Society for Testing and Materials, Philadelphia.

48. Quimby, P. C., Jr., Fulgham, F. E., Boyette, C. D., and Hoagland, R. E. 1988. New formulations nozzles boost efficacy of pathogens for weed control. Proc. Weed Sci. Soc. 28:52.

49. Quimby, P. C., Jr., and Walker, H. L. 1982. Pathogens as mechanisms for integrated weed management. Weed Sci. (Suppl.) 30:30–34.

50. Ridings, W. H. 1986. Biological control of stranglervine in citrus—a researcher's view. Weed Sci. 34s:31–32.

51. Ridings, W. H., Mitchell, D. J., and El-Gholl, N. E. 1975. Biological control of *Morrenia odorata* by *Phytophthora citropthora* in the field. Proc. Am. Phytopathol. Soc. 2:79.

52. Ridings, W. H., Mitchell, D. J., Schoulties, C. L., and El-Gholl, N. E. 1976. Biological control of milkweed vine in Florida citrus groves with a pathotype of *Phytophthora citropthora*. Pages 224–240, *in*: T. E. Freeman, ed., Proc. 4th Int. Symp. Biol. Control Weeds. University of Florida, Gainesville.

53. Riely, J. A. 1989. Personal communication. Mycogen Corporation, Ruston, LA.

54. Smith, R. J. 1982. Integration of microbial herbicides with existing pest management programs. Pages 189–203, *in*: R. Charudattan and H. L. Walker, eds., Biological Control of Weeds with Plant Pathogens. Wiley, New York.

55. Smith, R. J. 1986. Biological control of northern jointvetch (*Aexchynomene virginica*) in rice (*Oryza sativa*) and soybeans (*Glycine max*)—a researcher's view. Weed Sci. 34(Suppl. 1): 17–23.

56. TeBeest, D. O., and Templeton, G. E. 1985. Mycoherbicides: progress in the biological control of weeds. Plant Dis. 69:6–10.

57. Templeton, G. E., and Smith, R. J., Jr. 1977. Managing weeds with pathogens. Pages 167–176, *in*: J. G. Horsfall and E. B. Cowling, eds., Plant Disease: An Advanced Treatise, Vol. 1. Academic Press, New York.

58. Templeton, G. E., TeBeest, D. O., and Smith, R. J. 1979. Biological Weed Control with Mycoherbicides. Ann. Rev. Phytopathol. 17:301–310.

59. Toussoun, T. A., Nash, S. N., and Snyder, W. C. 1960. The effect of nitrogen sources and glucose on the pathogenesis of *Fusarium solani* f. *phaseoli*. Phytopathology 50:137–140.

60. Tuite, J. 1969. Plant Pathological Methods: Fungi and Bacteria. Burgess, Minneapolis.

61. Van Dyke, C. G., and Winder, R. S. 1985. *Bipolaris sorghicola*; a potential mycoherbicide to johnsongrass. Proc. South. Weed Sci. 38:373.

62. Walker, H. L. 1980. *Alternaria macrospora* as a potential biocontrol agent for spurred anoda: production of spores for field studies. U.S. Dep. Agric. Adv. Agric. Tech. South. Series (ISSN 0193–3728), No. 12, 5 pp.

63. Walker, H. L. 1981. *Fusarium lateritium*: a pathogen of spurred anoda (*Anoda cristata*), prickly sida (*Sida spinosa*), and velvetleaf (*Abutilon theophrasti*). Weed Sci. 29:629–631.

64. Walker, H. L. 1981. Granular formulations of *Alternaria macrospora* for control of spurred anoda (*Anoda cristata*). Weed Sci. 29:342–345.

65. Walker, H. L., and Boyette, C. D. 1985. Biological control of sicklepod (*Cassia obtusifolia*) in soybeans (*Glycine max*) with *Alternaria cassiae*. Weed Sci. 33:212–215.

66. Walker, H. L., and Connick, Jr., W. J. 1983. Sodium alginate for production and formulation of mycoherbicides. Weed Sci. 33:333–338.

67. Walker, H. L., and Riley, J. A. 1982. Evaluation of *Alternaria cassiae* for the biocontrol of sicklepod (*Cassia obtusifolia*). Weed Sci. 30:651–654.

68. Weidemann, G. J. 1988. Effects of nutritional amendments for conidial production of *Fusarium solani* f.sp. *cucurbitae* on sodium alginate granules and control of Texas gourd. Plant Dis. 72:757–759.

69. Weidemann, G. J., and Templeton, G. E. 1988. Control of Texas gourd, *Cucurbita texana*, with *Fusarium solani* f.sp. *cucurbitae*. Weed Technol. 2:271–274.

70. Weidemann, G. J., and Templeton, G. E. 1988. Efficacy and soil persistence of *Fusarium solani* f.sp. *cucurbitae* for control of Texas gourd (*Cucurbita texana*). Plant Dis. 72:36–38.

71. Winder, R. S., and Van Dyke, C. G. 1987. The effect of various adjuvants in biological control of johnsongrass [*Sorghum halepense* (L.). Pers.] with the fungus *Bipolaris sorghicola*. Proc. Weed Sci. Soc. 27:128.

72. Wymore, L. A., Poirier, C., Watson, A. K., and Gotlieb, A. R. 1988. *Colletotrichum coccodes*, a potential bioherbicide for control of velvetleaf (*Abutilon theophrasti*). Plant Dis. 72:534–538.

73. Wymore, L. A., and Watson, A. K. 1986. An adjuvant increases survival and efficacy of *Colletotrichum coccodes*, a mycoherbicide for velvetleaf (*Abutilon theophrasti*). Phytopathology 76:1115–1116.

Economic Aspects of
Biological Control

13

Submerged Fermentation of Biological Herbicides

Larry J. Stowell

Introduction

Development of production methods for commercial manufacture of bioherbicides is in its infancy. However, fermentation of fungi for use in industrial processes, such as the manufacture of wine and beer, bioconversion of organic molecules, and antibiotic production, is well-known and established technology (33,35,37,50). Interestingly, production of a bioherbicide differs from many conventional fermentation processes, because for bioherbicides the final product is the living biomass of the plant pathogen itself rather than a fermentation by-product. Not only is high biomass yield essential for commercialization of these products, but the biomass must also be viable, stable in a package on the distributor's shelf, simple for the farmer to apply, and effective under the widely varying environmental conditions encountered in production agriculture.

The same characters that make a biological control agent safe in the environment—high degree of specificity for the target and short residual activity—are the factors that dictate the need for inexpensive production systems (3). For this reason this chapter concentrates on the use of submerged fermentation methods, which are readily available and more cost effective than other production methods. Alternative methods of producing biological herbicides have been discussed elsewhere (39). The challenge is therefore to develop techniques for identifying the best available fermentation medium and conditions and downstream process. Future development of bioherbicide production systems should focus on the use of existing industrial submerged fermentation and downstream processing technology to provide low-cost, high-quality products.

Production Review

Bioherbicides have been produced by a variety of techniques. Three of these techniques have been cited in use patents for bioherbicides and will be described briefly. The first process was developed by Conway et al. (14) for production of

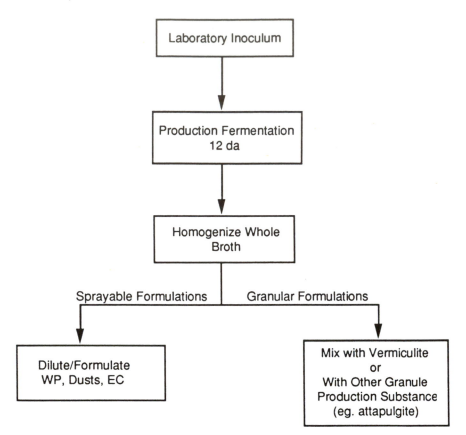

Figure 13.1. Process diagram for production of *Cercospora rodmanii* for control of water hyacinth (14).

Cercospora rodmanii (Figure 13.1) and utilizes surface culture of mycelium in Roux bottles followed by homogenization of the mycelium. After harvest, the mycelium is formulated into either granules or a sprayable material. Methods for stabilizing the mycelium were not described. However, this production strategy is simple and may be useful in early stages of product screening.

The second production process utilizes submerged fermentation of mycelium followed by harvesting and treating the mycelium to induce sporulation. This process was developed to circumvent the problems encountered in fermentation of fungi that do not readily sporulate in liquid culture (48,49) and is similar to spore production techniques for use in bioconversion of organic molecules (46). Following submerged fermentation, the whole broth is homogenized and either poured directly into trays or mixed with granules that immobilize the mycelium and provide a larger surface area for sporulation (Figure 13.2). In both cases, using whole broth or granules in shallow trays, the environment must be modified

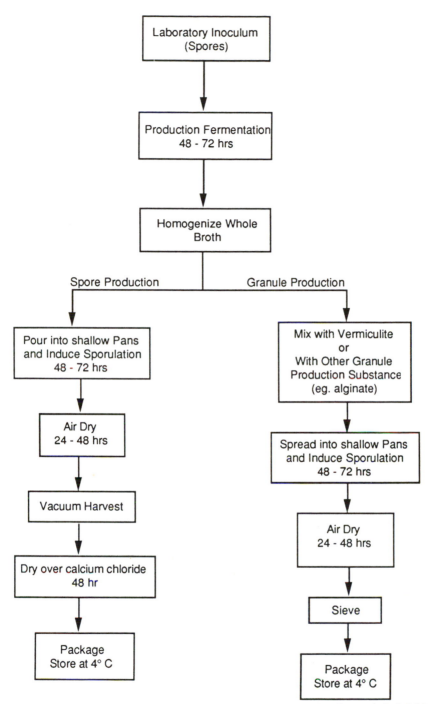

Figure 13.2. Process diagram for production of *Alternaria cassiae* for control of sickle-pod, showy crotalaria and coffee senna (48), and *Fusarium lateritium* (granules only) for control of prickley sida, velvetleaf, and spurred anoda (49).

to induce sporulation by providing light or removing CO_2 (16,39,48,49). Spores of *Alternaria cassiae* produced by these methods can be harvested by vacuum or the granular product can be applied directly to the field. The granular process has also been used to produce a *Fusarium*-based bioherbicide (49).

The third production scheme takes advantage of standard fermentation and downstream processing equipment (Figure 13.3). This technique has been used for production of several *Colletotrichum* sp. bioherbicides (12,18,41). Following submerged fermentation, the spores are separated from the mycelium by filtration, then concentrated by centrifugation. Freeze drying was used to stabilize spores. In a commercial production system a more common drying process (drum, spray, or fluid bed dryers) would probably replace freeze drying.

All of the preceding processes are valuable in discovery and early greenhouse and field evaluations of potential bioherbicides. However, a more economical process similar to that illustrated in Figure 13.4 is needed to bring bioherbicides to the commercial marketplace in the near future. In this case the number of steps in the process are minimized. The washing steps found in Figure 13.3 would be eliminated by modifying fermentation conditions to prevent production of unwanted by-products (e.g., germination inhibitors) or developing strains of the pathogen that do not produce by-products that require a washing step. In addition, protectants might be required to prevent loss of viability during the drying and rehydration of the bioherbicide prior to application in the field. Regardless of the nature of the organism, a successful bioherbicide product should conform to limitations imposed by existing industrial fermentation and downstream processing technology.

Production Alternatives

There are a variety of ways a bioherbicide might be produced, as evidenced earlier. However, only conventional fermentation and downstream processes are readily available for large-scale manufacture of these products. Fortunately, there is substantial evidence suggesting that most nonfastidious plant pathogenic fungi can be produced in submerged fermentation economically enough to be used as bioherbicides. In the following discussion the term *propagule* is used to describe any living portion of the pathogen that is capable of infecting and killing the target weed. For practical purposes (field application through 100-mesh screens), bioherbicide propagules should be less than 100 μm in length. Fortunately, only a few fungi that might be used for bioherbicides produce spores that are longer than 100 μm. Therefore, problems arise only when a bioherbicide candidate does not readily produce spores in liquid or when the fungus naturally produces large spores. In some cases (e.g., *Alternaria cassiae*), both reluctance to sporulate in liquid culture and large spore size (> 100 μm) can complicate development.

At least five different approaches can be identified for producing stable and

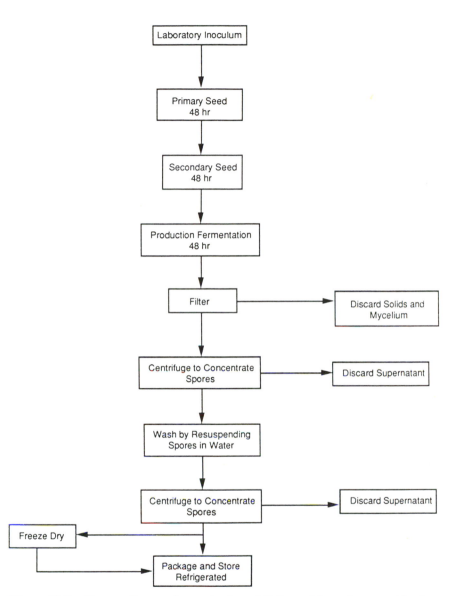

Figure 13.3. Process diagram for production of *Colleotorichum gloeosporoides* f.sp. *aeschynomene* for control of northern joint vetch (18), *Colletotrichum malvarum* for control of prickley sida, velvetleaf and other mallow species (41), and *Colletotrichum truncatum* for control of Florida beggarweed (12).

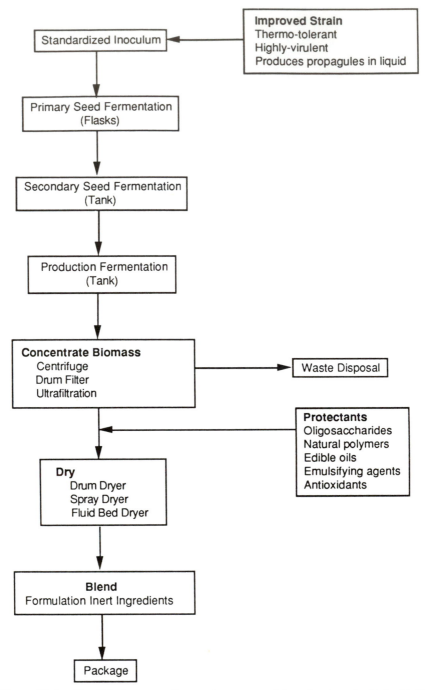

Figure 13.4. Simplified process diagram for commercial production of a bioherbicide using standard fermentation and downstream processing equipment.

small propagules (< 100 μm) of nonfastidious plant pathogens in submerged fermentation:

1. Submerged fermentation to produce mycelium followed by stabilization and milling to reduce propagule size.
2. Control of mycelium morphology during fermentation to yield small hyphal strands less than 100 μm long or mycelial pellets less than 100 μm in diameter.
3. Control of fungal dimorphism to induce yeastlike growth of filamentous fungi.
4. Induction of sporulation and control of spore size in submerged fermentation.
5. Strain improvement to alter morphology or sporulation characteristics in submerged fermentation.

Milling Mycelium

Production of mycelium in submerged fermentation has been evaluated for many fungi (5,10,11,35,24). Unfortunately, little information has been reported on the use of downstream processes to stabilize mycelium produced in submerged fermentation for use in biological control of plant pests. However, McCabe and Soper (28) developed the technique illustrated in Figure 13.5 for production of an insect pathogenic fungus. Mycelium is produced by liquid fermentation and harvested by filtration. Protectants are applied to the harvested mycelial cake to help maintain viability during drying. After drying, the mycelium is milled to an appropriate size. Similar techniques have been developed for *A. cassiae* and have provided equivalent activity to spores of the fungus on a viable propagule basis (39). Churchill also suggested that similar procedures might be valuable if the fungus can survive the process (13). Further research is needed to demonstrate the value of this and similar mycelium milling procedures.

Mycelial Morphology

Propagules might also be produced in submerged fermentation by altering the fermentation environment and medium to produce small mycelial fragments or pellets. Shear stress, initial inoculum load, and pH have been implicated in control of pellet morphology of *Aspergillus niger* (2,32). Pellets as small as 50 μm were reported at high shear. Short hyphal strands of *A. cassiae* have been produced in submerged fermentation using high agitation rates (unpublished). Likewise, a variety of morphological changes were reported by Borrow et al. working with *Gibberella fujikuroi* in stirred culture (5). The advantage of producing small mycelial pellets or short fragments is twofold. First, viable propagules

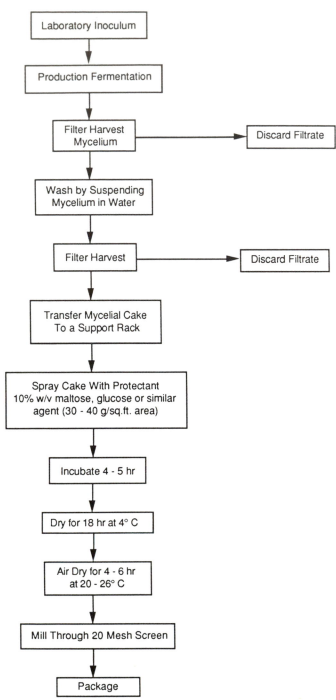

Figure 13.5. Process diagram for production of dried and milled mycelium of entomo-pathic fungal insect control agents (28).

would not be destroyed by the milling process, and second, the cost of downstream processing would be reduced by eliminating the milling process.

Dimorphism

Control of fungal dimorphism offers a promising opportunity in bioherbicide fermentation development and has been studied for a variety of fungi (5,19,25,33,38,51). *Dimorphism* refers to the ability of some fungi to grow as filamentous hyphae and also as a budding yeastlike form. Control of fungal dimorphism was first studied and understood by Pasteur more than a century ago (33). Nonetheless, the control of plant pathogen dimorphism has not been widely evaluated. However, production of COLLEGO (*Colletotrichum gloeosporioides* f. sp. *aeschoneme*) utilizes this technique. In this case 80–85% of the propagules are produced as yeastlike "fission spores," 8–10% conidia, and 5% blastospores and arthrospores (13). Unfortunately, the exact production conditions (medium, temperature, pH, aeration, agitation, downstream processing, etc.) developed for this bioherbicide remain proprietary. Further research is needed to determine whether dimorphism can be induced in other plant pathogenic fungi, particularly fungi that are reluctant to sporulate in liquid (e.g., *Alternaria*, *Drechslera*, etc.).

Sporulation in Liquid

Further research into induction of sporulation in liquid culture should not be overlooked. Many fungi have been induced to sporulate in liquid culture (1,47). For example, Vezina et al. (47) studied 19 species of filamentous fungi and were able to induce sporulation in liquid culture by 18 species. One *Fusarium* sp. yielded 2.3×10^{12} spores/l in a defined medium containing:

glucose	30g	
$NaNO_3$	3g	
K_2HPO_4	1g	
$MgSO_4 \cdot 7H_2O$	0.5g	
KCl	0.5g	
Zn	2000 μg	as $ZnSO_4 \cdot 7H2O$
Fe	200 μg	as $Fe_2(SO_4)_3 \cdot 6H_2O$
Mn	20 μg	as $MnCl_4 \cdot 4H_2O$
B	10 μg	as $Na_2B_4O_7$
Cu	100 μg	as $CuSO_4$
distilled water to:	1l	

pH adjusted to 5.5 with glacial acetic acid.

In this fermentation medium, optimum nutrition was needed for potassium phosphate, magnesium sulfate, and potassium chloride. Vitamins had no effect and both NaCl and CaCl were detrimental to spore production. In addition, the number of spores in the inoculum and temperatures between 25 and 32°C did not affect sporulation. The average spore yield for all 18 species of fungi that were induced to sporulate in liquid was 3×10^{11} spores/L. Vezina et al. concluded that a delicate balance between nutrition (carbon, nitrogen, and minerals) and environment (temperature, pH, agitation/aeration) controls sporulation of filamentous fungi in submerged fermentation. Once sporulation has been achieved in liquid, spore morphology might also be controlled by small alterations in nutrition (31).

Strain Improvement

Strain improvement is discussed last because the procedures needed to induce and select morphological variants that maintain virulence and ability to kill the target weed under field conditions may require years of costly research and development. This technique entails selection of natural variants or classical genetic manipulations in addition to more recent developments in mutagenesis (random or site directed) as well as rDNA technology. The growing opportunity for controlling morphology and differentiation of fungi was expressed by Allerman et al. when they wrote "The possibility now arises for identifying genes which may function as master switches for developmental processes" (1). For example, Cotty determined that CO_2 modulated sporulation in *Alternaria tagetica* and subsequently selected a mutant that was insensitive to CO_2. The mutant sporulated in the presence of CO_2 and was also capable of sporulation in submerged fermentation (15,16). Likewise, a simple process of selection might utilize liquid culture and passage through a 100-mesh screen to select small propagules. This selection process would be repeated until a stable small-propagule strain was selected. Although the use of rDNA technology for modification and control of fungal morphology is a more distant goal, the advances in understanding and manipulating plant pathogenic fungi holds great promise for the future (29).

Downstream Processing

Downstream processes include all steps in manufacture from the point the whole broth leaves the fermentation tank (including concentration, stabilization, and drying through packaging). Concentrating and stabilizing propagules produced in submerged fermentation may provide the most difficult challenge in production of bioherbicides. Fortunately, many lessons can be gleaned from industrial fermentation research, and, in particular, development of dried bakers' yeast. Since shortly after the turn of the century, bakers' yeast has been produced in aerobic deep-tank fermentations using incremental feeding techniques. The live biomass was concentrated by centrifugation to yield a cream with 15–20% dry matter. The initial concentration step was followed by a filtration or pressing procedure to bring the dry matter content up to 27–30%. The moist cake was then stored under refrigerated conditions. More recent developments incorporated drying the biomass using drum, spray, or fluidized bed driers. These techniques raise the dry matter content to 90–96%. Unfortunately, the first attempts to dry bakers' yeast products resulted in poor quality and low activity. However, development of yeast strains that were more tolerant to rehydration and use of protectants (fatty acid esters, polyols, antioxidants, etc.) led to modern dried bakers' yeast. The new products can be stored at temperatures reaching 30°C and are viable on the shelf for a longer period than the moist cake bakers' yeast. The disadvantages of drying the yeast include increased cost of production caused by the added drying step and reduction of activity to 65% of that of moist cake products on a dry weight basis (8,9). These disadvantages are far outweighed by the advantages of increased shelf life and elimination of refrigerated storage conditions. Similar approaches to development of bioherbicide products may lead to comparable successes.

Microorganisms are successfully stored following gentle removal of water by air drying or lyophilization. Anhydrobiosis is a natural method that many organisms display to survive or overwinter periods of adverse environmental conditions (e.g., spores, seeds, nematode galls of *Anguina tritici*). Removal of water to stabilize propagules is therefore a logical and also a demonstrated method of stabilizing bioherbicides (e.g., COLLEGO). However, other methods might be discovered that allow propagules to be stabilized without thorough drying. For example, conidia of some powdery mildews contain as much as 75% of their mass as water, compared to 6–25% for many other fungi (36). However, they do not usually germinate without an additional stimulus (e.g., the host surface or conducive environments). While discussing fungistasis, Sussman noted that low respiratory coefficients of dormant spores preserve endogenous energy reserves. He also suggested that antibiotics may have a positive role in nature in the survival of dormant spores by reversibly lowering the respiratory rate of cells and preserving their valuable nutrient reserves (40). Spore germination inhibitors have been described for many fungi, including *Colletotrichum gloeosporioides* (27,45). Therefore, induction of reversible fungistasis at the end of fermentation or during downstream processing may allow moist fermentation creams to be formulated and stored without drying, thereby reducing damage to the organism caused by the drying procedures and reducing the cost of production by eliminating the costly drying process.

Medium and Process Development

There are no specific rules to guide researchers in their attempts to develop and optimize new fermentation media. A medium that provides the needs of one organism may not provide the needs of others, even if they are closely related (47). This unpredictability of microorganisms has led some fermentation scientists to claim that medium and process development is as much an art as a science. However, the educated guesses that experienced fermentation scientists seem to pull out of thin air are based solidly on their past experiences—experiences that can only be gained by trial and many times error. The following sections describe a few important fermentation terms and experimental designs. They are meant to communicate the flavor of fermentation research and to be a guide to literature that might assist in development of industrial bioherbicide fermentation processes.

Specific Growth Rate

One of the most common terms used to communicate the results of a fermentation process is specific growth rate. It is a measure of biomass increase (either g dry weight (dwt)/L, or propagules/L) and is represented by the symbol μ.

$$\text{Specific growth rate} = (\text{change in biomass})/(\text{change in time})$$
$$\mu x = dx/dt$$

where $x \equiv$ biomass concentration (expressed in grams or cells per liter)

$t \equiv$ time expressed in hours

$\mu \equiv$ specific growth rate in hours^{-1}

Integrating the preceding relationship results in the familiar exponential growth equation:

$$x_{t2} = x_{t1} e^{\mu \Delta t}$$

where $x_{ty} \equiv$ biomass concentration at time y

$e \equiv$ the base of the natural system of logarithms

$\Delta t \equiv$ change in time between $t1$ and $t2$ in hours

Specific growth rate can be computed from experimental data by solving the equation for μ after taking the natural logarithm of each side of the equation:

$$\ln(x_{t2}) = \ln(x_{t1}) + \mu \Delta t$$
$$\mu = [\ln(x_{t2}) + \ln(x_{t1})]/\Delta t$$

Alternatively, regression analysis can be utilized to determine the slope of the growth curve to yield not only μ but also an estimate of the variation in the data. The following example illustrates the use of both of these techniques to determine μ from experimental data.

Specific growth rate can be calculated using two points on the exponential

Time (hr)	Yield (g dwt/l)	ln(g dwt/l)
0	0.0	
96	0.2	-1.6
120	2.0	0.7
126	3.2	1.2
132	5.4	1.7
141	11.0	2.4

Figure 13.6. Results of a test fermentation using *Sclerotinia sclerotiorum* in a 6% Nutrisoy, 4% Maltrin M-100 medium (50 ml working volume in a 500-ml nonbaffled flask, 150 rpm, 0.75-inch stroke).

portion of the growth curve. For example, Figure 13.6 lists the dry weight yield of a *Sclerotinia sclerotiorum* fermentation. Note the long lag period of 96 hours after inoculation before the fungus began to grow. Figure 13.7 graphically represents the data in Figure 13.6. To determine specific growth rate select two points from the exponential growth portion of the curve. The calculation of μ between 120 and 141 hours follows:

$$\mu = (\ln(x_{t141}) - \ln(x_{t120}))/\Delta t = (2.4 - 0.7)/21 \text{ hr} = 0.08 \text{ hr}^{-1}$$

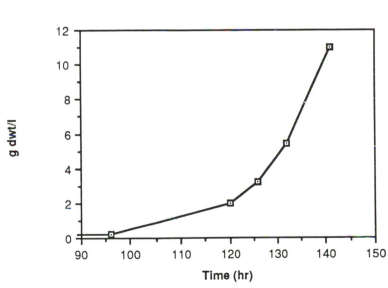

Figure 13.7 Linear plot of data listed in Table 13.1. Note the exponential growth after an initial lag of 96 hr.

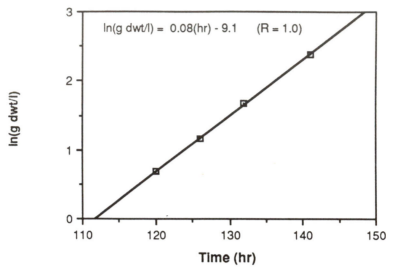

Figure 13.8.   Plot of natural log transformed yield (g dwt/L) data and linear regression analysis. The slope of the line is $\mu(0.08 \text{ hr}^{-1})$.

The alternative method of determining μ is illustrated in Figure 13.8. The slope of the regression equation is μ (0.08). The regression coefficient indicates the equation fits the line almost perfectly ($R = 1.0$). Low regression coefficients, below 0.8, may indicate that the organism is not growing exponentially or that there is a systematic error in the data.

Specific growth rate is widely used to communicate the results of fermentation experiments and the influence of nutritional and environmental conditions of a fermentation process on the organism. It should be monitored when evaluating or describing a fermentation process and may need to be slowed or controlled to produce a propagule that has the desired morphology, contains adequate energy reserves, and is viable and virulent under field conditions. Additionally, the time needed to double the biomass in a fermentation process, the doubling time (t_d), can be computed from the specific growth rate as follows:

$$t_d = \ln (2.00)/\mu$$
$$t_d = 0.69/0.08 \text{ hr}^{-1} = 8.63 \text{ hr}$$

In this example the fungus doubled in biomass every 8.6 hr. As is illustrated later, the doubling time is a critical factor in the cost of manufacturing any bioherbicide.

Mass Balance

Mass balance refers to the stoichiometry of a fermentation process. Ideally, the nutrients supplied in the fermentation medium can be accounted for at the end of the fermentation. Nutrients may be:

Incorporated into the fungal mycelium

Not utilized completely, the surplus remaining in the fermentation broth

Transformed into other molecules that remain in the fermentation broth

Utilized or broken down and expelled as gaseous fermentation by-products

The object of monitoring mass balance is to understand how the organism utilizes the nutrients supplied and to develop a fermentation medium that provides all the needed nutrients without large amounts of excess nutrients or undesired by-products at the end of the fermentation. Mass balance also identifies the needs of the organism to allow modification of the medium when batches of nutrients do not contain the desired composition or water supplies do not provide the needed minerals. Unfortunately, monitoring the composition of fermentation media and microorganisms usually requires time-consuming processes and expensive analytical equipment. However, major medium components, for example, glucose or other carbohydrates, can be monitored using simple analytical techniques (4,10,11,50,37). As a rule of thumb, gather as much information about each fermentation experiment as is technically possible under your own laboratory conditions.

Carbon and Nitrogen

Careful selection of carbon and nitrogen sources as well as mineral supplements is necessary for optimum growth of fungi (30). Chemical analysis of fungi reveals a range in carbon to nitrogen composition between 4:1 (carbon/nitrogen) to 9:1 (37). Therefore, a fermentation medium providing carbon and nitrogen within these ranges of ratios might be balanced for these elements. However, the medium that yielded the highest number of *Fusarium* spores per liter reported by Vezina et al. had a C/N ratio of 2.4:1 (47). In addition to the ratio of carbon to nitrogen, the sources of carbon and nitrogen may influence the virulence and stability of the bioherbicide.

The ratio of carbon to nitrogen can be estimated by determining the approximate amount of carbon delivered by carbohydrates in the medium and the approximate amount of nitrogen provided by proteins or inorganic compounds in the nitrogen source. For example, the highest yielding medium illustrated in Figure 13.9 contained 60 g/L Nutrisoy and 40 g/L Maltrin M–100. The total amount of carbohydrates provided in the medium was 40 g of Maltrin M–100 (a dextrin containing 53% carbon, see Figure 13.10) in addition to carbohydrates in the Nutrisoy which comprise 30% Figure (13.12) of the mass added, 18 g. In order to determine the contribution of the carbohydrates in the Nutrisoy to the carbon pool, the average value of 41% carbon composition for carbohydrates was used (Figure 13.11). The total number of grams of carbon added as carbohydrates is:

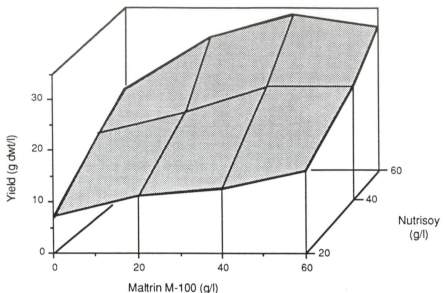

Figure 13.9. Response surface model for yield (g dwt/L) of *Sclerotinia sclerotiorum* cultured in various concentrations of Maltrin M-100 and Nutrisoy.

$$[(40 \text{ g Maltrin M–100}) \times (52\% \text{ carbon})] + [(18 \text{ g carbohydrate in Nutrisoy}) \times (41\% \text{ carbon})] = 28.2 \text{ g carbon}.$$

The nitrogen composition of Nutrisoy is estimated in Figure 13.12 at 8.2%, providing

$$(60 \text{ g Nutrisoy}) \times (8.2\% \text{ nitrogen}) = 4.9 \text{ g of nitrogen}$$

The C/N ratio is 28.2 g C to 4.9 g N, or 5.7:1.

Although the C/N ratio is helpful in keeping medium development within reasonable ranges, it is only one of many considerations in medium development.

The physiological role of the carbon source may be just as important as the balance or imbalance provided by the carbon source in the medium. For example, glucose serves a regulatory function (catabolite repression) in addition to its role as a carbon source. High glucose levels might be a signal to the organism that it is in a nutrient-rich environment and therefore does not need many of the inducible enzyme systems that break down complex molecules. Although these enzymes are not essential in a glucose-rich environment, they may be necessary for virulence of a pathogen on the host plant. For example, the synthesis and activation of pectolytic enzymes that are essential to the virulence of *Fusarium oxysporum* is retarded in the presence of glucose (52). In addition to catabolite repression of enzyme systems that are important in pathogenesis, catabolite

Figure 13.10. Some carbohydrates that might be used as a sole source of carbon by fungi (Garraway and Evans 1984), their molecular weights and elemental composition on a percent by weight basis.

Carbohydrate	MW	C	H	O	N	%C	%H	%O	%N
Adonitol	152.2	5	12	5		39.47	7.95	52.58	
Arabinose	150.1	5	10	5		40.00	6.71	53.29	
Arabinitol	152.2	5	12	5		39.47	7.95	52.58	
Cellobiose	342.3	12	22	11		42.10	6.48	51.42	
Chitin	Poly	8	13	5	1	47.29	6.45	39.37	6.89
Dextrin	Poly	6	10	5		52.56	3.65	43.80	
Dulcitol	182.2	6	14	6		39.56	7.75	52.70	
Erythritol	122.1	4	10	4		39.34	8.25	52.41	
Erythrose	20.11	4	8	4		40.00	6.71	53.29	
Fucose	164.2	6	12	5		43.90	7.37	48.73	
Fructose	180.2	6	12	6		40.00	6.72	53.29	
Galactose	180.2	6	12	6		40.00	6.72	53.29	
Gluconic acid	196.2	6	12	7		36.74	6.17	57.10	
Glucosamine	179.2	6	13	5	1	40.22	7.31	44.65	7.82
Glucose	180.2	6	12	6		40.00	6.72	53.29	
Glucuronic acid	194.1	6	10	7		37.12	5.19	57.69	
Glycerol	92.09	3	8	3		39.12	8.75	52.12	
Glycogen	Poly	6	10	5		52.56	3.65	43.80	
Inositol	180.2	6	12	6		40.00	6.71	53.29	
2-Ketogluconic acid	194.1	6	10	7		37.12	5.19	57.69	
Kojic acid	142.1	6	6	4		50.71	4.26	45.03	
Lactose	342.3	12	22	11		42.10	6.48	51.42	
Lyxose	150.1	5	10	5		40.00	6.71	53.29	
Maltose	342.3	12	22	11		42.10	6.48	51.42	
Mannitol	182.2	6	14	6		39.56	7.74	52.70	
Mannose	180.2	6	12	6		40.00	6.71	53.29	
Melibiose	342.3	12	22	11		42.10	6.49	51.42	
α-Methylglucoside	194.2	7	14	6		43.30	7.27	49.44	
Raffinose	504.5	18	32	16		42.86	6.39	50.72	
Rhamnose	164.2	6	12	5		43.90	7.37	48.73	
Ribose	150.1	5	10	5		40.00	6.71	53.29	
Salicin	286.3	13	18	7		54.54	6.34	39.12	
Sorbitol	182.2	6	14	6		39.56	7.75	52.70	
Sorbose	180.2	6	12	6		40.00	6.71	53.29	
Starch	Poly	6	10	5		52.56	3.65	43.80	
Sucrose	342.3	12	22	11		42.10	6.48	51.42	
Trehalose	342.3	12	22	11		42.10	6.48	51.42	
Turanose	342.3	12	22	11		42.10	6.48	51.42	
Xylose	150.1	5	10	5		40.00	6.71	53.29	
Average						41.35	6.80	51.45	

Figure 13.11.   Molecular weight and elemental composition (percent by weight) of some inorganic and amino nitrogen sources.

Nitrogen source	MW	C	H	O	N	P	S	K	Na	Ca	Cl
NaNO$_2$	69.00			43.38	20.30				33.32		
KNO$_2$	85.10			37.60	16.46			45.94			
Ca(NO$_2$)$_2$	132.10			48.45	21.21					30.34	
NaNO$_3$	85.01			56.47	16.48				27.05		
KNO$_3$	101.10			47.47	13.86			38.67			
Ca(NO$_3$)$_2$	164.10			58.50	17.07					24.42	
NH$_4$NO$_3$	80.05		5.04	59.96	35.00						
(NH$_4$)$_2$SO$_4$	132.14		6.10	48.43	21.20		24.27				
NH$_4$CL	53.50		7.54		26.18						66.28
Ala	89.09	40.44	7.92	35.92	15.72						
Arg	174.20	41.36	8.10	18.37	32.16						
Asn	132.12	36.36	6.10	36.33	21.20						
Asp	133.10	36.09	5.30	48.08	10.52						
Cys	121.16	29.74	5.82	26.41	11.56		26.47				
Gln	146.15	41.09	6.90	32.84	19.17						
Glu	147.13	40.18	6.16	43.50	9.52						
Gly	75.07	32.00	6.71	42.63	18.66						
His	155.16	46.44	5.85	20.26	27.08						
Ile	131.17	54.94	9.99	24.49	10.68						
Leu	131.17	54.94	9.99	24.40	10.67						
Lys	146.19	42.29	9.65	21.89	19.16						
Met	149.21	40.25	7.43	21.45	9.39		21.49				
Phe	165.19	65.43	6.71	19.37	8.48						
Pro	115.13	52.16	7.88	27.79	12.17						
Ser	105.09	34.28	6.71	45.67	13.33						
Thr	119.12	40.33	7.62	40.29	11.76						
Trp	204.22	64.69	5.92	15.67	13.72						
Tyr	181.19	59.66	6.12	26.49	7.73						
Val	117.15	51.26	9.46	27.32	11.96						

repression has also been implicated in control of dimorphism in *Mucor* (38). Accumulation of the critical storage and survival compounds trehalose (43) and lipids (21,22) may also be regulated by glucose and other fermentation conditions. Potential problems caused by catabolite repression do not rule out the use of glucose as an effective and inexpensive carbon source. However, the regulatory role of this carbon source and other nutrients should be considered during medium development.

Although carbon may be obtained from a variety of sources, some carbohydrates that have been reported as being used by fungi as a sole source of carbon are listed in Figure 13.10. To simplify medium design all the compounds have been listed with their respective percent composition of carbon, hydrogen, oxygen, and in two cases, nitrogen.

Figure 13.12. Nutritional composition of several complex nitrogen sources. All values are expressed in percent dry weight from manufacturers specifications. ADM: Archer Daniels Midland Co., Clinton, IA; Kalamazoo: Kalamazoo Paper Chemicals, Richland, MI; GPC: Grain Processing Corp., Muscatine, IA.

	ADM Toasted Nutrisoy 7B	Kalamazoo NPC Sonic Yeast	GPC 60% Corn Gluten meal	GPC Solulac Distillers Sol	Kalamazoo Potato Protein
Protein	52.00	50.00	60.00	26.00	80.00
N est.	8.15	7.84	9.40	4.08	12.54
Moisture	8.00	4.00			14.00
Fat	1.50	0.50	1.50	2.50	1.00
Fiber	3.00	1.50	4.00	8.00	
Ash	6.0000	6.5000	1.8000	7.0000	2.0000
Na	0.1500		0.0300	0.0500	
K	2.5000		0.4500	1.1200	
P	0.7000		0.7000	0.8800	
Ca	0.2500		0.0200	0.3500	
Fe	0.0070		0.0167		0.0040
Mg	0.2400		0.1500	0.3700	
Cl			0.1000		
Cu	0.0016		0.0022		
Zn	0.0051		0.0042		
Mn	0.0035				
S			0.8300		
Cr			0.0002		
Mo			0.0001		
Se			0.0001	0.0000	
Co			0.0000		
B					
Ala	2.22	7.00	5.20		4.10
Arg	3.79	6.10	1.90	1.04	4.90
Asn					
Asp	5.69	11.70	3.60		11.00
Cys	0.72	0.80	1.10	0.50	
Gln					
Glu	9.16	16.70	13.80		12.20
Gly	2.10	4.50	1.60	0.98	4.50
His	1.35	2.20	1.20	0.53	3.80
Ile	2.51	4.20	2.30	1.00	5.30
Leu	4.08	8.20	10.10	2.88	9.20
Lys	3.30	6.90	1.00	0.80	8.30
Met	0.72	1.40	1.90	0.50	2.10
Phe	2.60	4.20	3.80	1.21	5.20
Pro	3.04	4.50	5.50		9.00
Ser	2.52	5.80	3.10		4.20
Thr	2.17	5.00	2.00	0.91	4.80

(continued)

Figure 13.12.   (continued)

	ADM Toasted Nutrisoy 7B	Kalamazoo NPC Sonic Yeast	GPC 60% Corn Gluten meal	GPC Solulac Distillers Sol	Kalamazoo Potato Protein
Trp	0.74	0.70	0.30	0.25	
Tyr	1.70	3.80	2.90	0.94	4.70
Val	2.73	6.50	2.70	1.28	5.50
Vitamin A			0.0100		
Vitamin B$_{12}$				0.0000	
Vitamine D					
Vitamine E				0.0075	
Vitamine K					
Thiamine	0.0005	0.0100	0.0000	0.0006	
Riboflavin	0.0002	0.0050	0.0002	0.0017	
Ascorbic acid					
Biotin		0.0002	0.0000	0.0001	
Folic acid	0.0003	0.0005		0.0002	
Niacin	0.0027	0.0400	0.0015	0.0106	
Pantothenic acid	0.0025	0.0200	0.0003	0.0011	
Beta Carotene			0.0055		
Pyridoxine (B$_6$)	0.0006	0.0050	0.0006	0.0010	
Inositol	0.0025	0.4500	0.1890	0.6030	
Choline		0.2500	0.2200	0.4930	
NH3					1.20
Linoleic acid			3.20		
Carbohydrates	30.00				
Lactic acid				4.00	

Nitrogen is usually provided as a protein in complex plant meals or fermentation by-products in industrial fermentations. Inorganic compounds and amino acids may also be utilized to augment the nitrogen provided by a complex source or to study the physiology of the bioherbicide in defined media. Figure 13.11 lists several commonly used inorganic and amino nitrogen sources and their elemental composition. Although not comprehensive, this list may assist in medium design. Figure 13.12 lists the composition of several complex nutrient sources for comparison. All values have been converted to percent by weight to ease computations in medium design. A comprehensive summary of nutrient suppliers can be found in the *Manual of Industrial Microbiology and Biotechnology* (30).

Minerals and Vitamins

In addition to carbon and nitrogen, minerals and vitamins may be required for pathogen virulence or stability. For example, like other polyketides, alternariol (a toxin produced by some *Alternaria* sp.) synthesis is stimulated by transition metals, in particular Zn^{2+} and Mn^{2+}. Moreover, addition of both Zn^{2+} and Mn^{2+} results in a synergistic interaction and greatly enhanced production of alternariol by *Alternaria* sp. (17). Some *Fusarium* sp. require specific levels of certain minerals for sporulation (47). Garraway and Evans provide an introduction to the role of both minerals and vitamins in growth and development of fungi (24). Koser provides a sound overview of the role of vitamins in the growth and development of bacteria and yeast (26). Finally, some major and minor elements that have been used in culture of microorganisms are listed in Figure 13.13. As with the other tables, the elemental composition of each material is listed as a percentage by weight.

Environment

The environmental conditions during fermentation are equally as important as the nutrition provided by the medium. Fortunately, the physical principles that govern the environment of the bioreactor (flask or fermentor) are fairly well understood. The factors that are of primary concern include:

Temperature

pH

Shear (agitation, flask design, fermentor design, agitator tip speed)

Aeration (oxygen and carbon dioxide mass transfer rates, antifoam interactions)

Rheology (medium flow characteristics with and without mycelial growth)

All of these factors and their interaction with microorganism growth are discussed by Stanbury and Whitaker (37) and Wang et al (50). Oxygen requirements, measurement, and control of aeration in submerged culture has been reviewed by Brown (7). A discussion of the value of the flask bioreactor in bioengineering has been published by Freedman (23). The flask is the most valuable tool for evaluating fermentation media and conditions. Early field development can also be carried out using flask-produced inoculum. In fact, the yield assumptions that will be discussed next reveal that 1 L of fermentation broth should treat approximately 1 acre before production economics are highly profitable. At this application rate (1 L/A) only 10 L of fermentation broth are needed to treat the maximum field area permitted by the Environmental Protection Agency in the United States (10 acres) for application of pesticides for research purposes without an experimental use permit or full registration. Once a production system has been developed in a flask, it will probably scale up into a fermentor with only minor modifications. Therefore, costly fermentation equipment is not essential for early assessment of the commercial potential of bioherbicide products. However, control of fermentation conditions can be difficult using flasks as the only research

Figure 13.13. Molecular weight and elemental composition (percent by weight) of some major and minor mineral nutrient sources.

Chemical	MW				Composition
H_3BO_3	61.84	B 17.50	O 77.62	H 4.88	
$Na_2B_4O_7$	201.27	B 21.50	O 55.65	Na 22.84	
$Na_2B_4O_7 \cdot 10H_2O$	381.22	B 11.08	O 71.35	Na 12.06	H 5.25
$Ca(NO3)2 \cdot 4H2O$	246.08	Ca 16.26	O 65.04	N 11.38	H 7.32
$Ca(OH)_2$	74.10	Ca 54.09	O 43.19	H 2.72	
$CaCl_2$	110.99	Ca 36.11	Cl 36.89		
$CaCl_2 \cdot 2H_2O$	146.98	Ca 27.27	O 21.77	Cl 48.24	H 2.72
$CaCO_3$	100.09	Ca 40.04	O 47.96	C 12.00	
$CoCl_2$	129.85	Co 45.39	Cl 54.61		
$CoCl_2 \cdot 6H_2O$	237.83	Co 24.78	O 40.37	Cl 29.81	H 5.05
$CoSO_4$	155.00	Co 38.03	O 41.29	S 20.68	
$CoSO_4 \cdot 7H_2O$	280.99	Co 20.97	O 62.64	S 11.41	H 4.98
$Cu(NO_3)_2$	187.56	Cu 33.88	O 51.18	N 14.94	
$CuCl_2$	134.45	Cu 47.26	Cl 52.74		
$CuCO_3 \cdot Cu(OH)_2$	221.11	Cu 57.47	O 36.18	C 5.43	H 0.91
$CuSO_4$	159.61	Cu 39.81	O 40.10	S 20.09	
$CuSO_4 \cdot 5H_2O$	249.61	Cu 25.46	O 57.69	S 12.84	H 4.01
$Fe(NH_4)_2(SO_4)_2 \cdot 6H_2O$	391.97	Fe 14.25	O 57.15	N 7.14	S 16.36 H 5.10
$Fe(NO_3)_3$	241.87	Fe 23.09	O 59.54	N 17.37	
$Fe_2(SO_4)_3$	399.88	Fe 27.93	O 48.01	S 24.06	
$FeCl_2$	126.76	Fe 44.06	Cl 55.94		
$FeCl_3$	162.22	Fe 34.43	Cl 65.57		
$FeSO_4$	151.91	Fe 36.77	O 42.13	S 21.10	
$FeSO_4 \cdot 7H_2O$	277.91	Fe 20.10	O 63.33	S 11.54	H 5.04
K_2SO_4	174.26	K 44.87	S 18.40	O 36.73	
KCl	74.55	K 52.44	Cl 47.56		
KNO_3	101.10	K 38.67	O 47.47	N 13.86	
KOH	56.10	K 69.69	O 28.52	H 1.80	
$Mg(NO_3)_2$	148.32	Mg 16.39	O 64.73	N 18.88	
$MgCl_2$	95.23	Mg 25.54	Cl 37.08		

(continued)

tool, and fermentors will need to be used to assess more accurately the commercial potential of the bioherbicide.

Experimental Design

Many experimental design techniques can be implemented to evaluate the interactions between nutrients and the fermentation environment. A system with proven success utilizes response surface models or multiple regression analysis. For example, four levels of a dextrin carbon source (Maltrin M–100) and three levels of a nitrogen source (Nutrisoy) were evaluated in a dozen flasks (Figure 13.9). The results of the experiment were plotted in three dimensions and multiple regression analysis was used to provide a mathematical description of the interac-

Figure 13.13.   *(continued)*

Chemical	MW				Composition
MgCl2•6H₂O	203.32	Mg 11.95	O 47.22	Cl 19.48	H 5.96
MgSO₄	120.38	Mg 20.20	O 53.16	S 26.63	
MgSO₄•7H₂O	246.38	Mg 9.86	O 71.43	S 13.01	H 5.68
Mn(NO₃)₂	178.95	Mn 30.70	O 53.65	N 15.65	
MnCl₂	125.84	Mn 43.66	Cl 56.34		
MnCl₂•4H₂O	191.84	Mn 28.64	O 33.36	Cl 36.96	H 1.04
MnSO₄	151.00	Mn 36.38	O 42.38	S 21.23	
MnSO₄•4H₂O	223.00	Mn 24.64	O 57.40	S 14.38	H 3.59
MnSO₄•H₂O	169.00	Mn 32.51	O 47.34	S 18.97	H 1.18
MoO₃	143.95	Mo 66.66	O 33.34		
(NH₄)₆Mo₇N₆O₂₄	1163.89	Mo 57.71	O 32.99	N 7.22	H 2.08
(NH₄)₆Mo₇N₆O₂₄•4H₂O	1235.89	Mo 54.35	O 36.26	N 6.79	H 2.59
Na₂MoO₄	205.92	Mo 46.59	O 31.08	Na 22.33	
Na₂MoO₄•2H₂O	241.92	Mo 39.66	O 39.69	Na 19.01	H 1.65
Na₂SO₃	126.06	Na 36.49	O 25.44	S 25.44	
Na₂SO₄	142.06	Na 32.38	O 45.06	S 22.57	
NaCl	58.45	Na 39.34	Cl 60.66		
NaNO₂	69.00	Na 33.32	O 46.38	N 20.30	
NaNO₃	85.01	Na 27.05	O 56.47	N 16.48	
NaOH	40.01	Na 57.48	O 40.00	H 2.52	
H₃PO₄	98.00	P 31.61	O 65.31	H 3.09	
KH₂PO₄	136.09	P 22.76	K 28.73	O 47.03	H 1.48
K₂HPO₄	174.18	P 17.79	K 44.89	O 36.74	H 0.58
Na₂HPO₄	141.98	P 21.82	Na 32.39	O 45.08	H 0.71
NaHPO₄	119.98	P 25.81	Na 19.16	O 53.34	H 1.68
ZnCl₂	136.29	Zn 47.97	Cl 52.03		
ZnCO₃	125.38	Zn 52.14	O 38.28	C 9.58	
ZnSO₄	161.44	Zn 40.50	O 39.64	S 19.86	
ZnSO₄•7H₂O	287.44	Zn 22.75	O 61.23	S 11.15	H 4.87
ZnSO₄•H₂O	179.44	Zn 36.44	O 44.58	S 17.86	H 1.12

tion. The graph illustrated in Figure 13.9 is commonly referred to as a response surface model and is particularly valuable if the interaction demonstrates curvature or peaks and valleys. Multiple regression analysis also provides a mathematical description of the interaction in addition to an assessment of the variability of the system. The following equation describes the model illustrated in Figure 13.9:

$$\text{Yield (g dwt/L)} = 0.34(\text{g/L Nutrisoy}) + 0.17(\text{g/L Maltrin M--100}) + 0.5\ R = 0.94$$

A mathematical model based on experimental data such as that in the preceding equation allows a researcher to predict the yield of a fermentation from the medium composition. For example, if 6% Nutrisoy yielded the highest but is too viscous for proper aeration in a tank, a lower concentration of Nutrisoy might be used with extra Maltrin M--100 added to achieve the same yields. For example,

if the maximum allowable concentration of Nutrisoy was determined to be 4%
and the target yield is 30 g dwt/L, this yield could theoretically be attained by
addition of 9% Maltrin M–100. Of course, this theoretical medium would have
to be tested to confirm the model, in particular because 9% Maltrin M–100 falls
outside the limits of the experimental data.

Factorial Designs

Experimental designs employing factorial structure allow several parameters
to be evaluated simultaneously. There are few if any cases in the biological world
where knowledge of a single factor of the system allows prediction of how the
entire system will perform. For example, we would not expect to be able to
predict biomass yield from growth temperature alone. The interactions between
temperature, aeration, nutrient solubility, water potential, and so on, are all
crucial to the growth and development of a microorganism. By using factorial
designs, the impact of each factor and its interaction with the other factors in the
experiment are elucidated. The following example was developed to illustrate the
use of a 2^5 factorial design in fermentation development.

The 2^5 factorial design is used to determine whether a group of factors have
positive, negative, or no effect on the experimental system. The strategy incorpo-
rates selection of two widely (but reasonably) separated rates or conditions of
five factors. A "+" symbol usually designates the high rate or condition and
"−"indicates the low rate or condition of a factor. For example, the following
conditions and rates were selected to illustrate the use of a 2^5 factorial design:

	Level	
	(−)	(+)
Temperature	25.0	30.0
Glucose	1.0%	4.0%
Nutrisoy	2.0%	6.0%
Zinc	0.0 μg/ml	5.0 μg/ml
Cobalt	0.0 μg/ml	5.0 μg/ml

To evaluate all combinations of these treatments, $2^5 = 32$ flasks will be needed.
Figure 13.14 lists the temperature and medium treatments for each flask. The last
two columns, yield in g dwt/L and spores/L, list results fabricated for this
example. Although no single treatment is repeated in the 2^5 design, there is hidden
replication for each factor. Each treatment level for each factor is repeated 16
times. A simplified representation of the 2^5 factorial design is illustrated in Figure
13.15. This table can be used to assist in preparing media and placing flasks in
to proper environments.

Glancing down the two yield columns in Figure 13.14 (g dwt/L and spores $\times$
10^{10}/L) does not reveal many obvious interactions. Only treatments with Zn
appear to stand out to stimulate yield of spores/L. To identify significant interac-

Figure 13.14.   Treatment list for a 2^5 factorial experiment. Data was fabricated for this example.

Flask #	Temp °C	% Glucose	% Nutrisoy	Zn μg/ml	Co μg/ml	g dwt/l	Spores 10^{10}/1
1	25	1	2	0	0	25.9	5.9
2	30	1	2	0	0	31.9	2.5
3	25	4	2	0	0	29.1	5.7
4	30	4	2	0	0	33.9	2.3
5	25	1	6	0	0	29.1	8.9
6	30	1	6	0	0	33.9	5.5
7	25	4	6	0	0	30.9	8.7
8	30	4	6	0	0	36.1	5.3
9	25	1	2	5	0	32.0	15.9
10	30	1	2	5	0	36.8	12.5
11	25	4	2	5	0	34.2	15.7
12	30	4	2	5	0	38.9	12.3
13	25	1	6	5	0	33.9	18.9
14	30	1	6	5	0	39.0	15.5
15	25	4	6	5	0	33.9	18.7
16	30	4	6	5	0	39.1	15.3
17	25	1	2	0	5	26.2	6.1
18	30	1	2	0	5	31.8	2.3
19	25	4	2	0	5	28.8	5.9
20	30	4	2	0	5	34.0	2.1
21	25	1	6	0	5	28.9	9.1
22	30	1	6	0	5	34.1	5.3
23	25	4	6	0	5	31.2	8.9
24	30	4	6	0	5	35.8	5.1
25	25	1	2	5	5	31.8	16.1
26	30	1	2	5	5	37.2	12.3
27	25	4	2	5	5	33.8	15.9
28	30	4	2	5	5	39.1	12.1
29	25	1	6	5	5	34.2	19.1
30	30	1	6	5	5	38.8	15.3
31	25	4	6	5	5	34.2	18.9
32	30	4	6	5	5	38.9	15.1

tions, stepwise regression techniques were used to analyze the data. Other methods of analysis for factorial experiment designs are discussed by Box et al. (6). Stepwise regression analysis was chosen because of its simplicity and its availability in many statistical analysis packages for microcomputers.

Stepwise regression identifies linear relationships between independent variables (factors: temperature, % glucose, % Nutrisoy, Zn, Co) and dependent variables (yield in g dwt/L and spores $\times$ 10^{10}/L). The method evaluates one factor at a time to determine whether there is a correlation (positive or negative) with the dependent variable. If there is a significant correlation, the independent variable is added into the equation. As the analysis continues, the next factor is tested in the equation with the first factor and if the correlation is significant, the factor is included in the equation. If a significant correlation does not occur, the

Figure 13.15.   General design of a 2^5 factorial experiment illustrating the high "+" and low "−" levels for each factor. Notice the replication of each level for each factor.

TRT #	Factor 1	Factor 2	Factor 3	Factor 4	Factor 5
1	−	−	−	−	−
2	+	−	−	−	−
3	−	+	−	−	−
4	+	+	−	−	−
5	−	−	+	−	−
6	+	−	+	−	−
7	−	+	+	−	−
8	+	+	+	−	−
9	−	−	−	+	−
10	+	−	−	+	−
11	−	+	−	+	−
12	+	+	−	+	−
13	−	−	+	+	−
14	+	−	+	+	−
15	−	+	+	+	−
16	+	+	+	+	−
17	−	−	−	−	+
18	+	−	−	−	+
19	−	+	−	−	+
20	+	+	−	−	+
21	−	−	+	−	+
22	+	−	+	−	+
23	−	+	+	−	+
24	+	+	+	−	+
25	−	−	−	+	+
26	+	−	−	+	+
27	−	+	−	+	+
28	+	+	−	+	+
29	−	−	+	+	+
30	+	−	+	+	+
31	−	+	+	+	+
32	+	+	+	+	+

factor is removed from the equation. The process is repeated for each factor and the result is an equation containing only factors that demonstrate a significant influence on the dependent variables, in this example g dwt/L or spores/L.

The following analysis was carried out using a Macintosh Plus computer running StatView 512+ (BrianPower Inc., Calabasas, CA). Before the analysis can be carried out, the "F" value for the desired significance level must be determined. The following list has been provided to simplify the process.

Number of Factors	F to Enter	F to Remove
5	4.170	4.165
4	4.600	4.595
3	3.990	5.985

Stepwise Regression Y_1:g dwt/l   5 X variables

(Last Step) STEP NO. 4   VARIABLE ENTERED: X_2: % Glucose

R:	R-squared:	Adj. R-squared:	Std. Error:
.987	.973	.969	.655

Analysis of Variance Table

Source	DF:	Sum Squares:	Mean Square:	F-test:
REGRESSION	4	421.988	105.497	245.951
RESIDUAL	27	11.581	.429	
TOTAL	31	433.569		

STEP NO. 4   Stepwise Regression Y_1:g dwt/l   5 X variables

Variables in Equation

Parameter:	Value:	Std. Err.:	Std. Value:	F to Remove:
INTERCEPT	.4			
Temp. °C	1.015	.046	.689	480.364
% Glucose	.55	.077	.224	50.777
% Nutrisoy	.416	.058	.226	51.549
Zn µg/ml	.928	.046	.63	401.112

Variables Not in Equation

Parameter:	Par. Corr:	F to Enter:
Co ug/ml	.01	.003

Figure 13.16. Stepwise regression output for 2^5 factorial experiment data listed in Table 13.14. The dependent variable is yield in g dwt/L.

The F to enter refers to the level of significance level, 95% in our example, the correlation must demonstrate before the factor is added into the equation. The F to remove is the significance level that will cause a factor to be removed from the equation.

The results of the stepwise regressions are listed in Figures 13.16 and 13.17. The analysis indicates that the factors (parameters): temperature, % glucose, % Nutrisoy, and Zn all contributed positively to the yield in g dwt/L. On the other hand, Co was not added into the equation because it neither significantly increased nor decreased yield (Figure 13.16). Alternatively, Figure 13.17 illustrates the analysis of yield in spores/L. In this case, temperature and % glucose demonstrated a significant but negative effect upon yield in spores/L. However, both % Nutrisoy and Zn increased yield of spores/L. Again, Co had neither a positive nor negative influence on yield of spores/L.

The equations approximating these interactions are:

g dwt/L= 1.0 × (temperature °C) + 0.9 × (Zn µg/ml) + 0.6 × (% glucose) + 0.4 × (% Nutrisoy) + 0.4 Spores × 10^{10}/L= 2.0 × (Zn µg/ml) + 0.8 × (% Nutrisoy) −0.1 × (% glucose) −0.7 × (temperature °C) + 22.6

Stepwise Regression Y_1:spores x 10^10/l   5 X variables

(Last Step) STEP NO. 4   VARIABLE ENTERED: X_2: % Glucose

R:	R-squared:	Adj. R-squared:	Std. Error:
1	1	1	.109

Analysis of Variance Table

Source	DF:	Sum Squares:	Mean Square:	F-test:
REGRESSION	4	976	244	20587.5
RESIDUAL	27	.32	.012	
TOTAL	31	976.32		

STEP NO. 4   Stepwise Regression Y_1:spores x 10^10/l   5 X variables

Variables in Equation

Parameter:	Value:	Std. Err.:	Std. Value:	F to Remove:
INTERCEPT	22.567			
Temp. °C	-.72	.008	-.326	8748
% Glucose	-.067	.013	-.018	27
% Nutrisoy	.75	.01	.272	6075
Zn μg/ml	2	.008	.905	67500

Variables Not in Equation

Parameter:	Par. Corr:	F to Enter:
Co ug/ml	-1.041E-17	2.817E-33

Figure 13.17.   Stepwise regression output for 2^5 factorial experiment data listed in Table 13.14. The dependent variable is yield in spores $\times$ 10^{10}/L.

A note of caution must be injected at this point. Linear models do not accurately describe interactions with curvature or functions with peaks and valleys. Therefore, conclusions should not be based blindly upon statistical analyses. Improper application of experimental design and data analyses can easily mislead a researcher. Refer to *Statistics for Experimenters* (6) for further descriptions and use of various experimental designs and data analyses.

Costs of Goods

A rough estimate of the maximum allowable cost of goods (COG) for a bioherbicide can be determined to guide research toward commercial market demands. Cost of goods refers to the total cost of product manufacture from fermentation through formulation and packaging. During product development, the estimated COG will ultimately determine whether a project is worthy of further development. From the following rough estimate a researcher will be able to determine whether a production system is in the "ballpark." Caution should be used if production techniques greatly exceed the maximum value described, unless the product will command a much higher price than that assumed.

The values used here are estimations based upon literature citations and personal interviews. Many details have been omitted to allow evaluation of a complex problem using only the most critical parameters. This approach is a simplification of the use of fuzzy mathematics in bioengineering models proposed by Dohnal (20). Fuzzy mathematics was developed to help analyze complex problems based upon semiqualitative information.

Assumptions

The following assumptions for a bioherbicide product were used to develop a simple evaluation of production economics:

1. A niche market exists with limited or no competition expected in the near future.
2. The maximum that the consumer will readily pay for control of the weed is $37.00 per hectare ($14.97 per acre).
3. The product is effective using a single application per growing season.

These assumptions describe the potential bioherbicide product that will be used to determine an approximate maximum value for COG. If the product requires more than one application, competition enters the market, or the market will not bear the $37.00 per hectare price, then the COG will have to be reduced accordingly. However, if the product performs as assumed here and production costs fall within the limits to be described, the project might be successful.

A crude guidepost for determining the maximum value for COG is equivalent to 25–50% of the sale price of the product. (Joe Brumley, Mycogen Corporation, personal communication). However, the distributor will require approximately 20% profit to cover expenses. Therefore, a product that sells for $37.00 per hectare will provide $29.60 to the manufacturer, the sale price of the product. In addition to the COG, profit should range between 50% and 75% of the sale price to cover research, commercial development, regulatory expenses, and marketing costs. After all the costs are totaled, a company expects to make 15–20% profit.

Retail price per ha	$37.00	
Wholesale price per ha (retail, 20%)	$29.60	
Profit per ha (15 and 20%)	$4.44	$5.92
Research, development, and marketing (50% and 75%)	$13.45	$20.18

Based on these values, the following maximum COG matrix can be calculated and depends on high or low cost of research, development and marketing (50% or 75% of sale price) in addition to high or low profit requirements (15% or 20%

profit margin). Therefore, the maximum value for COG may range between $3.50 to $11.71 per hectare. Projects with an estimated COG exceeding $3.50 per ha should be approached cautiously. Moreover, projects with estimated COG exceeding $11.71 per hectare should be avoided.

		Research, development, and marketing costs	
Maximum Cost of goods (COGS)		Low	High
Profit	Low	$11.71	$4.98
Requirement	High	$10.23	$3.50

The second guidepost relates directly to the fermentation process and biomass yield. The cost of toll fermentations ranges from $4000 to $9000 per day for use of a fermentor with a working volume of 75,700 L (20,000 gallons), support facilities and operators, and, depending upon the fermentation requirements, some downstream processing needs (centrifuges, drum filters, driers, bulk mixers and loaders, etc.). These values do not include the cost of raw materials used in the fermentation medium and specialized downstream processes (e.g., special drying or particle-sizing procedures). For the sake of discussion, a value of $6000 per day will be used as a conservative estimate of cost of toll fermentation and downstream processing.

In addition to the daily cost of operating a fermentation facility, the amount of fermentation broth that will be used to treat each hectare (rate of application) must be considered. Figure 13.18 illustrates the relationship between cost of fermentation per hectare and rate of application (in liters of fermentation broth per hectare) as a function of fermentation duration. The importance of a short

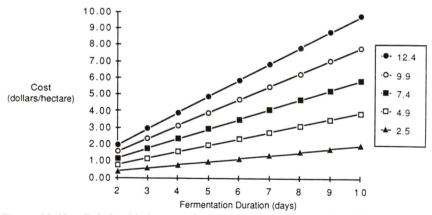

Figure 13.18.   Relationship between fermentation duration, rate of application in liters of whole broth per hectare, and cost of toll fermentation at $6,000/day.

fermentation process becomes critical as the rate of application increases. However, at a rate of application of 2.5 L of broth per hectare, the cost of the fermentation facility will add only about $1.50 per hectare even when the manufacturing process requires 10 days to complete at the toll fermentation facility. However, if 7.4 L are needed to treat each hectare, the toll fermentation alone will cost more than $1.00 per hectare even if the process could be completed in only 2 days.

Continuing along this path, the target yield of propagules per liter (PPL) can be estimated from the rate of application, expressed in liters of broth per hectare. As defined earlier, the term *propagule* will be used to describe any portion of the bioherbicide organism that is capable of infecting the target plant and resulting in control of the weed. In other words, a propagule could be a spore of the fungus, a small hyphal fragment produced in submerged fermentation or by mechanically reducing the size of long hyphal strands, or small budding yeastlike cells. For example, the recommended rates for application of the commercial bioherbicides COLLEGO and DeVine are 1.9×10^{11} propagules per hectare (PPHA) and 7.9×10^8 PPHA, respectively. However, the recommended rate of application for DeVine is considerably lower than effective rates of application reported for other bioherbicides, which average 5.5×10^{11} PPHA (12,18,41,48). Using the later value and the low rate (in liters of broth per acre) noted in Figure 13.18 (2.5 L broth/ha), the fermentation would have to yield 5.5×10^{11} propagules per 2.5 L, or 2.7×10^{11} PPL. This value is the target yield (PPL) for economical production of bioherbicides. It represents a generalization from reports of a variety of bioherbicides and should be used only as a guideline in evaluating processes.

The following "ballpark" statements summarize the preceding discussion:

1. Cost of goods should fall between $3.40 and $11.71 per hectare.
2. Yield of propagules in the tank should reach 2.7×10^{11} PPL. The average yield of spores reported by Vezina et al. for filamentous fungi was 3×10^{11} spores per liter (47).
3. Virulence should be high enough so that 5.5×10^{11} PPHA controls the weed.
4. When yield and virulence meet the limits described in statements 2 and 3, the cost of production (fermentation and downstream processing) is estimated to be less than $2.00 per hectare even when 10 days are required to complete the processes ($6000/days).

For more detailed analysis of economics of fermentation, refer to the book *Principles of Fermentation Technology* (37), which presents a clear analysis of the major components needed to prepare an economic analysis and cites a variety of recent reviews on this subject. For a more detailed description of economic analysis and an example of production of a biological insecticide refer to the report by Bartholomew and Reisman (3).

To illustrate the use of the "ballpark" values presented here, a fermentation process using the optimum medium for growth of *S. sclerotiorum* illustrated in Figure 13.9 and growth rate data reported in Figure 13.6 will be utilized. Cost of production and the yield will be estimated from the growth and yield information coupled with the process illustrated in Figure 13.4. In addition, the needed yield of propagules per liter of broth will be estimated. *Sclerotinia sclerotiorum* does not readily sporulate in liquid, requiring use of an alternative method of producing small propagules from mycelium or induction of sporulation as described earlier.

Sclerotinia sclerotiorum grew at $\mu = 0.08 \text{ hr}^{-1}$, doubling every 8.6 hr (Figure 13.8) in a medium containing 60/g/L Nutrisoy and 40/gL Maltrin M–100. The estimate of production costs will be based upon the process illustrated in Figure 13.4. A detailed scenario follows:

Ten 6-L flasks containing 600 ml of medium are each inoculated with 1g of freeze-dried standard inoculum (1.67 g dwt/L).

The primary seed will require a 96-hr lag before exponential growth begins and will require 34 hrs to reach 25 g dwt/L. The total primary seed fermentation time is 130 hr.

$t = [\ln(x_{t2}) - \ln(x_{t1})]/\mu = [\ln(25) - \ln(1.67)]/0.08 \text{ hr}^{-1} = 34 \text{ hr}$

The secondary seed fermentation tank contains a working volume of 2,271 l(3% of the production tank volume) and will be inoculated with the 10 primary seed flasks in exponential growth stage (25 g/L · 6l = 150 g dwt, 150 g dwt/ 2,271 = 0.07 g dwt/L). It will take 71 hr to reach 20 g dwt/L without a lag period.

$t = [\ln(x_{t2}) - \ln(2_{t1})]/\mu = [\ln(20) - \ln(0.07)]/0.08 \text{ hr}^{-1} = 71 \text{ hr}$

The production tank (75,700 L working volume) is inoculated with the secondary seed (2,271 l · 20 g dwt/L = 45 kg, 45 kg/75,700 L = 0.6 g/L) in exponential growth stage. The expected yield of the production fermentation process of 30 g dwt/L will be reached after 49 hr.

$t = [\ln(x_{t2}) - \ln (x_{t1})/\mu = [\ln(30) - \ln (0.6)]/0.08 \text{ hr}^{-1} = 49 \text{ hr}$

The total process from inoculation through production of 75,700 L of broth at 30 g dwt/L (2300 kg, 5000 lb total) will require 250 h (10 days). However, the fermentation tanks are only used for 120 hr (5 days). Additional time should be added for preparation and cleanup, 48 hr (2 days), and downstream processing,

48 hr (2 days). In this example the toll fermentation process would take 9 days. Using these simple estimates, production costs are roughly:

Toll Fermentation and Downstream Processing
(9 days @ $6,000/day):	$54,000
Nutrisoy ($18.80/cwt, 60 g/l):	$1,937
Maltrin M-100 ($29.00/cwt, 40 g/l):	$1,992
Total:	$57,929

The product would therefore cost $25.19/kg ($11.59/lb). Referring back to the earlier estimates of maximum COG of $11.71/ha, each kilogram of product will have to treat effectively at least 2.2 ha to meet our example market demands. The product would be more interesting if the lower value for COG of $3.50/ha could be achieved. In this case each kilogram of product would have to effectively treat 7.2 ha (140 g/ha, 0.12 lb/acre). As mentioned earlier, the average rate of bioherbicide application reported in the literature is about 5.5×10^{11} PPHA. Accordingly, 140 g of active ingredient would have to contain about 5.5×10^{11} propagules, or 3.9×10^{9} propagules per gram of active ingredient (dry propagules). This value is in the ballpark for the small-spored fungi *Penicillium* and *Aspergillus*, weighing in at 2×10^{10} spores per gram (34). The essential test is actual laboratory and field evaluations to insure that the product contains the necessary activity and kills the target weed.

The preceding example was prepared to illustrate the type of simple analysis a researcher can perform to evaluate program development. It is necessarily artificial because production estimates and processes like those above are usually proprietary information and carefully guarded by the manufacturer. However, this exercise illustrates the methods used by industry to determine whether a product should continue on its current development path, change course to another direction, or be canceled entirely. If all the numbers balance, and the production costs are below the limits described in the beginning of this section, a new biological control product may be on the horizon.

Streamlined Evaluation of Yield and Virulence

As a guideline to assessing yield of a bioherbicide, a baseline value of 2.7×10^{11} PPL has been proposed. Another expression of yield that included virulence was the value of activity equivalent to 2.5 L of broth per hectare. A simple method of assessing a fermentation process would therefore evaluate the biological activity of the diluted broth directly or centrifuge the propagules out of the broth and resuspend the propagules in an appropriate solution (e.g., sterile water, a nutrient suspension, or a buffering agent). The suspension of propagules could then be applied to plants to near runoff using an atomizer (approximately 1800 L/ha, 180 ml/m^2, or 17 ml/ft^2). At these rates of carrier per hectare, a dilution of 1:72 applied to runoff would need to kill the target weed to meet the yield–virulence requirement of 2.5 L of broth per hectare. To simplify the system a dilution of 1:50 could be used to evaluate fermentations during growth or at the end of a fermentation with little difficulty as follows:

1. Remove 0.5 ml fermentation broth aseptically from the test medium and transfer to a vial. If necessary, centrifuge the broth and wash with 1.0 ml water or appropriate solution, centrifuge again, and resuspend in 0.5 ml water or appropriate diluent.

2. Add 24.5 ml of diluent (water, nutrient suspension, adjuvant, etc.).

3. Mix and apply to test plants by atomization to near runoff (180 ml/m^2, or 17 ml/ft^2).

4. Incubate plants in a test environmental regime that simulates actual field conditions relative to temperature and dew duration.

5. Evaluate treatments for herbicidal effects after the required incubation period. An organism may not be commercially acceptable if it requires longer than 14 days to kill the target plant.

This process is a rapid screening procedure that eliminates costly propagule counting but still answers the key question of whether the fermentation process is yielding sufficient quantities of virulent propagules.

Conclusion

Submerged fermentation is the only method currently being used to produce the commercial bioherbicides COLLEGO and DeVine. The importance of developing submerged fermentation processes was stressed by Templeton, a pioneer in bioherbicide research, when he wrote: "from a practical standpoint, growth and sporulation by liquid culture technology may well be an essential requirement additional to the previously described attributes of activity, specificity and viability" (42). However, Templeton also suggested that solid culture or other unique production techniques may be suitable for public agency scientists to assess initial activity. With the growing interest in bioherbicide development, this is a fitting time to gain a consensus opinion that submerged fermentation is the most effective method of producing bioherbicides.

Small bioherbicide markets (usually less than $50 million) will probably require that the products be manufactured under contract at an existing fermentation facility on a toll basis if bioherbicides are to gain widespread use. Advances in fermentation and downstream processing are essential. The need for research in these areas extends beyond the private sector and should be pursued by public sector scientists to speed development and release of new bioherbicide products. It is hoped that the information presented in this chapter will spark public discussion of fermentation and downstream processing advances and encourage accelerated research into the production of biological control agents.

Literature Cited

1. Allerman, K, Olsen, J., and Smith, J.E. 1983. Asexual differentiation in the fungi. Pages 419–447, in: Fungal Differentiation: A Contemporary Synthesis. J. E. Smith, ed. Dekker, New York.

2. Ali Obaidi, Z.S., and Berry, D.R. 1980. cAMP concentration, morphological differentiation and citric acid production in *Aspergillus niger*. Biotechnol. Lett. 2:5–10.

3. Bartholomew, W.H., and Reisman, H.B. 1979. Economics of fermentation processes. Pages 464–496, in: Microbial Technology. H.J. Pepler and D. Perlman, eds. Academic Press, New York.

4. Birdson, E.Y., and Decker, A. 1970. Design and formulation of microbial culture media. Meth. Microbiol. 3:229–295.

5. Borrow, A., Jefferys, E.G., Kessell, R.H.J., Lloyd, E.C., Lloyd, P.B., and Nixon, I.S. 1961. The metabolism of *Gibberella fujikuroi* in stirred culture. Can. J. Microbiol. 7:227–276.

6. Box, G.E., Hunter, W.G., and Hunter, J.S. 1978. Statistics for Experimenters: An Introduction to Design, Data Analysis, and Model Building. Wiley, New York.

7. Brown, D.E. 1970. Aeration in the submerged culture of micro-organisms. Meth. Microbiol. 2:125–174.

8. Burrows, S. 1970. Bakers yeast. Pages 349–420, in: The Yeasts, Vol. 3.: Yeast Technology. A. H. Rose and J.S. Harrison, eds. Academic Press, New York.

9. Burrows, S. 1979. Bakers yeast. Pages 31–64, *in:* Economic Microbiology, Vol. 4.: Microbial Biomass. A.H. Rose, ed. Academic Press, New York.

10. Calam. C.T. 1969. The evaluation of mycelial growth. Meth. Microbiol. 1:567–591.

11. Calam, C.T. 1969. The culture of micro-organisms in liquid culture. Meth. Microbiol. 2:255–326.

12. Cardina, J., Littrell, R.H., and Stowell, L.J. 1987. Bioherbicide for Florida beggarweed. U.S. Patent No. 4,643,756.

13. Churchill, B.W. 1982. Mass production of microorganisms for biological control. Pages 134–156, in: Biological Control of Weeds with Plant Pathogens. R. Charudattan and H.L. Walker, eds. Wiley, New York.

14. Conway, K.E., Freeman, T.E., and Charudattan, R. 1978. Methods and compositions for controlling waterhyacinth. U.S. Patent No. 4,097,261.

15. Cotty, P.J. 1985. Carbon dioxide modulates the sporulation of *Alternaria* species. Phytopathology 75:1297.

16. Cotty, P.J. 1987. Modulation of sporulation of *Alternaria tagetica* by carbon dioxide. Mycologia 79:508–513.

17. Coupland, K., and Niehaus, W.G., Jr. 1987. Stimulation of alternariol biosynthesis by zinc and manganese ions. Exp. Mycol. 11:60–63.

18. Daniel, J.T., Templeton, G.E., and Smith, J., Jr. 1974. Control of aeschynomene sp. with *Colletotrichum gloeosporioides* penz. f.sp. *aeschynomene*. U.S. Patent No. 3,849,104.

19. Detroy, R.W., and Ciegler, A. 1971. Induction of yeastlike development in *Aspergillus parasiticus*. J. Gen. Microbiol. 65:259–264.

20. Dohnal, B. 1984. Fuzzy bioengineering models. Biotechnol. Bioeng. 27:1146–1151.

21. Evans, C.T., and Ratledge, C. 1984. Effects of nitrogen source on lipid accumulation in oleaginous yeasts. J. Gen. Microbiol. 130:1693–1704.

22. Fisher, D.J., Holloway, P.J., and Richmond, E.V. 1972. Fatty acid and hydrocarbon constituents of the surface and wall lipids of some fungal spores. J. Gen. Microbiol. 72:71–78.

23. Freedman, D. 1970. The shaker in bioengineering. Meth. Microbiol. 2:175–185.

24. Garraway, M.O., and Evans, R.C., eds. 1984. Fungal Nutrition and Physiology. Wiley, New York.

25. Kidd, G.H., and Wolf, F.T. 1973. Dimorphism in a pathogenic fusarium. Mycologia 65:1371–1375.

26. Koser, S.A. 1968. Vitamin Requirements of Bacteria and Yeasts. Charles C. Thomas, Springfield, IL.

27. Lax, A.R., Templeton, G.E., and Meyer, W.L. 1985. Isolation, purification and biological activity of a self-inhibitor from conidia of *Colletotrichum gloeosporioides*. Phytopathology. 75:386–390.

28. McCabe, D., and Soper, R.S. 1985. Preparation of an entomopathogenic fungal insect control agent. U.S. Patent No. 4,530,834.

29. Michelmore, R.W., and Hulbert, S.H. 1987. Molecular markers for genetic analysis of phytopathogenic fungi. Ann. Rev. Phytopathol. 25:383–404.

30. Miller, T.L., and Churchill, B.W., 1986. Substrates for large scale fermentations. Pages 122–136, *in:* Manual of Industrial Microbiology and Biotechnology. A.L. Demain and N.A. Solomon, eds. Amer. Soc. Microbiol. Cambridge, MA.

31. Misaghi, I.J., Grogan, R.G., Duniway, J.M., and Kimble, K.A. 1978. Influence of environment and culture media on spore morphology of *Alternaria alternata*. Phytopathology 68:29–34.

32. Mitard, A., and Riba, J.P. 1988. Morphology and growth of *Aspergillus niger* ATCC 26036 cultivated at several shear rates. Biotechnol. Bioeng. 32:835–840.

33. Pasteur, L. 1879. Studies on Fermentation: the Diseases of Beer, Their Causes, and the Means of Preventing Them. Macmillan, London. Reprinted in 1968 by Kraus Reprint Co., New York.

34. Sansing, G.A., and Ciegler, A. 1973. Mass propagation of conidia from several *Aspergillus* and *Penicillium* species. Appl. Microbiol. 26:830–831.

35. Smith, J.E., and Berry, D.R., eds. 1975. The Filamentous Fungi, Vol. 1: Industrial Mycology. Edward Arnold, London.

36. Sommers, E., and Horsfall, J.G. 1966. The water content of powdery mildew conidia. Phytopathology. 56:1031–1035.

37. Stanbury, P.F., and Whitaker, A. 1984. Principles of Fermentation Technology. Pergamon Press, New York.

38. Stewart, P.R., and Rogers, P.J. 1983. Fungal Dimorphism. Pages 267–313, *in:* Fungal Differentiation: A Contemporary Synthesis. J.E. Smith, ed. Dekker, Inc., New York.

39. Stowell, L.J., Nette, K., Heath, B., and Shutter, B. 1989. Fermentation alternatives for commercial production of a mycoherbicide. *in:* Topics in Industrial Microbiology. A. Demain, G. A. Somkuti, J.C. Hunter-Cevera, and H.W. Rossmore, eds. Elsevier, Amsterdam.

40. Sussman, A.S. 1965. Dormancy of soil microorganisms in relation to survival. Pages 99–109, *in:* Ecology of Soil-Borne Plant Pathogens: Prelude to Biological Control. K.F. Baker and W.C. Snyder, eds. University of California Press, Berkeley.

41. Templeton, G.E. 1976. *C. malvarum* spore concentrate, formulation, and agricultural process. U.S. Patent No. 3,999,973.

42. Templeton, G.E., Smith, R.J., and Komparens, W. 1980. Commercialization of fungi and bacteria for biological control. Biocontrol News Inf. 1:291–294.

43. Thevelein, J.M. 1984. Regulation of trehalose mobilization in fungi. Microbiol. Rev. 48:42–59.

44. Trinici, A.P.J. 1969. A kinetic study of the growth of *Aspergillus nidulans* and other fungi. J. Gen. Microbiol. 57:11–24.

45. Van Etten, J.L., Dahlberg, K.R., and Russo, G.M. 1983. Fungal spore germination. Pages 235–266, *in:* Fungal Differentiation: A Contemporary Synthesis. J.E. Smith, ed. Dekker, New York.

46. Vezina, C., and Singh, K. 1975. Transformation of organic compounds by fungal spores. Pages 158–192, *in:* The Filamentous Fungi. Vol. 1: Industrial Mycology. J.E. Smith and D.R. Berry, eds. Edward Arnold, London.

47. Vezina, C. Singh, K., and Sehgal, S.N. 1965. Sporulation of filamentous fungi in submerged culture. Mycologia 57:722–736.

48. Walker, H.L. 1983. Control of sicklepod, show crotalaria, and coffee senna with a fungal pathogen. U.S. Patent No. 4,390,360.

49. Walker, H.L. 1983. Control of Prickly sida, velvetleaf, and spurred anoda with fungal pathogens. U.S. Patent No. 4,419,120.

50. Wang, D.I.C., Cooney, C.L., Demain, A.L., Dunnill, P., Humphrey, A.E., Lilly, M.D. 1979. Fermentation and Enzyme Technology. Wiley, New York.

51. Wang, M.C., and Bartnicki-Garcia, S. 1970. Structure and composition of walls of the yeast form of *Verticllium albo-atrum*. J. Gen. Microbiol. 64:41–54.

52. Woltz, S.S., and Jones, J.P. 1981. Nutritional requirements of *Fusarium Oxysporum:* Basis for a Disease Control System. Pages 340–347, *in:* Fusarium: Diseases, Biology, and Taxonomy. P.E. Nelson, T.A. Tousson, and R.J. Cook, eds. Pennsylvania State University Press, University Park, PA.

14

Economic Aspects of Biological Weed Control with Plant Pathogens

Bruce A. Auld

Introduction

Economics is the study of the allocation of resources. Farmers are under pressure to allocate their resources in such a way as to maximize the difference between their revenues and their production costs. Scientists have also come under greater scrutiny in recent years to justify their allocation of research resources in order to maximize benefits, and biological weed control projects have not been immune from this.

The optimal allocation of resources varies depending on the point of view one is taking (i.e., whose benefits and costs are being measured). In some cases it may be appropriate to consider an individual farmer's viewpoint; in others, all farmers as a group, a commercial firm, or the whole of society.

In this chapter we consider the economics of classical and inundative control separately. (Possible augmentative approaches falling in between these two strategies are not discussed.)

Classical Biocontrol

Classical biological control is described by economists as a "pure public good." Once it is provided, because of the *spread of the control agent,* it is freely available to everyone irrespective of whether they have contributed to its provision. Although collective gain from the introduction of a classical biocontrol agent may far exceed the cost of its discovery and introduction, an individual alone would usually be unable to recover these costs from his own gain. Therefore, for it to be supplied optimally, it usually has to be provided by government. (Its provision is thus usually subject to whatever political shortcomings exist).

I would like to thank several colleagues for comments on a draft of this chapter: W. L. Bruckart, B. Gantotti, J. Harr, K. M. Menz, and H. L. Walker. The financial support of the Australian Wool Corporation is also gratefully acknowledged.

There are, in any case, more arguments for government involvement in weed control, and these are underpinned by the fact that *weeds spread*. The first is that a weed species spreading across farm boundaries imposes a cost that is external to individual farmers and thus becomes a social issue, which must be resolved by government (19). The second is that for maximum economic efficiency the measures needed to prevent reinfestation must be coordinated. Government intervention in control is usually necessary to achieve such coordination. Simultaneous attack on a weed species can be an important element in a successful weed control program, especially when the weed has a high capacity for rapid spread (5). Such coordination is usually administratively difficult and expensive. With classical biological weed control, coordination is a more or less automatic consequence of the technique and does not usually involve high additional administrative costs.

Initial Costs

Classical biocontrol programs have high initial fixed cost in searching, host range testing, and maintenance of material. Estimates are in the range of 12–24 scientist-years (2,14): $1200–1400 (U.S. dollars unless otherwise stated) per program if $100,000 is allowed per scientist-year. The biocontrol program for *Chondrilla juncea* L. by the rust fungus *Puccinia chrondrillina* Bubak & Syd. in Australia was estimated to have cost Aust. $2.6 million in 1980 (17).

Once an agent is released there are few additional costs, a characteristic of this form of control being that the agent spreads largely of its own accord and is self-perpetuating. This has several implications: The target weed does not have to be identified and located by humans; potential reinfection sources of the weed in inaccessible areas are controlled; there is no discrimination in control in different locations; and the control technique is irreversible and permanent.

Assessing Benefits

The benefits from a successful classical biocontrol program will continue indefinitely or at least for many years and the expected benefits may be great. However, care must be taken in assessing benefits as the value of money saved decreases in time by

$$\frac{\$x}{(1 + r)^n}$$

where **r**, the discount rate, indicates the difference in value between cash today and cash in the future and $x is the value received in *n* years. This is *discounting* (the reverse of *compounding*), a procedure to "weight" future benefits into present-value monetary terms

Beyond a certain point in time, benefits may be ignored because they approach zero in present-day terms. For instance, the present value of $100 at a discount rate of 10% is $2.22 in 40 years and 33 cents in 60 years.

However, in assessing future benefits of an apparently successful classical biocontrol program care must be taken to observe the consequences of removing the weed. Is the target weed simply replaced by another economically damaging weed? If so, the value of the project is seriously diminished. There appear to be few cases where data on species replacing biocontrol targets have been collected (see following discussion). In many cases there may be a need for an integrated approach to control of the target weed that includes sowing useful replacement species. This, of course, would be an additional control cost.

The Place for Classical Biological Control

The fact that there is no location specificity in the technique means that classical biocontrol costs do not vary with the size of weed infestations and are independent of area. Classical biocontrol of a weed may be economic from a collective point of view even when for an individual farmer control may be uneconomic (28).

In the usual crop loss–weed functions the *loss per weed* is greater as the density of weeds decreases (16). Thus, greater total loss of agricultural production may result when weeds are dispersed at low density over a large area than when the same population is concentrated over a small area. Most weed control costs are either proportional to the area covered or in some way dependent on area covered. Thus, in spite of considerable overall losses, control may be uneconomic by chemical or mechanical means. But with classical biological weed control costs per unit area fall with the size of the area being treated.

This implies a particular place for classical biocontrol in rangeland or broad-acre cropping land. Menz et al. (19) using a simulation model for a large area of southern Australia, the southern wheat–sheep zone, showed that an increase of only 1% in the naturalized pasture production in the region was valued at Aust. $14 million in (then) present-value terms (assuming a permanent benefit and a 5% discount rate). Marsden et al. (17) estimated the benefit–cost ratio of 200:1 for control of *Chondrilla juncea,* which occurs in this area as a weed of wheat. However, it is now thought that this may be an overestimate of the benefits because a form of *C. juncea* that is not attacked by the rust has expanded its range (12).

Conflicts of Interest

Lack of location specificity also may lead to conflicts of interest. In the broadest sense a weed is *a plant growing where it is not wanted,* and in the economic context of this chapter, that is where *its presence reduces economic output.* Many plant species are weeds in one location (e.g., a crop) and useful in another (e.g., pasture or rangeland). Cases of conflicts of interest in biocontrol of weeds have been discussed elsewhere (2,28,30).

Social cost–benefit analysis (SCBA) provides one method of resolving whether a project is beneficial from the point of view of society as a whole (29). In evaluating weed control it is often assumed that benefits and costs are limited to farmers. However, consumers may benefit through lower prices for farm products. A complication arises here if the farm products are exported (see 6). To evaluate the various costs and benefits from the point of view of the whole of society the *Kaldor–Hicks criterion* (16) may be used: if those gaining *could* compensate those losing, the project would be judged to be a social improvement. Note that there is no actual compensation paid under this test. If the income distribution consequences are socially unsatisfactory, we could ask if this disadvantage outweighs the benefits; if so, the change is socially unacceptable under a further test, the Little criterion (21).

SCBA was applied to the evaluation of biological control in the *Echium* species case in Australia. In this case, a conflict of interest arose between some farmers who wished to control the species and others as well as beekeepers who did not wish to control it (13,28,29). A government body, the Industries Assistance Commission (IAC) (15) investigated costs and benefits to various groups. The IAC assumed that the control of *Echium* in Australia would not have a significant impact on the prices of farm products. Therefore, consumers would be unaffected by the introduction of biocontrol agents and social gains, it was assumed, would be captured by farmers, some groups of whom would gain as a result of the biological control (e.g., wheat producers) and some of whom would lose (e.g., beekeepers) as the species are sources of nectar and pollen.

The IAC (15) concluded that the introduction of biological control on *Echium* in Australia would easily satisfy the Kaldor–Hicks test: Farmers as a *whole* could expect large net benefits and there would be no reduction in consumers' welfare. Although some primary producer groups (notably beekeepers) were anticipated to be worse off as a result of the project, the gains of others were expected to be much larger than the losses of the disadvantaged. The judgment was made that net gains to the community were more than adequate to make up for the income distribution consequences. It was concluded that those disadvantaged could be compensated for in "once-off" direct cash payments from consolidated revenue rather than levy the beneficiaries. As an exercise in SCBA the inquiry (15) has a number of limitations (26), even though it is useful in illustrating the application of the analysis to weed control projects. SCBA is a *single* objective criterion in that it requires all relevant costs and benefits to be expressed in money terms and the sum of net monetary benefits to be maximized. It is doubtful whether all relevant net benefits or outcomes from projects can be expressed solely in terms of money or by any other single measure.

Another approach to evaluating costs and benefits from society's point of view, involving multiple objectives, has been presented by Conway (11). This requires alternative farming technologies such as the use of pesticides versus biological control of pests to be evaluated in terms of four factors: (1) their impact on levels of yields or income (of farmers), (2) consequences for instability of yields or incomes, (3) effect on the equitability of income distribution, and (4) their consequences for sustainability of yields or income. In comparison to another technology, a technology that results in high income or yields, less instability of these, a more equitable distribution of income, and greater sustainability of income or yield is to be preferred.

Some difficulties involved in using Conway's approach to evaluation are the following: (1) It does not take account of consumers' surplus and therefore at this time is more applicable to subsistence economies than to market exchange economies. However, even in market economies it is relevant if farmers only are the focus of evaluation. (2) In the case where one technique is superior on the basis of one or more income characteristics and inferior on others to other techniques, there is no clear indication of how to rank it in relation to the other techniques. (3) Measurement of the characteristics poses some problems. For example, how are instability, equitability, and sustainability of income to be exactly specified and measured.

The irreversibility of classical biocontrol is another potential problem. Particularly with wind-borne pathogens such as rust fungi, the entire flora of a country or continent may be ultimately exposed to the biocontrol agent. Moreover, spread to neighboring nations is possible. It is not feasible to test an exotic organism on the entire flora of a country or continent, and apparent "jumps" in host range have occurred in microorganisms. For instance, a fungal pathogen, *Puccinia xanthii* Schw., apparently unintentionally introduced to Australia from the United States aroused interest because of its apparent specificity to *Xanthium* and *Ambrosia* species weeds in the United States and its potential for biological control of *Xanthium* species in Australia (1). However, since 1976 it has been recorded in the field on two sunflower (*Helianthus annuus*) cultivars (cv. Hysun 30, cv. Sunfola 68.2), even though it has not been recorded on *Helianthus* in the United States, and it has also been recorded on *Calendula officinalis* (31). Thus, some element of risk must be taken into account in evaluating classical biocontrol programs, especially with fungi. It should be borne in mind, however, that *P. xanthii* was not an official introduction to Australia and therefore not subject to a full testing program. In addition, its limited impact on the sunflower industry can be costed and weighed against benefits of *Xanthium* spp. control or suppression.

Mycoherbicides

Inundative biological control falls into the category described by economists as a "pure private good." It can be supplied to the market and by companies hoping to make a profit. Its acceptance will depend on the increased profits farmers perceive they make from its use. In this it is no different from conventional herbicides.

Although there is scope for the use of other microorganisms in an inundative approach to biocontrol of weeds, parasitic fungi, applied like conventional herbicides, as mycoherbicides (25) have received the greatest attention and success.

Commercial Considerations

For a commercial firm the development of a mycoherbicide would involve many considerations that would be identical for a conventional herbicide.

1.   What is the size and stability of the market?

   a. Is the weed annual or perennial?

   b. If annual, does it reoccur each year in similar amounts?

   c. What crops does it occur in? What value per unit area do the crops have? Over what area do the crops occur?

   d. Are there other forms of control available; what is their cost?

2.   Is the basic research protected by patents or secrecy?

3.   What is the projected cost of developing the basic research into a commercial product? In the case of a microorganism what is already known about the following factors?

   a. Identity

   b. Host range

   c. Pathogenicity

   d. Ecology

   e. Production in vitro

   f. Shelf life

   g. Efficacy under field conditions

   h. Compatibility with other herbicides

Up to this point it has been up to government research institutes or universities to provide most of the information in items 3a–h to stimulate the interest of private firms in developing a product and providing continued support during product development. Moreover, there has been a need for the research institutes and universities to look for pathogens of target weeds not only of major agricultural importance but where there might be a specific role for a biological agent (e.g., a vine growing over a horticultural crop, as in the case of DeVine).

Although agricultural chemicals have frequently been developed for specific insect species, commercial firms have rarely developed products for specific weed species. Market research in this area may be more difficult with firms unable to judge the *potential* market for new types of products and unaware of the potential economic returns.

Although the proportion of synthetic chemicals screened to those registered as pesticides is of the order of 10,000:1 (22), according to Bowers (9) the road to the discovery of chemical pesticides is more straightforward and more attractive to chemically oriented private firms than that for finding microbial pesticides. This is presumably because the infrastructure for the former and not the latter already exists.

For some companies with in-house expertise mainly in microbiology, the development of a mycoherbicide may be technically attractive but the company may lack experience in agricultural marketing.

In most large firms the development of a biological product for weed control would require integrated activity between divisions in which fermentation research and development is done (usually pharmaceutical) and an agricultural division. This could require organizational changes.

Although host specificity may limit the market size of a product (9), mycoherbicides have advantages that may become increasingly attractive to the manufacturers of agricultural chemicals. These include reduced costs in toxicity testing requirements and (usually) little carryover effect. The advantages particularly apply where indigenous pathogens are used as they have been in COLLEGO and DeVine. If exotic pathogens were used, most of the risk and uncertainty factors associated with classical biocontrol would apply.

It is likely that some mycoherbicides will be sold for use with chemical herbicides to broaden the spectrum of weed species controlled and increase efficacy on the target species. In addition, it would be feasible to produce mycoherbicides that are mixtures of microbial agents.

The cost of research and development of COLLEGO, one of the first mycoherbicides, was estimated to be approximately $2 million (24), whereas the cost of research and development for a conventional herbicide was approximately $30 million at that time. Given that the development of the first mycoherbicides provides the knowledge and technological basis for future projects, the relative cost advantage is likely to improve in the future. A recent example demonstrates the current costs of a mycoherbicide project to the stage of providing a basis for commercial research and development. Research on the anthracnose disease *Colletotrichum orbiculare* of the weed *Xanthium spinosum*, Bathurst burr or spiny clotburr (7), by NSW Agriculture & Fisheries, Australia, reached the stage in May 1988 where expressions of interest for commercial development of the fungus into a product were sought and obtained. Costs for the project up to that time were estimated to be approximately $500,000, of which about 10% was for new equipment; about half of the total funds were obtained from extramural primary industry–based funding bodies.

Patents obtained during such research projects have been licensed to commercial firms, who have funded subsequent research via royalty payments or grants.

Benefits to Society

From society's point of view other important advantages mycoherbicides have include reduced damage to nontarget organisms, no mammalian toxicity, and no contamination of soil or groundwater.

Combellack (10) estimated that only 0.5–2.0% of herbicide applied to weeds in cereals is actually used to control the weeds. This is exacerbated by uneven distribution of weeds in crops. The remainder of the herbicide either misses the weeds (impacting on the ground or crop); is lost as airborne drift or blown or washed outside the target area; or contacts the weed but is in excess of the amount required to achieve the desired level of control. Drift assessment is usually made by measuring off-target deposits. However, droplets too small to deposit on any natural object are lost to the atmosphere and collected or precipitated in an unknown place and at an unknown time; Bals (8) has estimated that in the UK 72 million liters of spray liquid every year are contributing to long-distance drift, that is, general environmental contamination. There may therefore be a case for society to subsidize the development of mycoherbicides and, to the extent that much of the basic research is conducted in public institutions, it already does.

Strong economic arguments exist for government funding of and involvement in basic research (20), including research into weed control. Private firms can be expected to be motivated in their activities by *their* resulting profit. Private firms are as a rule able to appropriate only a small fraction of community gains from the results of basic research. Basic research is, therefore, likely to be undersupplied from a social point of view if it is left to private firms.

As research becomes of a more applied character and the likelihood increases of its resulting in an identifiable product, private firms can be expected to appropriate a larger share of social gains from it. Therefore, if inventions can be protected by patents or similar means, private companies have a greater incentive to carry out the requisite research.

A Farmer's Viewpoint

From an individual farmer's economic point of view the same factors would be taken into account as with a conventional herbicide (see ref. 6; Chs. 4 and 5). These include the fundamental factors: anticipated loss caused by weeds and cost of control (Figure 14.1). Other factors that may have to be considered with conventional herbicide use (e.g., phytotoxicity to crop, drift problems, applicator

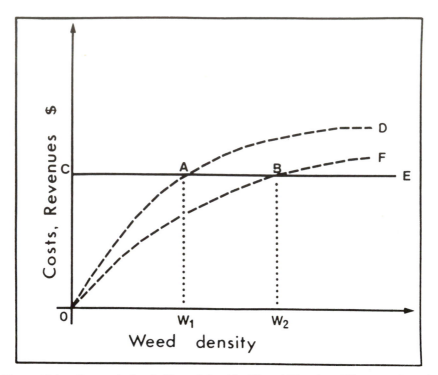

Figure 14.1.   Economic threshold weed densities W_1 and W_2, where CE represents costs of control and OAD and OBF represent revenues from weed control (see text).

exposure, effects on nontarget organisms, and residue effects on subsequent crops) would not usually be relevant.

The economics of treating a weed depends not only on the gain in yield of the crop inferred from the yield loss function but also on the monetary value of the extra yield and the cost of weed treatment. The increase in value of the product can be obtained by multiplying the price per unit of yield by the increased yield brought about by the elimination of the weed.

In Figure 14.1 the economic threshold for a basic case is demonstrated. CE represents the cost of control and OAD the value of additional yield obtained from weed control, say by using a mycoherbicide. W_1 is the threshold weed density. Let us assume that a conventional chemical herbicide is also available to control the weed but that it is slightly phytotoxic to the crop, and in this case the value from treatment will be reduced to OBF and the threshold weed density will rise to W_2; that is, greater potential yield loss must be accepted before control is economic. Note also that because of their specificity mycoherbicides may also be able to be applied earlier in the crop's life than many conventional herbicides, thus increasing potential yield.

In this simple case (Figure 14.1) we have considered a single-field, single-year situation, but farmers have to consider their actions in a whole-farm framework. Use of certain conventional herbicides may prevent use of the treated field for certain crops in the following season, and drift onto neighboring susceptible crops may occur.

Farmers (and herbicide producers) guided only by their individual profits may tend to use (or promote the use of) herbicides to the extent that unfavorable spillovers for society occur (e.g., groundwater contamination). In addition for most unirrigated crops, accurate prediction of yield is difficult and farmers are faced with variation in potential yield. Consider, for example, in Figure 14.1 that curves OAD and OBF encompass the range of possible revenues from weed control using one herbicide treatment.

The economic threshold weed density may be between W_1 and W_2. Given that farmers tend to be risk averse (23), treatment at low weed density (W_1) is more likely, even though it is not economic if the revenue function is OBF (see ref. 4 for a fuller discussion). This indicates that, overall, there will be greater weed control and herbicide use than is economically necessary. Therefore, given that there are unfavorable spillovers from herbicide use, the social desirability of biological controls such as mycoherbicides is further strengthened.

Conclusion

Classical biocontrol and inundative biocontrol using microbes have an established place in weed control technology. Although there are more potential risks and uncertainties with classical biocontrol than with inundative approaches using native organisms, successful projects may produce a stream of benefits over many years for one initial fixed cost. Inundative biocontrol techniques such as mycoherbicides generally have a number of cost advantages over contentional herbicides in terms of reduced registration requirements and cheap production— for example, by submerged culture fermentation. (However if, for instance, a potential mycoherbicide microorganism cannot be produced by submerged culture fermentation, this may ultimately limit commercial development.) Notwithstanding whatever benefits may be obtained by individual farmers or commercial firms, there are compelling economic reasons from the whole of society's point of view that biological weed control be fostered and financially supported by government.

Literature Cited

1. Alcorn, J.L. 1976. Host range of *Puccinia xanthii*. Trans. Br. Mycol. Soc. 66:365–367.

2. Andres, L.A. 1976. The economics of biological control of weeds. Aquat. Bot. 3:111–123.

3. Andres, L.A. 1980. Conflicting interests and the biological control of weeds. Pages 11–20, *in:* Proc. 5th Int. Symp. Biol. Cont. Weeds, Brisbane, Australia.

4. Ault, B.A., and Tisdell, C.A. 1987. Economic thresholds and response to uncertainty in weed control. Agric. Syst. 25:219–227.

5. Auld, B.A., Menz, K.M., and Monaghan, N.M. 1979. Dynamics of weed spread: implications for policies of public control. Prot. Ecol. 1:141–148.

6. Auld, B.A., Menz, K.M., and Tisdell, C.A. 1987. Weed Control Economics. Academic Press, London.

7. Auld, B.A., McRae, C.F., and Sayn M.M. 1988. Possible control of *Xanthium spinosum* by a fungus. Agric. Ecosystems Environ. 21:219–223.

8. Bals, E.J. 1984. Where have all the droplets gone? Pages 81–85, *in:* Proc. 7th Aust. Weeds Conf. Vol. 2. Perth, Australia.

9. Bowers, R.C. 1982. Commercialisation of microbial biological control agents. Pages 157–173, *in:* Biological Control of Weeds with Plant Pathogens, R. Charudattan and H.L. Walker, eds. Wiley, New York.

10. Combellack, J.H. 1981. Herbicide application techniques—a review of ground practices in Australia. Pages 153–166, *in:* Proc. 6th Austral. Weeds Conf. Vol. 2. Gold Coast, Australia.

11. Conway, G.R. 1985. Agroecosystems analysis. Agric. Admin. 20:31–55.

12. Cullen, J.M. 1985. Bringing the cost benefit analysis of biological control of *Chondrilla juncea* up to date. Pages 145–152, *in:* Proc. VI Int. Symp. Biol. Contr. Weeds, Vancouver, Canada.

13. Delfosse, E.S. 1986. *Echium plantagineum* in Australia: effects of a major conflict of interest. Pages 293–299, *in:* Proc. VI Int. Symp. Biol. Contr. Weeds, Vancouver.

14. Harris, P. 1979. Cost of biological control of weeds by insects in Canada. Weed Sci. 27:242–250.

15. IAC. 1985. Biological control of *Echium* species (including Paterson's Curse/Salvation Jane). Industries Assistance Commission Report No. 371, Australian Government Publishing Service, Canberra.

16. Little, I.M.D. 1957. A Critique of Welfare Economics, 2nd ed. Oxford University Press, London.

17. Marsden, J.S., Martin, G.E., Parham, D.J., Risdell Smith, T.J., and Johnston, B.G. 1980. Returns on Australian Research. CSIRO, Canberra.

18. Menz, K.M., and Auld, B.A. 1977. Galvanised burr, control and public policy towards weeds. Search 8:281–287.

19. Menz, K.M., Auld, B.A., and Tisdell, C.A. 1984. The role for biological weed control in Australia. Search 15:208–210.

20. Nelson, R.R. 1959. The single economics of basic scientific research. J. Polit. Econ. 67:297–306.

21. Ng, Y. 1979. Welfare Economics. Macmillan, London.

22. Oelhaf, R.C. 1978. Organic Agriculture, Economic and Ecological Comparisons with Conventional Methods. Allunheld, Osmun, Montclair, NJ.

23. Reichelderfer, K.H. 1980. Economics of integrated pest management: discussion. Am. J. Agric. Econ. 62:1012–1013.

24. Templeton, G.E. 1983. Department of Plant Pathology, University of Arkansas, Fayetteville. Personal communication.

25. Templeton, G.E., TeBeest, D.O., and Smith, R.F., Jr., 1979. Biological weed control with mycoherbicides. Annu. Rev. Phytopathol. 17:301–310.

26. Tisdell, C.A. 1987. Economic evaluation of biological weed control. Plant Prot. Quart. 2:10–12.

27. Tisdell, C.A., Auld, B.A., and Menz, K.M. 1984a. On assessing the value of biological control of weeds. Prot. Ecol. 6:169–179.

28. Tisdell, C.A., Auld, B.A., and Menz, K.M. 1984. Crop loss elasticity in relation to weed density and control. Agric. Syst. 13:161–166.

29. Tisdell, C.A., and Auld, B.A. 1988. Evaluation of biological control projects. Proc. VII Symp. Biol. Contr. Weeds, Rome, *in press*.

30. Turner, C.E. 1986. Conflicting interests and biological control of weeds. Pag̱ ₃ 203–225, *in:* Proc. VI International Symp. Biol. Contr. Weeds, Vancouver.

31. Walker, J. 1982. NSW Agriculture & Fisheries, Biological and Chemical Research Institute, Rydalmere NSW, Australia. Personal communication.

Summary

The authors of the chapters in this volume have discussed a considerable body of information accumulated from investigations of a relatively modest number of plant pathogens regarded as potential biological herbicides. The focus of many of these investigations was clarification of the interactions of the host, pathogen and environment to assess the potential use of these pathogens in agricultural ecosystems. Some of these projects also investigated sporulation in culture and inoculum production and the results led to the development of commercially acceptable inoculum as a product. The development of a viable product from spores and demonstration of effectiveness maintained continued interest in biological herbicides.

The validity of the concept that plant pathogens can be used successfully in agriculture has been established with a few pathogens which have naturally served as models for later investigations. A few additional pathogens may soon be commercialized pending completion of regulatory and registration requirements. However, many plant pathogens of weeds were "discarded" from serious study after investigations proved that they "failed" to meet the criteria established by "successful" models such as *Colletotrichum gloeosporioides f.sp aeschynomene*. However, these "unsuccessful" pathogens, lacking the ability to control weeds adequately in the field or greenhouse, have prompted discussions relative to modification of the genetics of host-parasite interactions.

The use of molecular biological techniques has been proposed to increase the virulence, and presumably, therefore the ability of certain "unsuccessful" plant pathogens to control weeds. Currently, only a very small effort is being directed toward these studies. This research takes on a special importance since it redirects research toward more fundamental questions relative to the intentional large-scale release of these new virulent strains. Genetic manipulation of plant pathogens as biological control agents rightfully raises questions concerning the use of these genetically engineered organisms in the ecosystem. It challenges our concepts and understanding of gene transfer within a single species and between related species and the ecology of transformed isolates in the environment.

The intent of this book was to bring together some of the various aspects of biological control of weeds to show the uniqueness of this area of study and to highlight areas in need of additional work. Much still remains to be done and it will be exciting to watch its development and progress.

David O. TeBeest
June, 1990

Index

Contributors

Dr. Bruce A. Auld
Agricultural Research and Veterinary
  Centre
Orange, Australia

Dr. C. D. Boyette
USDA–ARS
Southern Weed Science Laboratory
Stoneville, MS

Dr. W. L. Bruckart
USDA
Foreign Disease-Weed Science Research Unit
Ft. Detrick
Frederick, MD

Dr. R. Charudattan
Department of Plant Pathology
University of Florida
Gainesville, FL

Dr. William J. Connick, Jr.
USDA–ARS
Southern Regional Research Center
New Orleans, LA

Dr. Donald J. Daigle
USDA–ARS
Southern Regional Research Center
New Orleans, LA

Dr. Floyd E. Fulgham
USDA–ARS
Jamie Whitten Delta States Research Center
Stoneville, MS

Dr. Dean Gabriel
Department of Plant Pathology
University of Florida
Gainesville, FL

Dr. Gary Harman
Department of Horticultural Science
Cornell University
Geneva, NY

Dr. S. Hasan
C.S.I.R.O.
Biological Control Unit
Montpellier, France

Dr. H. Corby Kistler
Department of Plant Pathology
University of Florida
Gainesville, FL

Dr. George Lacy
Department of Plant Pathology and Physiology
Virginia Polytechnic Institute and
  State University
Blacksburg, VA

Dr. Paul E. Parker
USDA–ARS
Mission Biological Control Lab
Mission, TX

Dr. C. Quimby
USDA–ARS
Montana State University
Bozeman, MT

Dr. R. J. Smith, Jr.
USDA–ARS
Rice Research and Extension Center
Stuttgart, AK

Dr. T. Stasz
Agricultural Research
Life Sciences Research Lab
Eastman Kodak Company
Rochester, NY

Mr. Larry Stowell
PACE Consulting
San Diego, CA

Dr. Alan K. Watson
Department of Plant Science
McGill University
Quebec, Canada

Dr. G. J. Weidemann
Department of Plant Pathology
University of Arkansas
Fayetteville, AK